Fuzzy Geometric Programming

Applied Optimization

Volume 76

Series Editors:

Panos M. Pardalos
University of Florida, U.S.A.

Donald Hearn
University of Florida, U.S.A.

The titles published in this series are listed at the end of this volume.

Fuzzy Geometric Programming

by

Bing-Yuan Cao

Department & Institute of Mathematics,
Shantou University,
Guangdong, P.R. China

SPRINGER-SCIENCE+BUSINESS MEDIA, B.V.

A C.I.P. Catalogue record for this book is available from the Library of Congress.

ISBN 978-1-4613-4884-9 ISBN 978-1-4615-0009-4 (eBook)
DOI 10.1007/978-1-4615-0009-4

Printed on acid-free paper

Contents

List of Figures

List of Tables

Preface

Fuzzy geometric programming was originated by the author in the Proceeding of the second IFSA conferences, 1987(Tokyo) 14 years ago. Later, the paper was invited for formal publication in the International Journal of Fuzzy Sets and Systems. From then on, more and more papers have been written by scholars all over the world who have been interested in its research. So this programming method has been acknowledged by experts and has gradually formed a new branch of fuzzy mathematics.

Inspired by Zadeh's fuzzy sets theory, fuzzy geometric programming emerges from the combination of fuzzy sets theory with geometric programming, where models are built in the fuzzy posynomial and the reverse geometric programming. The present book is intended to discuss fuzziness of objective function and constraint conditions, a variety of fuzzy numbers in coefficients and variables and problems about multi-objective fuzzy geometric programming. It establishes and rounds out an entire theory system, showing that there exist conditions of fuzzy optimal or most satisfactory solutions in fuzzy geometric programming, and it develops some effective algorithms.

In order to introduce this new branch, the book aims at the exposition of three points: encompassing ideas and conception, theory and methods, and diffusion and application. It lays more emphasis on the second point than the first one, and less on the third. Besides, it introduces some knowledge of classical geometric programming and of fuzzy sets theory and application examples of fuzzy geometric programming in electric power systems as well.

The book is based on 26 papers by the author himself, of which some have been published in major international journals and indexed in SCI and in EI up to 1997.

The research is supported by the National Natural Science Foundation of China (1997–1999), and also supported by the Scientific and Technical Committees Project of Hunan Province and " 211" Project Foundation of Shantou

University and Li Jia-cheng Science Development Foundation of Shantou University P. R. China.

The author has complied this book by sorting out his research achievements over 14 years, offering a few commonplace remarks by way of introduction to fuzzy geometric programming so that others may formulate opinions about it and mane valuable contribution to it.

CAO BING-YUAN

SHANTOU UNIVERSITY

P.R.CHINA

June 2001

To My Wife Wang Pei-hua

Acknowledgements

This book was written while the author was being supported by the National Natural Science Foundation of P.R.China (No.79670012), Scientific and Technical Committees Project of Hunan Province, "211" Project Foundation and Science Advance Foundation of Shantou University by Li Jiacheng.

The author is grateful for the constant encouragement as well as support by Prof. M. Sugeno, Prof. H.J.Zimmermann and Prof. D.Dubois and Prof. H.Prade after he posed the idea of fuzzy geometric programming.

He wishes to thank Mrs.Wang Peihua, an associate professor of Shantou University, for great contribution to the writing of the book in English and its revised version as well.

The author wishes to express his deep appreciation for Kluwer Academic Publisher B.V providing him with the opportunity of revealing the first of his study to scholars all over the world and to Doctor Editorial Angela Quilici, Doctor Language editor Edwin Beschlev, the director and editor, and anonymous referees of the book, who have giving great help.

Finally, he is thankful to the author's Master Ms.Liu Meijiao and typist Cao Fanghua for much work in the proofreading and typesetting.

Introduction
of Author

Cao Bing-yuan was born in Hunan Changde, P. R. China in Oct. 1951. He graduated from Mathematics Department of Hunan Normal University and spent one year in Teachers' Class for Advanced Studies. His current positions are: Chairman of the Mathematics Department, Professor of the Mathematics Department & Institute of Shantou University (one of the leading Universities in China). Member of Gesellschaft fur Angewandte Mathematic und Mechanik. Director of China Operation Research Society. Permanent Member of Council in Chinese Matter Element Analysis Society. President of Fuzzy Mathematics and Systems Chapter in South China. Associate President of Guangdong Mathematics Society Industrial and Applied.

In 1987, he developed fuzzy geometric programming, which has become a new branch in fuzzy mathematics, and published more than 60 papers, and six books including research works, teaching materials and proceedings papers. Now, he concentrates on research in fuzzy geometric programming, fuzzy statistics and mathematical economics. In 1997, he was awarded the Zen Xianzi National Education Foundation Prize for his achievements in both excellent teaching and research work, and an award in Advanced Science and Technology by the Hunan Education Committee, among other honors. Since 1997, he has taken charge of projects supported by the National Natural Science Foundation of China and the Science Project of Hunan Province. The project directed by him is acknowledged by experts to embody the most advanced standards at home and abroad. His research achievements have been included in Who's Who in the World, World Celebrity Books.

Chapter 1

MATHEMATICAL PRELIMINARIES

This chapter is devoted to presentation of some necessary mathematical background for the sake of further discussion of fuzzy geometric programming (GP).

Section 1 introduces fuzzy sets and their operations.

Section 2 provides decomposition theorems for fuzzy sets and convex fuzzy sets.

Section 3 presents fuzzy matrices and fuzzy vectors.

Section 4 develops a geometric inequality and properties of posynomial geometric programming (PGP).

Section 5 relates an extension principle and a representation theorem.

Section 6 exhibits fuzzy functions.

1.1 Fuzzy Sets and Their Operations

Let X be an ordinary set, whose generic elements are denoted by x.

Definition 1.1.1 A mapping

$$\tilde{A}: \quad \begin{array}{l} X \to [0,1] \\ x \to \tilde{A}(x) \end{array}$$

is called a fuzzy collection or fuzzy sets [Zad65a] for short, $\tilde{A}(x)$ a membership x corresponding to a fuzzy set $\tilde{A}$, and $\tilde{A}(\cdot)$ a membership function with respect to a fuzzy set $\tilde{A}$.

$\tilde{A}$ and $\tilde{A}(x)$ mean respectively a fuzzy subset and its membership function, a distinction that is necessary in order to avoid misunderstanding.

Definition 1.1.2 Let $\tilde{A}, \tilde{B} \in \mathcal{F}(X)$. Then $\forall x \in X$,

$$\tilde{A} = \tilde{B} \Longleftrightarrow \tilde{A}(x) = \tilde{B}(x), \qquad \text{(equalities)}$$

$$\tilde{A} \subseteq \tilde{B} \Longleftrightarrow \tilde{A}(x) \leqslant \tilde{B}(x). \qquad \text{(inclusions)}$$

If ϕ represents fuzzy sets in which a membership function identically equals 0, and X represents fuzzy sets where it identically equals 1, then there exist properties as follows:

1^0 $\phi \subset \tilde{A} \subset X$ (existence of maximum and minimum fuzzy sets)
2^0 $\tilde{A} \subset \tilde{A}$ (reflection)
3^0 If $\tilde{A} \subset \tilde{B}, \tilde{B} \subset \tilde{A}$, then $\tilde{A} = \tilde{B}$ (symmetry)
4^0 If $\tilde{A} \subset \tilde{B}, \tilde{B} \subset \tilde{C}$, then $\tilde{A} \subset \tilde{C}$ (transitivity)

Since relation " $\subseteq$ " constitutes an order relation on $\mathcal{F}(X), (\mathcal{F}(X), \leqslant)$ stands for a partially ordered set.

Definition 1.1.3 Suppose $\tilde{A}, \tilde{B} \in \mathcal{F}(X)$ and we define the following:

1^0 Union of $\tilde{A}$ and $\tilde{B}$ means $\tilde{A} \cup \tilde{B}$, whose membership function is
$(\tilde{A} \cup \tilde{B})(x) = \tilde{A}(x) \vee \tilde{B}(x) = \max\{\tilde{A}(x), \tilde{B}(x)\}$
2^0 Intersection of $\tilde{A}$ and $\tilde{B}$ denotes $\tilde{A} \cap \tilde{B}$, whose membership function is
$(\tilde{A} \cap \tilde{B})(x) = \tilde{A}(x) \wedge \tilde{B}(x) = \min\{\tilde{A}(x), \tilde{B}(x)\}$
3^0 Complement of $\tilde{A}$ is $\tilde{A}^c$ and its membership function means
$\tilde{A}^c(x) = 1 - \tilde{A}(x)$

Obviously, if $\tilde{A}, \tilde{B} \in \mathcal{F}(X)$, there exist $\tilde{A} \cup \tilde{B}, \tilde{A} \cap \tilde{B} \in \mathcal{F}(X)$ and $\tilde{A}^c \in \mathcal{F}(X)$.

Theorem 1.1.1 $(\mathcal{F}(X), \cup, \cap, c)$ *has properties as follows:*

1^0 $\tilde{A} \cup \tilde{B} = \tilde{B} \cup \tilde{A}, \tilde{A} \cap \tilde{B} = \tilde{B} \cap \tilde{A}$ *(commutative law)*
2^0 $(\tilde{A} \cup \tilde{B}) \cup \tilde{C} = \tilde{A} \cup (\tilde{B} \cup \tilde{C})$ *(associative law)*
$(\tilde{A} \cap \tilde{B}) \cap \tilde{C} = \tilde{A} \cap (\tilde{B} \cap \tilde{C})$
3^0 $\tilde{A} \cup (\tilde{B} \cap \tilde{C}) = (\tilde{A} \cup \tilde{B}) \cap (\tilde{A} \cup \tilde{C})$ *(distributive law)*
$\tilde{A} \cap (\tilde{B} \cup \tilde{C}) = (\tilde{A} \cap \tilde{B}) \cup (\tilde{A} \cap \tilde{C})$
4^0 $\tilde{A} \cup (\tilde{A} \cap \tilde{B}) = \tilde{A}, \tilde{A} \cap (\tilde{A} \cup \tilde{B}) = \tilde{A}$ *(absorptive law)*
5^0 $\tilde{A} \cup \tilde{A} = \tilde{A}, \tilde{A} \cap \tilde{A} = \tilde{A}$ *(idempotent law)*
6^0 $(\tilde{A}^c)^c = \tilde{A}$ *(involution law)*
7^0 $X \cap \tilde{A} = \tilde{A}, \quad X \cup \tilde{A} = X$ $(0 - 1\ law)$
$\phi \cap \tilde{A} = \phi, \phi \cup \tilde{A} = \tilde{A}$
8^0 $(\tilde{A} \cup \tilde{B})^c = \tilde{A}^c \cap \tilde{B}^c, (\tilde{A} \cap \tilde{B})^c = \tilde{A}^c \cup \tilde{B}^c$ *(DeMorgan's laws)*
9^0 $(\tilde{A}^c \cup \tilde{B}) \cap (\tilde{A} \cup \tilde{B}^c) = (\tilde{A}^c \cap \tilde{B}^c) \cup (\tilde{A} \cap \tilde{B})$ *(equivalence)*
10^0 $(\tilde{A}^c \cap \tilde{B}) \cup (\tilde{A} \cap \tilde{B}^c)$
$= (\tilde{A}^c \cup \tilde{B}^c) \cap (\tilde{A} \cup \tilde{B})$ *(symmetrical difference)*

◊

Proof We prove only the first formula of 8^0 and we can obtain the rest by testing them directly.

Because $\forall x \in X$ we have

$$\begin{aligned}(\tilde{A} \cup \tilde{B})^c(x) &= 1 - (\tilde{A} \cup \tilde{B})(x) \\ &= 1 - \max\{\tilde{A}(x), \tilde{B}(x)\} \\ &= \min\{1 - \tilde{A}(x), 1 - \tilde{B}(x)\} \\ &= \min\{\tilde{A}^c(x), \tilde{B}^c(x)\} \\ &= (\tilde{A}^c \cap \tilde{B}^c)(x),\end{aligned}$$

It follows that

$$(\tilde{A} \cup \tilde{B})^c = \tilde{A}^c \cap \tilde{B}^c.$$

□

We point out that fuzzy sets no longer satisfy the excluded-middle law, i.e.,

$$\tilde{A} \cup \tilde{A}^c \neq X, \tilde{A} \cap \tilde{A}^c \neq \phi.$$

Example 1.1.1 Let $X = [0, 1], \tilde{A}(x) = x$. Then $\tilde{A}^c(x) = 1 - x$, while

$$(\tilde{A} \cup \tilde{A}^c)(x) = \begin{cases} 1 - x, & x \leqslant \frac{1}{2}, \\ x, & x > \frac{1}{2}, \end{cases}$$

$$(\tilde{A} \cap \tilde{A}^c)(x) = \begin{cases} x, & x \leqslant \frac{1}{2}, \\ 1 - x, & x > \frac{1}{2}, \end{cases}$$

hence

$$\tilde{A} \cup \tilde{A}^c \neq X, \tilde{A} \cap \tilde{A}^c \neq \phi.$$

Especially,

$$(\tilde{A} \cup \tilde{A}^c)\left(\frac{1}{2}\right) = (\tilde{A} \cap \tilde{A}^c)\left(\frac{1}{2}\right) = \frac{1}{2}.$$

A fuzzy set operation does not satisfy the excluded-middle law, which complicates the study of fuzzy sets. Just because of this, fuzzy sets can exhibit more objective properties than classical sets.

1.2 Decomposition Theorems of Fuzzy Sets and Convex Fuzzy Sets

1.2.1 α–cut Set and Decomposition Theorem

The decomposition theorem of fuzzy sets is a bridge between fuzzy sets and ordinary ones.

Definition 1.2.1 Let $\tilde{A} \in \mathcal{F}(X), \forall\alpha \in [0,1]$, written

$$(\tilde{A})_\alpha = A_\alpha = \{x|\tilde{A}(x) \geqslant \alpha\}.$$

A_α is said to be an α–cut set of a fuzzy set $\tilde{A}$. Again, we write

$$(\tilde{A})_{\underset{\cdot}{\alpha}} = A_{\underset{\cdot}{\alpha}} = \{x|\tilde{A}(x) > \alpha\}.$$

$A_{\underset{\cdot}{\alpha}}$ is called a strong α–cut set of a fuzzy set $\tilde{A}$, α a confidence level.

Definition 1.2.2 Let $\tilde{A} \in \mathcal{F}(X)$, written

$$A_{\underset{\cdot}{0}} = \{x|\tilde{A}(x) > 0\} = \mathrm{supp}\tilde{A}.$$

$A_{\underset{\cdot}{0}}$ is called a support of a fuzzy set $\tilde{A}$. Ker $\tilde{A}$, written

$$\mathrm{Ker}\ \tilde{A} = \{x|\tilde{A}(x) = 1\},$$

is called a kernel of fuzzy set $\tilde{A}$ and $\tilde{A}$ is a normal fuzzy set for Ker $\tilde{A} \neq \phi$.

Theorem 1.2.1 *We have the following properties for α–cut and strong α–cut sets:*

$$\begin{array}{ll} 1^0 & (\tilde{A} \cup \tilde{B})_\alpha = A_\alpha \cup B_\alpha, \quad (\tilde{A} \cap \tilde{B})_\alpha = A_\alpha \cap B_\alpha; \\ 2^0 & (\tilde{A} \cup \tilde{B})_{\underset{\cdot}{\alpha}} = A_{\underset{\cdot}{\alpha}} \cup B_{\underset{\cdot}{\alpha}}, \quad (\tilde{A} \cap \tilde{B})_{\underset{\cdot}{\alpha}} = A_{\underset{\cdot}{\alpha}} \cap B_{\underset{\cdot}{\alpha}}. \end{array}$$

◊

Proof We prove only the first formula in 1^0.

$$\begin{aligned} (\tilde{A} \cup \tilde{B})_\alpha &= \{x|(\tilde{A} \cup \tilde{B})(x) \geqslant \alpha\} = \{x|\tilde{A}(x) \vee \tilde{B}(x) \geqslant \alpha\} \\ &= \{x|\tilde{A}(x) \geqslant \alpha\} \cup \{x|\tilde{B}(x) \geqslant \alpha\} = A_\alpha \cup B_\alpha. \end{aligned}$$

Proofs of the other formulas are similar. □

1.2.2 Decomposition Theorems

Definition 1.2.3 Suppose $\alpha \in [0,1], \tilde{A} \in \mathcal{F}(X)$, whose product is

$$(\alpha\tilde{A})(x) = \alpha \wedge \tilde{A}(x).$$

Theorem 1.2.2 (decomposition theorem I) *As for an arbitrary $\tilde{A} \in \mathcal{F}(X)$, we have*

$$1^0 \quad \tilde{A} = \bigcup_{\alpha\in[0,1]} \alpha A_\alpha,$$

$$2^0 \quad \tilde{A} = \bigcup_{\alpha\in[0,1)} \alpha A_{\underset{\cdot}{\alpha}}.$$

◊

Proof We prove only 1^0; the other proofs are similar.

Because

$$A_\alpha(x) = \begin{cases} 1, & x \in A_\alpha, \\ 0, & x \notin A_\alpha, \end{cases}$$

then

$$\begin{aligned} \Big(\bigcup_{\alpha\in[0,1]} \alpha A_\alpha\Big)(x) &= \sup\nolimits_{0\leqslant\alpha\leqslant 1} \alpha A_\alpha(x) \\ &= \sup\nolimits_{x\in A_\alpha} \alpha = \sup\nolimits_{\alpha\leqslant \tilde{A}(x)} \alpha \\ &= \tilde{A}(x). \end{aligned}$$

□

Theorem 1.2.3 (decomposition theorem II) *Let $\tilde{A} \in \mathcal{F}(X)$. If there exists a set value mapping*

$$\begin{aligned} H : [0,1] &\Longrightarrow \mathcal{F}(X), \\ \alpha &\longmapsto H(\alpha), \end{aligned}$$

such that $\forall \alpha \in [0,1], A_{\underset{\cdot}{\alpha}} \subseteq H(\alpha) \subseteq A_\alpha$, *then*

$$1^0 \quad \tilde{A} = \bigcup_{\alpha\in[0,1]} \alpha H(\alpha),$$

$$2^0 \quad \alpha_1 < \alpha_2 \Longrightarrow H(\alpha_1) \supset H(\alpha_2),$$

$$3^0 \quad A_\alpha = \bigcap_{\lambda<\alpha} H(\lambda), \alpha \neq 0,$$

$$A_{\underset{\cdot}{\alpha}} = \bigcup_{\lambda>\alpha} H(\lambda), \alpha \neq 1.$$

◊

Proof 1^0 $A_{\underset{\cdot}{\alpha}} \subseteq H(\alpha) \subseteq A_\alpha$

$$\begin{aligned} &\Longrightarrow \alpha A_{\underset{\cdot}{\alpha}} \subseteq \alpha H(\alpha) \subseteq \alpha A_\alpha \\ &\Longrightarrow \tilde{A} = \bigcup_{\alpha\in[0,1)} \alpha A_{\underset{\cdot}{\alpha}} \subseteq \bigcup_{\alpha\in[0,1]} \alpha H(\alpha) \subseteq \bigcup_{\alpha\in[0,1]} \alpha A_\alpha = \tilde{A} \\ &\Longrightarrow \tilde{A} = \bigcup_{\alpha\in[0,1]} \alpha H(\alpha). \end{aligned}$$

2^0 Because $\forall x \in X$, we have

$$x \in A_{\alpha_2} \Rightarrow \tilde{A}(x) \geqslant \alpha_2 > \alpha_1 \Rightarrow x \in A_{\underset{\cdot}{\alpha_1}},$$

therefore

$$\alpha_1 < \alpha_2 \Rightarrow H(\alpha_1) \supseteq A_{\alpha_1} \supseteq A_{\alpha_2} \supseteq H(\alpha_2).$$

3^0 We prove only the first formula

$$\forall \lambda < \alpha, H(\lambda) \supseteq A_\lambda \supseteq A_\alpha \Rightarrow \bigcap_{\lambda<\alpha} H(\lambda) \supseteq A_\alpha, \alpha \neq 0.$$

Again

$$\bigcap_{\lambda<\alpha} H(\lambda) \subseteq \bigcap_{\lambda<\alpha} A_\lambda = A_{(\bigwedge_{\lambda<\alpha} \lambda)} = A_\alpha, \alpha \neq 0,$$

hence

$$A_\alpha = \bigcap_{\lambda<\alpha} H(\lambda).$$

□

From the knowledge of decomposition theorem II, a collection of sets $\{H(\alpha)| \alpha \in [0,1]\}$ is replaced one by one with variations of α, i.e., a collection sleeve. Then we have

Definition 1.2.4 If a set-valued mapping $H : [0,1] \longrightarrow \mathcal{F}(X)$ satisfies

$$\begin{aligned} &\forall\, \alpha_1, \alpha_2 \in [0,1], \\ &\alpha_1 < \alpha_2 \Rightarrow H(\alpha_1) \supseteq H(\alpha_2), \end{aligned}$$

then H is called a collection sleeve on X.

A set composed of all the collection sleeves on X is written as $\mathcal{U}(X)$.

1.2.3 Convex Fuzzy Sets

Definition 1.2.5 Let $X = \Re^m$ be an $m-$dimensional Euclidean space and A be an ordinary subset of X. If $\forall x_1, x_2 \in A$ and $\forall \lambda \in [0,1]$,

$$\lambda x_1 + (1-\lambda)x_2 \in A,$$

Then A is a convex set.

Definition 1.2.6 Suppose $\tilde{A} \in \mathcal{F}(X)$. If $\forall x_1, x_2 \in X$,

$$\tilde{A}(\lambda x_1 + (1-\lambda)x_2) \geqslant \tilde{A}(x_1) \wedge \tilde{A}(x_2), \lambda \in [0,1],$$

then $\tilde{A}$ is a convex fuzzy set.

Theorem 1.2.4 *$\tilde{A} \in \mathcal{F}(X)$ is a convex fuzzy set $\Longleftrightarrow \forall \alpha \in [0,1]$, A_α is a convex set.* ◊

Proof Let $\tilde{A}$ be a convex fuzzy set. Since

$$A_\alpha = \{x|\tilde{A}(x) \geqslant \alpha\}, \quad \alpha \in [0,1],$$

we have

$$\tilde{A}(\lambda x_1 + (1-\lambda)x_2) \geqslant \tilde{A}(x_1) \wedge \tilde{A}(x_2) \geqslant \alpha, \lambda \in [0,1]$$

for $x_1, x_2 \in A_\alpha$. Therefore, $\lambda x_1 + (1-\lambda)x_2 \in A_\alpha$, i.e., A_α is a convex set.

Conversely, if $\forall \alpha \in [0,1]$, A_α is a convex set, then $\forall x_1, x_2 \in X$, we suppose $\alpha = \tilde{A}(x_1) \wedge \tilde{A}(x_2)$. So $x_1 \in A_\alpha, x_2 \in A_\alpha$.

Because A_α is a convex set $\forall \alpha, \lambda \in [0,1]$, we have $\lambda x_1 + (1-\lambda)x_2 \in A_\alpha$, so that

$$\tilde{A}(\lambda x_1 + (1-\lambda)x_2) \geqslant \tilde{A}(x_1) \wedge \tilde{A}(x_2);$$

therefore, $\tilde{A}$ is proved to be a convex fuzzy set. □

Theorem 1.2.5 *If $\tilde{A}$ and $\tilde{B}$ are convex sets, so is $\tilde{A} \cap \tilde{B}$.* ◊

Proof $\forall x_1, x_2 \in X$, $\forall \lambda \in [0,1]$,

$$\begin{aligned}
&\big(\tilde{A} \cap \tilde{B}\big)\big(\lambda x_1 + (1-\lambda)x_2\big) \\
&\qquad = \tilde{A}\big(\lambda x_1 + (1-\lambda)x_2\big) \wedge \tilde{B}\big(\lambda x_1 + (1-\lambda)x_2\big) \\
&\qquad \geqslant \big(\tilde{A}(x_1) \wedge \tilde{A}(x_2)\big) \wedge \big(\tilde{B}(x_1) \wedge \tilde{B}(x_2)\big) \\
&\qquad = \big(\tilde{A}(x_1) \wedge \tilde{B}(x_1)\big) \wedge \big(\tilde{A}(x_2) \wedge \tilde{B}(x_2)\big) \\
&\qquad = (\tilde{A} \cap \tilde{B})(x_1) \wedge (\tilde{A} \cap \tilde{B})(x_2).
\end{aligned}$$

Overall, $\tilde{A} \cap \tilde{B}$ denotes convex fuzzy sets. □

Definition 1.2.7 Let $\tilde{A}, \tilde{B}, \tilde{\wedge} \in \mathcal{F}(X)$; a convex combination with respect to $\tilde{\wedge}$ of $\tilde{A}$ and $\tilde{B}$ is a fuzzy set, denoted $(\tilde{A}, \tilde{B}; \tilde{\wedge})$, with its membership function being

$$(\tilde{A}, \tilde{B}; \tilde{\wedge}) = \tilde{\wedge}(x)\tilde{A}(x) + \big(1 - \tilde{\wedge}(x)\big)\tilde{B}(x), \quad \forall x \in X.$$

Generally, if $\tilde{A}_i, \tilde{\wedge}_i \in \mathcal{F}(X)(1 \leqslant i \leqslant m)$ and $\sum\limits_{i=1}^{m} \tilde{\wedge}_i(x) = 1(\forall x \in X)$, then a convex combination with respect to $\tilde{\wedge}$ of $\tilde{A}_i$ is written as

$$(\tilde{A}_1, \tilde{A}_2, \ldots, \tilde{A}_m; \tilde{\wedge}_1, \tilde{\wedge}_2, \ldots, \tilde{\wedge}_m)(x) = \sum_{i=1}^{m} \tilde{\wedge}_i(x)\tilde{A}_i(x), \quad \forall x \in X.$$

Definition 1.2.8 Let $\tilde{A} \in \mathcal{F}(X)$. If $\forall \alpha \in [0,1]$, then all of A_α represent bounded sets in X and $\tilde{A}$ is called a bounded fuzzy set in X.

It is easy to prove the following by Theorem 1.2.5.

Theorem 1.2.6 *Both union and intersection of two bounded fuzzy sets are bounded fuzzy sets.* ◊

1.3 Fuzzy Matrices and Vectors

Fuzzy matrices and vectors play the same role in fuzzy mathematical programming as matrices and vectors do in the classical mathematical programming. If all of the objects discussed under a fuzzy environment are discrete, the use of fuzzy matrices and vectors makes our expression and application brief and convenient, respectively.

1.3.1 Fuzzy Matrices

(1) Notion of a fuzzy matrix

Definition 1.3.1 If $x = \{x_1, x_2, \ldots, x_m\}, y = \{y_1, y_2, \ldots, y_n\}$, then a membership function value of a fuzzy relation $\tilde{R}$ can be exhibited by means of an $m \times n$ matrix in $X \times Y$, written $\tilde{R}$, i.e.,

$$\tilde{R} = \begin{pmatrix} r_{11} & r_{12} & \cdots & r_{1n} \\ r_{21} & r_{22} & \cdots & r_{2n} \\ \cdots & \cdots & \cdots & \cdots \\ r_{m1} & r_{m2} & \cdots & r_{mn} \end{pmatrix}$$

where $r_{ij} = \mu_{\tilde{R}}(x_i, y_j) \in [0,1](1 \leqslant i \leqslant m; 1 \leqslant j \leqslant n)$ is a matrix element, and $\tilde{R}$ is called a fuzzy matrix. If $r_{ij} \in \{0,1\}$(i.e., all entries in the matrices are 0 or 1), then the matrix is called Boolean matrix, written R. Such three such matrices have specific names.

$\tilde{O}$—called a fuzzy-zero matrix, and all of its elements are 0.

$\tilde{I}$—called a fuzzy-unit matrix, its diagonal elements are 1 and the others are 0.

$\tilde{E}$—called a fuzzy complete matrix and all of its elements are 1.

These three matrices are identical to ordinary zero, unit and complete matrices.

Definition 1.3.2 As for $\forall \alpha \in [0,1]$, we call $T_\alpha = (t_{ij}^\alpha)$ an $\alpha-$cut matrix in a fuzzy matrix $\tilde{T} = (t_{ij})(t_{ij} \in [0,1])$, where

$$t_{ij}^\alpha = \begin{cases} 1, & t_{ij} \geqslant \alpha, \\ 0, & t_{ij} < \alpha. \end{cases}$$

Obviously, T_α behaves as a Boolean matrix.

(2) Relations and operations in a fuzzy matrix

The discussion is limited to square matrices for the sake of convenience. All m order fuzzy square matrices are written as $\mathcal{M}_{m \times m}$.

Definition 1.3.3 Let $\tilde{R} = (r_{ij})_m, \tilde{S} = (s_{ij})_m \in \mathcal{M}_{m \times m}, r_{ij}, s_{ij} \in [0,1]$. Then

1^0 Equality: $\tilde{R} = \tilde{S} \Longleftrightarrow r_{ij} = s_{ij},\ 1 \leqslant i, j \leqslant m.$
2^0 Inclusion: $\tilde{R} \subset \tilde{S} \Longleftrightarrow r_{ij} \leqslant s_{ij},\ 1 \leqslant i, j \leqslant m.$
3^0 Union: If $T_{ij} = r_{ij} \vee s_{ij}$, then $\tilde{T} = \tilde{R} \cup \tilde{S}$ is a union of $\tilde{R}$ and $\tilde{S}$.
4^0 Intersection: If $T_{ij} = r_{ij} \wedge s_{ij}$, then $\tilde{T} = \tilde{R} \cap \tilde{S}$ is an intersection of $\tilde{R}$ and $\tilde{S}$.
5^0 Complement: $\tilde{R}^c = (1 - r_{ij})$ is called a complement matrix of $\tilde{R}$.

Properties 1.3.1

1^0 $\tilde{R} \cup \tilde{S} = \tilde{S} \cup \tilde{R};\ \tilde{R} \cap \tilde{S} = \tilde{S} \cap \tilde{R}$ *(commutative law)*

2^0 $(\tilde{R} \cup \tilde{S}) \cup \tilde{T} = \tilde{R} \cup (\tilde{S} \cup \tilde{T})$ *(associative law)*
$(\tilde{R} \cap \tilde{S}) \cap \tilde{T} = \tilde{R} \cap (\tilde{S} \cap \tilde{T})$

3^0 $\tilde{R} \cup (\tilde{S} \cap \tilde{T}) = (\tilde{R} \cup \tilde{S}) \cap (\tilde{R} \cup \tilde{T})$ *(distributive law)*
$\tilde{R} \cap (\tilde{S} \cup \tilde{T}) = (\tilde{R} \cap \tilde{S}) \cup (\tilde{R} \cap \tilde{T})$

4^0 $\tilde{R} \cup (\tilde{R} \cap \tilde{S}) = \tilde{R}$ *(absorptive law)*
$\tilde{R} \cap (\tilde{R} \cup \tilde{S}) = \tilde{R}$

5^0 $\tilde{R} \cup \tilde{R} = \tilde{R}, \tilde{R} \cap \tilde{R} = \tilde{R}$ *(idempotent law)*

6^0 $(\tilde{R}^c)^c = \tilde{R}$ *(involution law)*

7^0 $(\tilde{R} \cup \tilde{S})^c = \tilde{R}^c \cup \tilde{S}^c$ *(dual law)*
$(\tilde{R} \cap \tilde{S})^c = \tilde{R}^c \cup \tilde{S}^c$

8^0 $(\tilde{R}^m)^n = \tilde{R}^{mn}$ *(power)*

9^0 $\tilde{R}^T$ *(transitive fuzzy matrices)*

10^0 $(\tilde{R}^T)^T = \tilde{R}$

11^0 $\tilde{R} \subseteq \tilde{S} \Longleftrightarrow \tilde{R}^T \subseteq \tilde{S}^T$

12^0 $(\tilde{R} \cup \tilde{S})^T = \tilde{R}^T \cup S^T, (\tilde{R} \cap \tilde{S})^T = \tilde{R}^T \cap \tilde{S}^T$

13^0 $\tilde{R} \subseteq \tilde{S} \Longleftrightarrow R_\alpha \subseteq S_\alpha$

14^0 $(\tilde{R} \cup \tilde{S})_\alpha = R_\alpha \cup S_\alpha, (\tilde{R} \cap \tilde{S})_\alpha = R_\alpha \cap S_\alpha$

◊

1.3.2 Fuzzy Vector

Definition 1.3.4 A vector

$$\overrightarrow{a} = (a_1, a_2, \ldots, a_m)$$

is called an $m-$dimensional fuzzy vector, if an arbitrary component satisfies

$$0 \leqslant a_i \leqslant 1\ (1 \leqslant i \leqslant m).$$

And if only values of 0 or 1 are selected for all of the components, then, the vector is called a Boolean (or Binary) vector.

If $\overrightarrow{\alpha} \in \mathcal{U}_{1\times m}$, the transpose of $\overrightarrow{\alpha}$ is $\overrightarrow{\alpha}^{\mathrm{T}} \in \mathcal{U}_{m\times 1}$, called an $m-$dimensional column vector.

A fuzzy vector has two characterizations:

(1) It denotes a fuzzy set $\tilde{A}$ on a finite universe $X = \{x_1, x_2, \ldots, x_m\}$

$$a_i \triangleq \mu_{\tilde{A}}(x_i)\ (1 \leqslant i \leqslant m).$$

(2) It denotes a fuzzy relation as well as a fuzzy matrix.

Analogous to an ordinary matrix made up of binary vectors, A fuzzy matrix is made up of binary fuzzy vectors. An $m \times n$ matrix is composed of m fuzzy vectors. The difference between fuzzy vectors and ordinary ones is that each component of a fuzzy vector is $a_i \in [0, 1]$.

A fuzzy set on a universe X denotes a relation from that set to X and a collection of fuzzy vector, $\overrightarrow{\alpha}$ from a matrix representing that relation.

1.4 Geometric Inequality and Properties of PGP

This section introduces only some basic results, as for the content in details we refer to References [Eck80][DP73][WB67] and [WY82].

1.4.1 Geometric Inequality

Theorem 1.4.1 (geometric inequality) *Let* $v_k > 0, w_k > 0$ $(1 \leqslant k \leqslant J)$ *and* $\sum_{k=1}^{J} w_k = 1$. *Then*

$$\sum_{k=1}^{J} v_k^{w_k} \leqslant \sum_{k=1}^{J} w_k v_k. \tag{1.1}$$

The equal sign in the formula holds iff $v_1 = v_2 = \cdots = v_J$. ◊

Proof On the basis of convex function theory, since

$$\frac{d^2(\log v)}{dv^2} = -\frac{1}{v^2} < 0,$$

$\log v$ is a strictly concave function for v on the open interval $(0, +\infty)$, such that

$$\sum_{k=1}^{J} w_k \log v_k \leqslant \log \sum_{k=1}^{J} w_k v_k$$

with the equal sign holding iff $v_1 = v_2 = \cdots = v_J$. Obviously, it is equivalent to (1.1). □

Corollary 1.4.1 *Suppose* $w_k \geqslant 0, v_k > 0 (1 \leqslant k \leqslant J)$*; then*

$$(\sum_{k=1}^{J} v_k)^{w_0} \geqslant \prod_{k=1}^{J} (\frac{v_k}{w_k})^{w_k} w_0^{w_0}. \tag{1.2}$$

Here

$$w_0 = \sum_{k=1}^{J} w_k \tag{1.3}$$

and we stipulate $(v_k/w_k)^{w_k}|_{w_k=0} = 1$. *Again, the equal sign holds in (1.2) iff*

$$w_k \sum_{k=1}^{J} v_k = v_k w_0 \quad (1 \leqslant k \leqslant J_0). \tag{1.4}$$

◊

1.4.2 Geometric Programming

Definition 1.4.1 We call

$$(P) \quad \begin{aligned} &\min \ g_0(x) \\ &\text{s.t. } g_i(x) \leqslant 1 (1 \leqslant i \leqslant p) \\ &\quad x > 0 \end{aligned} \tag{1.5}$$

a primal PGP [Duf62a,b][DP66][DPZ67], where

$$g_i(x) = \sum_{k=1}^{J_i} v_{ik}(x) = \sum_{k=1}^{J_i} c_{ik} \prod_{l=1}^{m} x_l^{\gamma_{ikl}} (0 \leqslant i \leqslant p)$$

is posynomial of x, while we say a programming in a reverse direction to be inequality constraint and

$$(P^-) \quad \begin{aligned} &\min \ g_0(x) \\ &\text{s.t. } g_i(x) \leqslant 1 (1 \leqslant i \leqslant p') \\ &\quad g_i(x) \geqslant 1 \ (p' + 1 \leqslant i \leqslant p) \\ &\quad x > 0 \end{aligned} \tag{1.6}$$

a reverse PGP [AW70][DP72], where all of $g_i(x) (0 \leqslant i \leqslant p)$ is posynomial of x, with

$$g_i(x) = \sum_{k=1}^{J_i} v_{ik}(x)$$

where

$$v_{ik}(x) = \begin{cases} c_{ik} \prod\limits_{l=1}^{m} x_l^{\gamma_{ikl}}, & (0 \leqslant i \leqslant p \,; 1 \leqslant k \leqslant J_i) \quad \text{for a primal PGP,} \\ \begin{cases} c_{ik} \prod\limits_{l=1}^{m} x_l^{\gamma_{ikl}}, & (1 \leqslant i \leqslant p' \,; 1 \leqslant k \leqslant J_i) \\ c_{ik} \prod\limits_{l=1}^{m} x_l^{-\gamma_{ikl}}, & (p' + 1 \leqslant i \leqslant p \,; 1 \leqslant k \leqslant J_i) \end{cases} & \text{for a reverse PGP.} \end{cases}$$

Here $x = (x_1, x_2, \dots, x_m)^{\mathrm{T}}$ represents a positive variable vector, coefficient c_{ik} represents positive real numbers and exponent γ_{ikl} $(1 \leqslant k \leqslant J_i, 0 \leqslant i \leqslant p, 1 \leqslant l \leqslant m)$ means an arbitrary real number.

Definition 1.4.2 We call

$$(D) \quad \begin{aligned} &\max d(w) \\ &\text{s.t. } w_{00} = 1 \\ &\quad \Gamma^{\mathrm{T}} w = 0 \\ &\quad w \geqslant 0 \end{aligned}$$

a dual programming of the primal PGP (P), where $\Gamma = (\gamma_{ikl})$ an exponent matrix, and

$$d(w) = \prod_{i=0}^{p}\prod_{k=1}^{J_i}\left(\frac{c_{ik}}{w_{ik}}\right)^{w_{ik}}\prod_{i=1}^{p}(w_{i0})^{w_{i0}}.$$

Here $w = (w_{01}, \ldots, w_{0J_0}, \ldots, w_{p1}, \ldots, w_{pJ_p})^{\mathrm{T}}$ denotes a $J-$dimensional variable vector $(J = J_0 + J_1 + \cdots + J_p)$, $w_{i0} = w_{i1} + \cdots + w_{iJ_i} (0 \leqslant i \leqslant p)$ is exactly the sum of each dual variable corresponding to an objective function $\tilde{g}_0(x)$ (if $i = 0$) or constraint functions $\tilde{g}_i(x)$ (if $i \neq 0$).

When

$$d(w) = \prod_{i=0}^{p'}\prod_{k=1}^{J_i}\left(\frac{c_{ik}}{w_{ik}}\right)^{w_{ik}}\prod_{i=p'+1}^{p}\prod_{k=1}^{J_i}\left(\frac{c_{ik}}{w_{ik}}\right)^{-w_{ik}}\prod_{i=1}^{p'}(w_{i0})^{w_{i0}}\prod_{i=p'+1}^{p}(w_{i0})^{-w_{i0}},$$

(D) is written as (D^-), representing a dual programming of reverse PGP (P^-).

1.4.3 Basic Properties

Theorem 1.4.2 [WY82] (basic lemma) *For any feasible solution in x and w in the primal PGP (P) and a dual program (D), we have*

$$g_0(x) \geqslant g_0(x)\prod_{i=1}^{p}\big(g_i(x)\big)^{w_{i0}} \geqslant d(w) \tag{1.7}$$

with $g_0(x) = d(w)$ iff

$$w_{ik} = \begin{cases} v_{0k}(x)/g_0(x), & (i = 0; 1 \leqslant k \leqslant J_0) \\ w_{i0}v_{ik}(x), & (i \neq 0; 1 \leqslant k \leqslant J_i) \end{cases} \tag{1.8}$$

holds; here x and w stand for optimum solutions of (P) and (D), respectively.

◊

Theorem 1.4.3 [WY82] (the first dual theorem) *Suppose that the primal PGP (P) is super-consistent, with an optimum solution x^*; then there must exist a Lagrange multiplier $\mu^* = (\mu_1^*, \mu_2^*, \ldots, \mu_p^*)^T \geqslant 0$, such that*

$$\nabla g_0(x^*) + \sum_{i=1}^{p}\mu_i^* \nabla g_i(x^*) = 0, \tag{1.9}$$

$$\mu_i^*\big(g_i(x^*) - 1\big) = 0 \quad (1 \leqslant i \leqslant p), \tag{1.10}$$

while w^* defined by

$$w_{ik}^* = \begin{cases} v_{0k}(x^*)/g_0(x^*), & (i = 0; 1 \leqslant k \leqslant J_0) \\ \mu_i^* v_{ik}(x^*)/g_0(x^*), & (i \neq 0; 1 \leqslant k \leqslant J_i) \end{cases} \tag{1.11}$$

means an optimum solution to a dual program (D), with

$$g_0(x^*) = d(w^*).$$

◊

Theorem 1.4.4 [WY82] (the second dual theorem) *If (P) is viewed as consistent and (D) maintains a feasible solution with components being positive, then an optimum solution exists in (P).* ◊

For a reverse PGP (P^-), we have the following [WY82]

Definition 1.4.3 Let us find a feasible solution x^* to (P^-) and $\mu^* = (\mu_1^*, \mu_2^*, \dots, \mu_p^*)^{\mathrm{T}} \geqslant 0$ satisfying $\mu_i^*(g_i(x^*) - 1) = 0 (1 \leqslant i \leqslant p)$, such that a Lagrange function

$$L(x, \mu) = g_0(x) + \sum_{i=1}^{p'} \mu_i(g_i(x) - 1) + \sum_{i=p'+1}^{p} \mu_i(1 - g_i(x))$$

satisfies $\nabla_x L(x^*, \mu^*) = 0$, which is called a Lagrange problem in (P^-).

Theorem 1.4.5 *Let x^* be a feasible solution of reverse PGP (P^-), written $\mathcal{I} = \{i | g_i(x^*) = 1, 1 \leqslant i \leqslant p\}$ as a subscript set of fuzzy effective constraints at x^*, $\exists : \mu^*$, which enables (x^*, μ^*) to be a solution to the Lagrange problem $\Longleftrightarrow$ to all vectors $x > 0$ satisfying*

$$\sum_{l=1}^{m} \Gamma_{il}(\log x_l - \log x_l^*) \leqslant 0 \ (i \in \mathcal{I}),$$

we have

$$g_0(x^*) \leqslant g_0(x).$$

Here

$$\Gamma_{il} = \sum_{k=1}^{J_i} \gamma_{ikl} v_{ik}(x^*) \ (i \in \mathcal{I}, 1 \leqslant l \leqslant m).$$

◊

1.5 Extension Principle and Representation Theorem

1.5.1 Extension Principle

Theorem 1.5.1 (extension principle I, L.A.Zadeh)[Zad65a,b] *Let $f : X \to Y$ be an ordinary point function, $\tilde{A} \in \mathcal{F}(X)$. Two mappings can be induced by f as follows:*

$$f : \mathcal{F}(X) \to \mathcal{F}(Y), \qquad\qquad f^{-1} : \mathcal{F}(Y) \to \mathcal{F}(X),$$
$$\tilde{A} \mapsto f(\tilde{A}) \in \mathcal{F}(Y), \qquad\qquad \tilde{B} \mapsto f^{-1}(\tilde{B}) \in \mathcal{F}(X),$$

whose membership functions are denoted by

$$f(\tilde{A})(y) = \begin{cases} \bigvee\limits_{f(x)=y} \tilde{A}(x), & f^{-1}(y) \neq \phi, \\ 0, & f^{-1}(y) = \phi, \end{cases}$$
$$f^{-1}(\tilde{B})(x) = \tilde{B}(f(x)),\ y = f(x),$$

respectively. $f(\tilde{A})$ is called an image of $\tilde{A}$ under f and $f^{-1}(\tilde{B})$ an inverse image of $\tilde{B}$. ◊

The form for an extension principle represents a cut-set.

Theorem 1.5.2 (extension principle II): *Let $f : X \to Y, \tilde{A} \in \mathcal{F}(X), \tilde{B} \in \mathcal{F}(Y)$. Then $\forall \alpha \in [0, 1]$, we have*

1^0 $f(\tilde{A})_{\underset{\cdot}{\alpha}} = f(A_{\underset{\cdot}{\alpha}})$

2^0 $f^{-1}(\tilde{B})_{\underset{\cdot}{\alpha}} = f^{-1}(B_{\underset{\cdot}{\alpha}})$

3^0 $f^{-1}(\tilde{B})_{\alpha} = f^{-1}(B_{\alpha})$

Here, $f(\tilde{A})_\alpha$ is simplified for $(f(\tilde{A}))_\alpha$. ◊

Proof Only 1^0 is proved and the others can be verified in a similar way.

$$\begin{aligned} & y \in f(\tilde{A})_{\underset{\cdot}{\alpha}} \\ \Longleftrightarrow\ & f(\tilde{A})(y) > \alpha \\ \Longleftrightarrow\ & \bigvee_{f(x)=y} \tilde{A}(x) > \alpha \\ \Longleftrightarrow\ & \exists : x \in X, \text{ corresponding with } f(x) = y, \text{ such that } \tilde{A}(x) > \alpha \\ \Longleftrightarrow\ & \exists : x \in X, \text{ corresponding with } f(x) = y, \text{ such that } x \in A_{\underset{\cdot}{\alpha}} \\ \Longleftrightarrow\ & y \in f(A_{\underset{\cdot}{\alpha}}); \end{aligned}$$

therefore

$$f(\tilde{A})_{\underset{\cdot}{\alpha}} = f(A_{\underset{\cdot}{\alpha}}).$$

□

Corollary 1.5.1 (extension principle III) *Let* $f : X \to Y, \tilde{A} \in \mathcal{F}(X), \tilde{B} \in \mathcal{F}(Y)$. *Then*

$$\begin{aligned} f(\tilde{A}) &= \bigcup_{\alpha \in [0,1)} \alpha f(A_{\underset{\cdot}{\alpha}}), \\ f^{-1}(\tilde{B}) &= \bigcup_{\alpha \in [0,1]} \alpha f^{-1}(B_{\alpha}), \\ f^{-1}(\tilde{B}) &= \bigcup_{\alpha \in [0,1)} \alpha f^{-1}(B_{\underset{\cdot}{\alpha}}). \end{aligned}$$

◊

1.5.2 Representation Theorem

Theorem 1.5.3 (representation theorem I) *Given* $X \in \mathcal{U}(X)$,

$$\bigcup_{\alpha \in [0,1]} \alpha H(\alpha)$$

denotes a fuzzy set on X, *written* $\tilde{A}$, *with* $\forall \lambda, \alpha \in [0,1]$,

$$\begin{aligned} &1^0\ A_{\alpha} = \bigcap_{\lambda < \alpha} H(\lambda), \quad \alpha \neq 0, \\ &2^0\ A_{\underset{\cdot}{\alpha}} = \bigcup_{\lambda > \alpha} H(\lambda), \quad \alpha \neq 1. \end{aligned}$$

◊

Proof According to a number-and-set-product definition, $\forall \alpha \in [0,1], H(\alpha) \in \mathcal{F}(X)$, we have that $\alpha H(\alpha) \in \mathcal{F}(X)$, such that $\bigcup\limits_{\alpha \in [0,1]} \alpha H(\alpha) \in \mathcal{F}(X)$, written

$$\tilde{A} = \bigcup_{\alpha \in [0,1]} \alpha H(\alpha);$$

hence, $\forall \alpha \in [0,1]$,

$$\begin{aligned} x \in A_{\underset{\cdot}{\alpha}} &\Longrightarrow \tilde{A}(x) > \alpha \\ &\Longrightarrow (\bigcup_{\lambda \in [0,1]} \lambda H(\lambda))(x) > \alpha \\ &\Longrightarrow \bigvee_{\lambda \in [0,1]} \lambda \wedge H(\lambda)(x) > \alpha \\ &\Longrightarrow \exists \alpha_0 \in [0,1], \text{ such that } \alpha_0 \wedge H(\alpha_0)(x) > \alpha \\ &\Longrightarrow \alpha_0 > \alpha \text{ and } H(\alpha_0)(x) = 1 \\ &\Longrightarrow x \in H(\alpha_0) \subseteq H(\alpha)(\alpha \neq 1), \end{aligned}$$

$$\begin{aligned} x \in H(\alpha) &\Longrightarrow H(\alpha)(x) = 1 \\ &\Longrightarrow \bigvee_{\lambda \in [0,1]} \lambda \wedge H(\lambda)(x) \geqslant \alpha \wedge H(\alpha)(x) = \alpha \\ &\Longrightarrow \tilde{A}(x) \geqslant \alpha \\ &\Longrightarrow x \in A_{\alpha}. \end{aligned}$$

Therefore, we have $A_{\underset{\cdot}{\alpha}} \subseteq H(\alpha) \subseteq A_\alpha$ and by decomposition theorem II, 1^0 and 2^0 hold. □

Corollary 1.5.2 (representation theorem II) *Assume* $H \in \mathcal{U}(X)$ *and write*

$$\tilde{A} = \bigcup_{\alpha \in [0,1]} \alpha H(\alpha);$$

then

1^0 $\forall \alpha \in [0,1], A_{\underset{\cdot}{\alpha}} \subseteq H(\alpha) \subseteq A_\alpha.$

2^0 $\tilde{A}(x) = \sup\{\alpha | x \in H(\alpha), \alpha \in [0,1]\}.$ ◊

The representation Theorem I provides a convenient structure of fuzzy sets and lays a foundation for both study of the theory and application of fuzzy GP.

1.6 Fuzzy Functions

A fuzzy-valued function is one of the most important conceptions in a fuzzy optimum problem. The discussion is divided into two parts below [DP80]

1.6.1 Fuzzy-valued functions from universe X to another one Y

Definition 1.6.1 Let $\mathcal{F}(X)$ and $\mathcal{F}(Y)$ represent all fuzzy sets on universe X and Y, respectively. If there exists an ordinary mapping $f : \mathcal{F}(X) \to \mathcal{F}(Y)$, then we call f a fuzzy-valued function from X to Y, writing $\tilde{f} : X \sim\!\!\to Y$.

Definition 1.6.2 Let $\tilde{f} : X \sim\!\!\to Y, \tilde{g} : Y \sim\!\!\to Z$ be two fuzzy-valued functions. Then

$$\tilde{g} \circ \tilde{f} : \mathcal{F}(X) \to \mathcal{F}(Z),$$

i.e.,

$$\forall \tilde{A} \in \mathcal{F}(X), (\tilde{g} \circ \tilde{f})(\tilde{A}) \in \mathcal{F}(Z)$$

is called a compound fuzzy function of $\tilde{f}$ and $\tilde{g}$.

Proposition 1.6.1 *If $f : X \to Y, g : Y \to Z$ denote two ordinary mappings, two fuzzy functions $\tilde{f} : X \sim\!\!\to Y$ and $\tilde{g} : Y \sim\!\!\to Z$ can be obtained by means of the extension principle. Their compound under Definition 1.6.2 consists with a fuzzy function of compound $g \circ f : X \to Z$ from ordinary mappings $f : X \to Y$ and $g : Y \to Z$ by means of the extension principle.* ◊

Proof As for $\forall \tilde{A} \in \mathcal{F}(X)$, the image

$$\tilde{f}(\tilde{A})(y) = \begin{cases} \sup\limits_{x \in f^{-1}(y)} \tilde{A}(x), & f^{-1}(y) \neq \phi, \\ 0, & f^{-1}(y) = \phi, \end{cases}$$

obtained with fuzzy functions $\tilde{f} : X \sim\!\!\to Y$, is extended from $f : X \to Y$ for $\forall y \in Y$.

$\forall \tilde{B} \in \mathcal{F}(Y)$, the image

$$\tilde{g}(\tilde{B})(z) = \begin{cases} \sup\limits_{y \in g^{-1}(z)} \tilde{B}(y), & g^{-1}(z) \neq \phi, \\ 0, & g^{-1}(z) = \phi \end{cases}$$

is achieved by a fuzzy function $\tilde{g} : Y \sim\!\!\to Z$ for $\forall z \in Z$, such that their compound denotes $\tilde{g} \circ \tilde{f} : X \sim\!\!\to Z$. From Definition 1.6.2, $\forall \tilde{A} \in \mathcal{F}(X), \forall z \in$

Z,

$$\tilde{g} \circ \tilde{f}(\tilde{A})(z) = \tilde{g}(\tilde{f}(\tilde{A}))(z)$$

$$= \begin{cases} \sup\limits_{y \in g^{-1}(z)} f(\tilde{A})(y), & g^{-1}(z) \neq \phi \\ 0, & g^{-1}(z) = \phi \end{cases}$$

$$= \begin{cases} \sup\limits_{y \in g^{-1}(z)} \begin{cases} \sup\limits_{x \in f^{-1}(y)} \tilde{A}(y), & f^{-1}(y) \neq \phi, \\ 0, & f^{-1}(y) = \phi, \end{cases} & g^{-1}(z) \neq \phi \\ 0, & g^{-1}(z) = \phi \end{cases}$$

$$= \begin{cases} \sup\limits_{y \in g^{-1}(z)} \sup\limits_{x \in f^{-1}(y)} \tilde{A}(x), & f^{-1}(y) \neq \phi, g^{-1}(z) \neq \phi \\ 0, & f^{-1}(y) = \phi \text{ or } g(z) = \phi; \end{cases}$$

therefore, $\forall z \in Z$,

$$\tilde{g} \circ \tilde{f}(\tilde{A})(z) = \begin{cases} \sup\limits_{x \in f^{-1}(g^{-1}(x))} \tilde{A}(x), & f^{-1}(g^{-1}(z)) \neq \phi \\ 0, & f^{-1}(g^{-1}(z)) = \phi \end{cases}$$

$$= \begin{cases} \sup\limits_{x \in (y \circ f)^{-1}(x)} \tilde{A}(x), & (g \circ f)^{-1}(z) \neq \phi \\ 0, & (g \circ f)^{-1}(z) = \phi. \end{cases} \tag{1.12}$$

Formula (1.12) on the right is a fuzzy function gained from the ordinary compound mapping $g \circ f$ by means of the extension principle. □

1.6.2 Fuzzy-valued Functions in Fuzzy Sets from Universe X to Universe Y

Definition 1.6.3 Let $f : X \to Y$ be an ordinary mapping. If for fuzzy sets $\tilde{A}$ and $\tilde{B}$ defined on X and Y, respectively, we have $\tilde{B}(f(x)) = \tilde{A}(x)$ for $\forall x \in X$, then we call $\tilde{f}$ a fuzzy-valued function from fuzzy set $\tilde{A}$ to $\tilde{B}$, writing

$$\tilde{f} : \tilde{A} \sim\to \tilde{B}.$$

Let $\tilde{A} \in \mathcal{F}(X), \tilde{B} \in \mathcal{F}(Y), \tilde{C} \in \mathcal{F}(Z)$, and let $f : \tilde{A} \to \tilde{B}$ and $g : \tilde{B} \to \tilde{C}$. Then a composite mapping $g \circ f : X \to Z$ is a fuzzy function from $\tilde{A}$ to $\tilde{C}$, i.e, $\tilde{g} \circ \tilde{f} : \tilde{A} \sim\to \tilde{C}$.

In fact, $\forall x \in X, (\tilde{g} \circ \tilde{f})(x) = \tilde{g}(\tilde{f}(x))$, therefore,

$$\begin{aligned}(\tilde{C} \circ (\tilde{g} \circ \tilde{f}))(x) &= \tilde{C}((\tilde{g} \circ \tilde{f})(x)) \\ &= \tilde{C}\Big(\tilde{g}(\tilde{f}(x))\Big) \\ &= (\tilde{C} \circ \tilde{g})(\tilde{f}(x)) \\ &= \tilde{B}(\tilde{f}(x)) \\ &= \tilde{A}(x).\end{aligned}$$

1.6.3 Fuzzy-valued Function Constantly Used

We introduce some fuzzy-valued functions constantly used for the sake of discussion [Cao93a][Cao94b][Caoupd][DP80].

Definition 1.6.4 $\forall x \in X$, $g(x)$ is a real bounded function defined on X, and its infimum and supremum are written as $\inf(g)$ and $\sup(g)$, respectively, such that we define

$$\tilde{M}(x) = \left[\frac{g(x) - \inf(g)}{\sup(g) - \inf(g)}\right]^n \tag{1.13}$$

calling $\tilde{M} : X \to [0,1]$ a maximal set of g, where $\tilde{M}(x) \neq 0$.

Definition 1.6.5 c_i^1, c_i^2 are left and right endpoints of an interval, therefore, for $\tilde{c}_i$ freely fixed in a closed value interval $[c_i^1, c_i^2]$, the degree of accomplishment is determined by

$$\tilde{\phi}_i(\tilde{c}_i) = \begin{cases} 0, & \text{if} \quad c_i \leqslant c_i^1 \\ \left(\dfrac{c_i - c_i^1}{c_i^2 - c_i^1}\right)^n, & \text{if} \quad c_i^1 < c_i \leqslant c_i^2 \\ 1, & \text{if} \quad c_i > c_i^2 \end{cases} \tag{1.14}$$

where n denotes a natural number.

For fuzzy constraint sets and fuzzy objective sets, we have

Definition 1.6.6 If

$$\tilde{D}_i = \{x \in \Re^m | g_i(x) \lesssim 1\} \ (1 \leqslant i \leqslant p)$$

is a fuzzy constraint set corresponding to fuzzy constraint inequations $g_i(x) \lesssim 1$, then the membership functions of $\tilde{D}_i$ are

$$\tilde{B}_i(x) = \begin{cases} 0, & \text{if} \quad g_i(x) \geqslant 1 + d_i \\ (1 - t_i/d_i)^n, & \text{if} \quad g_i(x) = 1 + t_i, 0 \leqslant t_i \leqslant d_i \\ 1, & \text{if} \quad g_i(x) \leqslant 1 \end{cases} \tag{1.15}$$

where $d_i \in \Re$ ($\Re$ is a real number set) denotes a maximum flexible index of $g_i(x)$.

Definition 1.6.7 Regard

$$\tilde{D}_0 = \{x \in \Re^m | g_0(x) \gtrsim z_0\}$$

as a fuzzy objective set and assume a membership function of $\tilde{D}_0$ as follows:

$$\tilde{B}_0(x) = \begin{cases} 0, & \text{if} \quad g_0(x) \geqslant z_0 - d_0 \\ (1 - t_0/d_0)^n, & \text{if} \quad g_0(x) = z_0 - t_0, 0 \leqslant t_0 \leqslant d_0 \\ 1, & \text{if} \quad g_0(x) \leqslant z_0 \end{cases} \tag{1.16}$$

where $d_0 \geqslant 0$ is a maximum flexible index of $g_0(x)$ and z_0 an objective value.

Definition 1.6.8 Let fuzzy sets $\tilde{D}_i (1 \leqslant i \leqslant p)$ be

$$\tilde{D}_i = \{x \in \Re^m | g_i(x) \lesssim 1\} \quad (0 \leqslant i \leqslant p')$$

and

$$\tilde{D}_i = \{x \in \Re^m | g_i(x) \gtrsim 1\} \quad (p' + 1 \leqslant i \leqslant p).$$

Then their membership functions are defined as

$$\tilde{B}_i(x) = \begin{cases} 1, & g_i(x) \leqslant 1 \\ e^{-\frac{1}{d_i}\left(g_i(x) - 1\right)}, & 1 < g_i(x) \leqslant 1 + d_i \end{cases} \tag{1.17}$$

for $1 \leqslant i \leqslant p'$, and

$$\tilde{B}_i(x) = \begin{cases} 0, & g_i(x) \leqslant 1 \\ 1 - e^{-\frac{1}{d_i}\left(g_i(x) - 1\right)}, & 1 < g_i(x) \leqslant 1 + d_i \end{cases} \tag{1.18}$$

for $p' + 1 \leqslant i \leqslant p$, where $d_i \geqslant 0$ is a maximum flexible index of i-th function $g_i(x)$.

Definition 1.6.9 Let

$$\tilde{D}_i = \{x \in \Re^m | \tilde{g}_i(x) \lesssim 1\} \quad (0 \leq i \leq p)$$

be fuzzy sets corresponding to fuzzy inequality $\tilde{g}_i(x) \lesssim 1$, and we assume the existence of membership function $\tilde{B}_i(x)$ in $\tilde{D}_i (0 \leqslant i \leqslant p)$ defined as one of (1.15)–(1.18). Then definition symbol" $\lesssim$ " is a flexible version of $\leqslant$ at a 'certain degree' [Ver84][LL01], or approximately less than or equal to.

We introduce the possibility grade of dominance of $\tilde{1}$ over $\tilde{g}_i(x)$, a concept introduced by Dubois and Prade in 1980 which represents the fuzzy extension for $g_i(x) \leqslant 1$.

Definition 1.6.10 The degree of possibility of $\tilde{g}_i(x) \leqslant \tilde{1}$ is defined as

$$v(\tilde{g}_i(x) \leqslant \tilde{1}) = \sup_{x,y:x \geqslant y} \min\left(\mu_{\tilde{1}}(x), \mu_{\tilde{g}_i(x)}(y)\right).$$

This formula is an extension of the inequality $x \geqslant y$ according to the extension principle. When pair (x, y) exists, such that $x \geqslant y$ and $\mu_{\tilde{1}}(x) = \mu_{\tilde{g}(x)}(y) = 1$, then $v(\tilde{g}_i(x) \leqslant \tilde{1}) = 1$.

When $\tilde{g}_i(x)$ and $\tilde{1}$ are convex fuzzy numbers, we have

$$v(\tilde{g}_i(x) \leqslant \tilde{1}) = 1, \text{ iff } \tilde{g}_i(\bar{x}') \leqslant 1,$$
$$v(\tilde{g}_i(x) \leqslant \tilde{1}) = \text{hgt}\,(\tilde{g}_i(x) \cap \tilde{1}) = \mu_{\tilde{g}_i(x)}(d),$$

where *d* is the ordinate of the highest intersection point between $\mu_{\tilde{1}}$ and $\mu_{\tilde{g}_i(x)}$.

Chapter 2

FUZZY POSYNOMIAL GEOMETRIC PROGRAMMING

In economic management, we often meet with a problem as follows:

Example 2.1 Suppose that we manufacture a case for transporting cotton. The case is Vm^3 in volume and has a bottom, but no top. The bottom and two sides are made from Cm^2 of a special flexible material with negligible cost. The material for the other two sides cost more than A yuan $/m^2$ (yuan means Renmingbi) and it costs about k yuan to transport the case. What is the cost at least to ship one case of cotton?

Such a problem can be posed for solution by a Geometric Program. Since a classical GP cannot give a good account of the problem, or obtain a practical solution to it, in 1987 we advanced a fuzzy GP, which is effective to solve the problem.

Now, we introduce some basic knowledge before introducing a fuzzy PGP and its algorithms.

2.1 Basic Knowledge and Properties

Definition 2.1.1 [Cao93a] Let $\tilde{B}_i(0 \leqslant i \leqslant p)$ be a fuzzy-valued function on $X \subset \Re^m$. If there exists a fuzzy function $\tilde{\phi}_i = \tilde{c}_{ik} \circ \tilde{\gamma}_{ikl}(1 \leqslant k \leqslant J_i, 0 \leqslant i \leqslant p, 1 \leqslant l \leqslant m)$ on X, such that $c_{ik} \geqslant \tilde{c}_{ik}^{-1}(\beta), \gamma_{ikl} \geqslant \tilde{\gamma}_{ikl}^{-1}(\beta)$, then we define a fuzzy compound function as $\tilde{A}_i = \tilde{B}_i \circ \tilde{\phi}_i(0 \leqslant i \leqslant p)$ on X, and their crisp form as $\tilde{A}_i(\bar{x}')$, where " $\circ$ " denotes a compound operation, $\bar{x}' = (x, \beta)$.

For instance, we define $\tilde{A}_i(\bar{x}')$ as

$$\tilde{A}_i(\bar{x}') = (M' - \sum_{k=1}^{J_i} \tilde{c}_{ik}^{-1}(\beta) \prod_{l=1}^{m} x_l^{\tilde{\gamma}_{ikl}^{-1}(\beta)})/(M' - N),$$

where N and M' are the infimum and supremum of

$$g_i(\bar{x}') = \sum_{k=1}^{J_i} \tilde{c}_{ik}^{-1}(\beta) \prod_{l=1}^{m} x_l^{\tilde{\gamma}_{ikl}^{-1}(\beta)},$$

respectively.

Definition 2.1.2 Let $\tilde{A}_i(0 \leqslant i \leqslant p)$ be a convex fuzzy set, $\tilde{A}_i(\bar{x}') = \tilde{B}_i(g_i(\bar{x}'))$ be fuzzy-valued sets defined on X and for $\alpha, \beta \in [0,1]$,

$$A_{i\alpha} = \{x | \tilde{A}_i(\bar{x}') \leqslant \alpha\}$$
$$(\text{ resp. } A'_{i\alpha} = \{x | \tilde{A}_i(\bar{x}') \geqslant \alpha\})$$

represent a convex set. Then $g_i(x)$ is called a fuzzy-valued convex (resp. concave) function with respect to $\tilde{A}_i$.

Definition 2.1.3 If we define $\tilde{A}_i(x)(0 \leqslant i \leqslant p)$ in Definition 2.1.2 as a strongly quasi-convex (resp. quasi-concave) function of x, $g_i(x)$ stands for a strongly fuzzy-valued convex (resp. concave) function about $\tilde{A}_i$.

Definition 2.1.4 Suppose that $\tilde{A}_i(0 \leqslant i \leqslant p)$ is a strictly quasi-convex (resp. quasi-concave) function of x, with $\tilde{A}_i(\bar{x}') = \tilde{B}_i(g_i(\bar{x}'))$; then for all of $\alpha, \beta \in [0,1]$,

$$A_{i\dot{\alpha}} = \{x | \tilde{A}_i(\bar{x}') < \alpha\}$$
$$(\text{resp. } A'_{i\dot{\alpha}} = \{x | \tilde{A}_i(\bar{x}') > \alpha\})$$

represents a convex set and $g_i(x)$ is called a strictly fuzzy-valued convex (resp. concave) function with respect to $\tilde{A}_i$.

If $-g_i(x)$ means a convex (resp. strictly quasi-convex, or strongly quasi-convex) function, then $g_i(x)$ is a concave (resp. strictly quasi-concave, or strongly quasi-concave) function.

Definition 2.1.5 If $\tilde{A}_i(x)(0 \leqslant i \leqslant p)$ denotes a strongly quasi-convex (resp. strongly quasi-concave) function, then $\forall x^1, x^2 \in X, x^1 \neq x^2$ and $\lambda, \beta \in [0,1]$,

$$\tilde{A}_i(\lambda x^1 + (1-\lambda)x^2, \beta) < \max\{\tilde{A}_i(x^1,\beta), \tilde{A}_i(x^2,\beta)\}$$
$$(\text{resp. } \tilde{A}_i(\lambda x^1 + (1-\lambda)x^2, \beta) > \min\{\tilde{A}_i(x^1,\beta), \tilde{A}_i(x^2,\beta)\}).$$

Theorem 2.1.1 *$\tilde{g}_i(x)(0 \leqslant i \leqslant p)$ is a strictly fuzzy convex (resp. concave) function with respect to a strictly nondecreasing function $\tilde{A}_i$ iff $\tilde{A}_i(x) = \tilde{B}_i(g_i(x))$ is a strictly quasi-convex (resp. quasi-concave) one valued on $[0,1]$.* ◊

Proof We prove only the fuzzy concave case with the others being similar. "$\Longrightarrow$" From the condition, we know $g_i(\bar{x}')$ is a strictly fuzzy concave function of $\tilde{A}_i$, and for all $x^1, x^2 \in X, x^1 \neq x^2, g_i(x^1,\beta) \neq g_i(x^2,\beta)$ and $\lambda, \beta \in [0,1]$,

$$\begin{aligned}\tilde{B}_i(g_i(\lambda x^1 + (1-\lambda)x^2, \beta) &> \tilde{B}_i(\lambda g_i(x^1,\beta) + (1-\lambda) g_i(x^2,\beta)) \\ &> \min\{(\tilde{B}_i(g_i(x^1,\beta)), \tilde{B}_i(g_i(x^2,\beta))\},\end{aligned}$$

i.e.

$$\tilde{A}_i(\lambda x^1 + (1-\lambda)x^2, \beta) > \min\{\tilde{A}_i(x^1, \beta), \tilde{A}_i(x^2, \beta)\}.$$

Therefore, $\tilde{A}_i(\bar{x}') = \tilde{B}_i\big(g_i(\bar{x}')\big)$ is a strictly quasi-concave function valued on $[0,1]$, which is equivalent to the validity of this theorem.

"$\Longleftarrow$" For all $\lambda, \alpha, \beta \in [0,1]$, if $x^1, x^2 \in A_{i\alpha}$, then

$$\tilde{A}_i(x^1, \beta) = \tilde{B}_i\big(g_i(x^1, \beta)\big) > \alpha,$$
$$\tilde{A}_i(x^2, \beta) = \tilde{B}_i\big(g_i(x^2, \beta)\big) > \alpha,$$

hence

$$\tilde{B}_i\big(g_i(\lambda x^1 + (1-\lambda)x^2, \beta)\big) > \alpha$$

so $g_i\big(\lambda x^1 + (1-\lambda)x^2, \beta\big) \in A_{i\alpha}$, i.e., $\tilde{g}_i(x)$ represents a strictly fuzzy convex function with respect to $\tilde{A}_i$. □

Theorem 2.1.2 *$\tilde{g}_i(x)(0 \leqslant i \leqslant p)$ is a fuzzy (resp. strongly fuzzy) convex (resp. concave) function about a strictly nondecreasing function $\tilde{A}_i$ iff $\tilde{A}_i(x) = \tilde{B}_i\big(g_i(\bar{x}')\big)$ is a quasi-convex (resp. quasi-concave) or strongly quasi-convex (resp. strongly quasi-concave) function valued on [0,1].* ◊

Proof Due to the quasi-convex (resp. quasi-concave) or strong quasi-convex (resp. quasi-concave) properties of the chosen function, this theorem proves true analogous to the proof of Theorem 2.1.1. □

Theorem 2.1.3 *Assume that $\tilde{g}_i(x)(0 \leqslant i \leqslant p)$ is a quasi-convex (resp. strongly quasi-convex) function on $X \subset \Re^m$; then*

(i) If $\tilde{B}_i(\tilde{g}_i)$ is a nondecreasing function with respect to $\tilde{g}_i$, then $\tilde{g}_i(x)$ is a fuzzy (resp. strongly fuzzy) convex function about $\tilde{A}_i$.

(ii) If $\tilde{B}_i(\tilde{g}_i)$ is a nonincreasing function with respect to $\tilde{g}_i$, then $\tilde{g}_i(x)$ is a fuzzy (resp. strongly fuzzy) concave function about $\tilde{A}_i$. ◊

Proof We prove only a quasi-convex state in (i).

From our knowledge of the condition, for all $x^1, x^2 \in X, \lambda, \beta \in [0,1]$,

$$g_i\big(\lambda x^1 + (1-\lambda)x^2, \beta\big) \leqslant \max\{g_i(x^1, \beta), g_i(x^2, \beta)\}.$$

From the nondecrease of $\tilde{A}_i$, we can get that of $\tilde{A}_i(\bar{x}')$, so

$$\begin{aligned} \tilde{B}_i\big(g_i(\lambda x^1 + (1-\lambda)x^2, \beta)\big) &\leqslant \tilde{B}_i\{\max\big(g_i(x^1, \beta), g_i(x^2, \beta)\big)\} \\ &= \max\{\tilde{B}_i\big(g_i(x^1, \beta)\big), \tilde{B}_i\big(g_i(x^2, \beta)\big)\}, \end{aligned}$$

i.e.

$$\tilde{A}_i\big(\lambda x^1 + (1-\lambda)x^2)\big) \leqslant \max\{\tilde{A}_i(x^1), \tilde{A}_i(x^2)\}.$$

Therefore, $\tilde{A}_i(x)(0 \leqslant i \leqslant p)$ represents a quasi-convex function about x.

From the knowledge of Theorem 2.1.1, $\tilde{g}_i(x)$ is a fuzzy convex function about $\tilde{A}_i$. □

Theorem 2.1.4 *Suppose that $\tilde{g}_i(x)(0 \leqslant i \leqslant p)$ stands for a quasi-concave (resp. strongly quasi-concave) function on $X \subset \Re^m$; then*

(i) If $\tilde{B}_i(\tilde{g}_i)$ is a nondecreasing function with respect to $\tilde{g}_i$, then $\tilde{g}_i(x)$ is a fuzzy (resp. strongly fuzzy) quasi-concave function about $\tilde{A}_i$.

(ii) If $\tilde{B}_i(\tilde{g}_i)$ is a nonincreasing function with respect to $\tilde{g}_i$, then $\tilde{g}_i(x)$ is a fuzzy (resp. strongly fuzzy) quasi-convex function about $\tilde{A}_i$. ◊

Note 2.1: A fuzzy convex (resp. concave) or strongly fuzzy quasi-convex (resp. quasi-concave) function is fuzzy quasi-convex (resp. quasi-concave) and also is strictly fuzzy quasi-convex (resp. quasi-concave).

Definition 2.1.6 Let $\tilde{g}_i(x)(0 \leqslant i \leqslant p)$ be a fuzzy-valued function about $\tilde{A}_i$. As for all $\beta \in [0,1]$, if $g_i(\bar{x}')$ has an nth partial derivative at an ordinary point $\bar{x}' = (x_1, x_2, \ldots, x_m; \beta)$, then $\tilde{g}_i(x)$ contains an n$th(n > 0)$ one at point $x = (x_1, x_2, \ldots, x_m)$, writing $\tilde{g}_i^{(n)}(x)$, i.e.

$$\tilde{g}_i^{(n)}(x) \triangleq \left[\tilde{B}_i(g_i(\bar{x}'))\right]_x^{(n)}.$$

Again, since

$$\tilde{B}_i(g_i(\bar{x}')) \geqslant \alpha \Longleftrightarrow g_i(\bar{x}')/\tilde{B}_i^{-1}(\alpha) \leqslant 1,$$

writing $G_i(\overline{x}) = g_i(\bar{x}')/\tilde{B}_i^{-1}(\alpha)$, then

$$\tilde{g}_i^{(n)}(x) \triangleq (G_i)_x^{(n)}(\overline{x})$$

where $\overline{x} = (x; \alpha, \beta)$.

2.2 Fuzzy PGP and Its Properties

In reality, constraint conditions need softening in classical GP (P); coefficients, exponents, even variables in it are regarded as fuzzy parameters or fuzzy variables in objective functions as well as in constraint conditions, in order to deal with a large number of problems, which involves extending GP (P). In this chapter, we consider problem of PGP.

Example 2.2 Consider Example 2.1 at the beginning of this chapter.

Now, suppose that the length, width and height are x_1, x_2 and x_3, respectively, which is induced to find a fuzzy GP:

$$\begin{aligned} &\widetilde{\min}\ g_0(x) \\ &\text{s.t. } g_i(x) \lesssim C \\ &\qquad x_1, x_2, x_3 > 0 \end{aligned}$$

where $g_i(x) = \sum\limits_{i=1}^{J_i} c_{ik} \prod\limits_{l=1}^{3} x_l^{\gamma_{ikl}} (i = 0, 1, 2)$.

Definition 2.2.1 We consider mathematical programming for variables $x_l (1 \leqslant l \leqslant m)$ to be a positive, objective function $\tilde{g}_0(x)$ and constraint conditions $\tilde{g}_i(x)(1 \leqslant i \leqslant p)$ to be a fuzzy posynomial

$$\begin{aligned} (\tilde{P})\quad &\widetilde{\min}\ \tilde{g}_0(x) \\ &\text{s.t. } \tilde{g}_i(x) \lesssim 1 (1 \leqslant i \leqslant p) \\ &\qquad x > 0. \end{aligned} \tag{2.1}$$

We call it a fuzzy PGP, where $x = (x_1, x_2, \ldots, x_m)^{\mathrm{T}}$ represents a variable vector, and $\tilde{g}_i(x) = \sum\limits_{k=1}^{J_i} \tilde{v}_{ik}(x) = \sum\limits_{k=1}^{J_i} \tilde{c}_{ik} \prod\limits_{l=1}^{m} x_l^{\tilde{\gamma}_{ikl}} (0 \leqslant i \leqslant p)$ is a posynomial in x, its coefficients $\tilde{c}_{ik}$ and exponents $\tilde{\gamma}_{ikl}$ may be freely fixed in closed value intervals $[c_{ik}^1, c_{ik}^2]$ and $[\gamma_{ikl}^1, \gamma_{ikl}^2](1 \leqslant k \leqslant J_i, 0 \leqslant i \leqslant p, 1 \leqslant l \leqslant m)$, respectively, with $c_{ik}^1, c_{ik}^2, \gamma_{ikl}^1, \gamma_{ikl}^2$ being the left and right endpoints in the intervals and with all of them being real numbers, stipulating $0 < c_{ik}^1 \leqslant c_{ik}^2, \gamma_{ikl}^1 \leqslant r_{ikl}^2$.

Symbol $\lesssim$ represents a flexible version of $\leqslant$ at a 'certain degree' [Ver84] [LL01], say approximately less than or equal to. Let fuzzy sets $\tilde{D}_i = \{x \in \Re^m | \tilde{g}_i(x) \lesssim 1\}\quad (0 \leq i \leq p)$ be fuzzy constraint sets corresponding to fuzzy constraints $\tilde{g}_i(x) \lesssim 1$. We assume the existence of membership function in $\tilde{D}_i(0 \leqslant i \leqslant p)$ defined as

$$\tilde{B}_i(x) = \begin{cases} 1, & \tilde{g}_i(x) \leq 1, \\ e^{-\frac{1}{d_i}\left(\tilde{g}_i(x)-1\right)}, & 1 < \tilde{g}_i(x) \leq 1 + d_i (1 \leq i \leq p), \end{cases}$$

or

$$\tilde{B}_i(x)=\begin{cases}0, & \text{if } \ g_i(x)\geqslant 1+d_i\\ \left(1-t_i/d_i\right)^n, & \text{if } \ g_i(x)=1+t_i, 0\leqslant t_i\leqslant d_i\\ 1, & \text{if } \ g_i(x)\leqslant 1\end{cases}$$

where $d_i \geq 0 (0 \leq i \leq p)$ denotes a maximum flexible index of $\tilde{g}_i(x)$.

We define the fuzzy objective as a fuzzy set of the objective set. This intuitive definition agrees with the Fuzzification Principle of [Gog67], and implies that, given (2.1), with $g_0 : \Re^n \to \Re$, the fuzzy optimization is shown by the symbol $\sim$. Regard $\tilde{D}_0 = \{x \in \Re^m | \tilde{g}_0(x) \gtrsim z_0\}$ as a fuzzy objective set and we assume the existence of a membership function about $\tilde{D}_0$ as follows

$$\tilde{B}_0(x)=\begin{cases}0, & \tilde{g}_i(x)\leq z_0-d_i,\\ 1-e^{-\frac{1}{d_i}\left(\tilde{g}_i(x)-z_0\right)}, & \tilde{g}_i(x)>z_0,\end{cases}$$

or

$$\tilde{B}_0(x)=\begin{cases}0, & \text{if } \ g_0(x)\geqslant z_0-d_0\\ \left(1-t_0/d_0\right)^n, & \text{if } \ g_0(x)=z_0-t_0, 0\leqslant t_0\leqslant d_0\\ 1, & \text{if } \ g_0(x)\leqslant z_0\end{cases}$$

where $d_0 \geq 0$ denotes the maximum flexible index of $\tilde{g}_0(x)$ and z_0 an objective value. $\widetilde{\min}$ as an extended min operation, increasing in $\Re$. $\widetilde{\min}\tilde{g}_0(x)$ denotes the minimum of the fuzzy objective function $\tilde{g}_0(x)$, which is a fuzzy number and is obtained by considering the nonfuzzy minimum of the membership function at x.

In the present book, 0 and 1 are regarded as special fuzzy numbers.

We rewrite the form of (2.1) for the sake of discussion.

Definition 2.2.2 Let $\tilde{F}(\bar{x}') = \wedge_{1\leqslant i\leqslant p}\tilde{A}_i^*(\bar{x}')$, where $\tilde{A}_i^*(\bar{x}') = \tilde{B}_i(g_i^*(\bar{x}'))$ be a constraint condition defined on X which is a fuzzy feasible solution set for $\tilde{A}_i^*$. And let $\tilde{A}_0^*(\bar{x}') = \tilde{B}_0\left(g_0^*(\bar{x}')\right)$ be an objective function defined on X, which is a fuzzy optimum value set for $\tilde{A}_0^*$, with $\tilde{Y}^* = \tilde{A}_0^* \cap \tilde{F}$. Then (2.1) is written as

$$\max_{x\in\Re^+}\{\tilde{Y}^*(\bar{x}')=\min_{0\leqslant i\leqslant p}\tilde{A}_i^*(\bar{x}')\}. \tag{2.2}$$

(2.2) is said to be a fuzzy PGP having parameter variables with respect to $\tilde{Y}^*$ of $g_0^*(\bar{x}')$, where

$$g_0^*(\bar{x}')=g_i(\bar{x}')-b_i', \quad b_i'=\begin{cases}z_0, & \text{for } i=0\\ 1, & \text{for } 1\leqslant i\leqslant p.\end{cases}$$

Definition 2.2.3 Let $X \subset \Re^m$ be a convex fuzzy set. If $g_0^*(\bar{x}')$ and $g_i^*(\bar{x}')$ denote fuzzy convex (resp. strongly fuzzy convex) functions with respect to $\tilde{A}_0^*$ and $\tilde{A}_i^*(1 \leqslant i \leqslant p)$, respectively, then (2.2) represents a fuzzy convex (resp. a strongly fuzzy convex) programming problem with respect to $g_0^*(\bar{x}')$.

Definition 2.2.4 Let $\tilde{g}_i^{*(n)}(x) = \left(\tilde{g}_i(x) - b_i'\right)^{(n)} (0 \leqslant i \leqslant p)$ be a fuzzy-valued derived function. If $G_i^{*(n)}(\overline{x}) = \left(G_i(\overline{x}) - b_i'\right)^{(n)}$ continues on X for all $\alpha, \beta \in [0,1]$, then $\tilde{g}_i^{*(n)}(x)$ is called continuation on X, where $\bar{x} = (x, \alpha, \beta)$.

Theorem 2.2.1 *Let a fuzzy-valued function $\tilde{B}_i$ and membership functions $\tilde{c}_{ik}$, $\tilde{\gamma}_{ikl}(1 \leqslant k \leqslant J_i, 0 \leqslant i \leqslant p, 1 \leqslant l \leqslant m)$ be continuous and strictly monotone (CSM) functions. Then (2.1) is equivalent to*

$$\begin{aligned} &\max \alpha = \min(-\alpha), \\ &\text{s.t. } g_i(\overline{x}) \leqslant 1, \alpha, \beta \in [0,1], \\ &\qquad x_l > 0, (0 \leq i \leq p, 1 \leq l \leq m). \end{aligned} \tag{2.3}$$

◊

Proof Suppose $\tilde{B}_i$ as in (1.15) and (1.16) (or (1.17) and (1.18)), with $\tilde{c}_{ik}$ and $\tilde{\gamma}_{ikl}$ represented as in (1.14). From the assumption, we know there exist $\tilde{B}_i^{-1}, \tilde{c}_{ik}^{-1}$ and $\tilde{\gamma}_{ikl}^{-1}$; then (2.1) is equivalent to

$$\begin{aligned} &\max \quad \alpha \\ &\text{s.t. } \tilde{B}_0\Big(\sum_{k=1}^{J_0} c_{0k} \prod_{l=1}^{m} x_l^{\gamma_{0kl}}\Big) \geqslant \alpha, \\ &\qquad \tilde{\phi}_{0k}(c_{0k}) \geqslant \beta, \tilde{\phi}_{0kl}(\gamma_{0kl}) \geqslant \beta, \\ &\qquad \tilde{B}_i\Big(\sum_{k=1}^{J_i} c_{ik} \prod_{l=1}^{m} x_l^{\gamma_{ikl}}\Big) \geqslant \alpha, (1 \leqslant i \leqslant p), \\ &\qquad \tilde{\phi}_{ik}(c_{ik}) \geqslant \beta, \tilde{\phi}_{ikl}(\gamma_{ikl}) \geqslant \beta, (0 \leqslant i \leqslant p), \\ &\qquad \alpha, \beta \in [0,1], c_{ik}, \gamma_{ikl} \in \Re, \\ &\qquad x > 0, \end{aligned}$$

where $\beta = 1 - \alpha$.

For all of $c, \gamma \in \Re^m$, $c = (c_{i1}, c_{i2}, \ldots, c_{iJ_i}), \gamma = (\gamma_{i1l}, \gamma_{i2l}, \ldots, \gamma_{iJ_il})$, $\tilde{c} = \inf_k \tilde{c}_{ik}, \tilde{\gamma} = \inf_k \tilde{\gamma}_{ikl}, \tilde{c}(c) \geqslant \beta, \tilde{\gamma}(\gamma) \geqslant \beta$ iff $\tilde{c}_{ik}(c_{ik}) \geqslant \beta, \tilde{\gamma}_{ikl}(\gamma_{ikl}) \geqslant \beta$, which means

$$c_{ik} \geqslant \tilde{c}_{ik}^{-1}(\beta), \gamma_{ikl} \geqslant \tilde{\gamma}_{ikl}^{-1}(\beta)(1 \leqslant k \leqslant J_i, 0 \leqslant i \leqslant p, 1 \leqslant l \leqslant m).$$

The problem is equivalent to

$$\begin{aligned} &\max \alpha = \min(-\alpha), \\ &\text{s.t. } \sum_{k=1}^{J_0} c_{0k} \prod_{l=1}^{m} x_l^{\gamma_{0kl}} \leqslant \tilde{B}_0^{-1}(\alpha), \\ &\qquad \sum_{k=1}^{J_i} c_{ik} \prod_{l=1}^{m} x_l^{\gamma_{ikl}} \leqslant \tilde{B}_i^{-1}(\alpha)(1 \leqslant i \leqslant p), \\ &\qquad c_{ik} = \tilde{c}_{ik}^{-1}(\beta), \gamma_{ikl} = \tilde{\gamma}_{ikl}^{-1}(\beta) \in \Re\ (0 \leqslant i \leqslant p), \alpha, \beta \in [0,1], \\ &\qquad x > 0, \end{aligned} \tag{2.4}$$

and also to problem (2.3). □

Lemma 2.2.1 *Suppose $\tilde{g}_i^*(x)(0 \leqslant i \leqslant p)$ has a 2nd fuzzy derivative on an open interval $X \subset \Re^m$. Then $\tilde{g}_i^*(x)$ is a fuzzy convex (resp. strictly fuzzy convex) function on $x \in X$ iff, for each $x \in X$, we have*

$$\frac{d^2\tilde{g}_i^*(x)}{dx^2} \geqslant 0 \quad (\text{resp. } \frac{d^2\tilde{g}_i^*(x)}{dx^2} > 0).$$

◊

Proof From the condition of the lemma, we know

$$\left(g_i^*(x+\Delta x,\beta) - g_i^*(x,\alpha,\beta)\right)/\Delta x \geqslant 0 (> 0)$$

and

$$\frac{dg_i^*(x'',\beta)}{dx} - \frac{dg_i(x',\beta)}{dx} = \frac{d^2g_i^*(\xi,\beta)}{dx^2},$$

$$\frac{d^2g_i^*(\xi,\beta)}{dx^2} \geqslant 0(> 0)(x' \leqslant \xi \leqslant x'', x' < x'').$$

From the convexity of an ordinary parameter variable function, we know $g_i^*(x,\alpha,\beta)$ is convex (resp. strictly convex), such that $\tilde{g}_i^*(x)$ is fuzzy convex on X.

Similarly, the condition of fuzzy (resp. strictly fuzzy) concavity can be obtained as follows:

$$\frac{d^2\tilde{g}_i^*(x)}{dx^2} \leqslant 0 \quad (\text{resp. } \frac{d^2\tilde{g}_i^*(x)}{dx^2} > 0).$$

□

Proposition 2.2.1 *If a constraint set $\tilde{A}_i^*(x)(0 \leqslant i \leqslant p)$ is strictly convex fuzzy on $X \subset \Re^m$, i.e.,*

$$\tilde{A}_i^*(\lambda x + (1-\lambda)x^2) > \tilde{A}_i^*(x^1) \wedge \tilde{A}_i^*(x^2)$$

for all $\lambda \in (0,1), x^1, x^2 \in X \subset \Re^m, x^1 \neq x^2$, then

$$G_0(\alpha) = sup_{x\in F_\alpha}\tilde{Y}^*(x)$$

is called continuation, where $F_\alpha(\alpha \in [0,1])$ denotes an α-cut set in constraint sets $\tilde{F}$. ◊

Theorem 2.2.2 *As for CSM functions $\tilde{A}_i^*(0 \leqslant i \leqslant p)$, a fuzzy PGP (2.1) can be turned into a fuzzy convex GP.* ◊

Proof Let $x_l = e^{z_l}(1 \leqslant l \leqslant m)$. Then

$$\begin{aligned} g_i^*(\overline{x}) &= \tilde{B}_i\big(g_i^*(\overline{x}')\big) = \sum_{k=1}^{J_i} \tilde{c}_{ik}^{-1}(\beta)/\tilde{B}_i^{-1}(\alpha)\prod_{l=1}^{m} x_l^{\gamma_{ikl}^{-1}(\beta)} \\ &= \sum_{k=1}^{J_i} \tilde{c}_{ik}^{-1}(\beta)/\tilde{B}_i^{-1}(\alpha)\exp\{\sum_{l=1}^{m}\tilde{\gamma}_{ikl}^{-1}(\beta)z_l\} \\ &= G_i^*(\overline{z}) \quad (0 \leqslant i \leqslant p), \end{aligned}$$

where $\overline{z} = (z, \alpha, \beta)$.

Because

$$\frac{d^2}{dz^2}\left(\exp\{\sum_{l=1}^{m} \tilde{\gamma}_{ikl}^{-1}(\beta) z_l\}\right) > 0,$$

$\exp\{\sum\limits_{l=1}^{m} \tilde{\gamma}_{ikl}^{-1}(\beta) z_l\}$ denotes a strictly convex function about z_l, such that for $\lambda, \beta \in [0, 1]$, we have

$$\begin{aligned}&\exp\{\sum_{l=1}^{m} \tilde{\gamma}_{ikl}^{-1}(\beta)[(1-\lambda)\bar{u}_l + \lambda\bar{v}_l]\}\\ &\leqslant (1-\lambda)\exp\{\sum_{l=1}^{m} \tilde{\gamma}_{ikl}^{-1}(\beta)\bar{u}_l\} + \lambda\exp\{\sum_{l=1}^{m} \tilde{\gamma}_{ikl}^{-1}(\beta)\bar{v}_l\}\end{aligned}$$

where $\bar{u}_l = u_l(\beta), \bar{v}_l = v_l(\beta)$.

Therefore, $\exp\{\sum\limits_{l=1}^{m} \tilde{\gamma}_{ikl}^{-1}(\beta) z_l\}$ is a convex function of z.

Again, because of $(0, c_i^2) \subset [c_i^1, c_i^2]$, and by applying (1.14), (1.15) in Section 1.6.3 in Chapter 1, we know $\tilde{c}_{ik}^{-1}(\beta) = c_i^1 + (s_i - b_i)\beta^{1/l}$, while

$$\tilde{B}_i^{-1}(\alpha) = kd_i - d_i\beta^{1/l} > 0 (k > 0),$$

hence

$$c_{ik}(\alpha, \beta) = \tilde{c}_{ik}^{-1}(\beta)/\tilde{B}_i^{-1}(\alpha) > 0$$

Therefore, $G_i^*(\overline{z})$ denotes a convex function of z. From the arbitrary choice of α, β on [0,1], we know $\tilde{G}_i^*(z)$ denotes a fuzzy convex programming with respect to z.

Again, from the Definition 2.2.3, we know (2.1) can be changed into a fuzzy convex programming. □

Theorem 2.2.3 *Suppose that $\tilde{A}_i^*(0 \leqslant i \leqslant p)$ is a CSM function, then a fuzzy local minimum solution to fuzzy PGP (2.1) denotes its fuzzy global minimum one.* ◊

Proof From Theorem 2.2.2, we know (2.1) can be deformed into a fuzzy convex GP

$$\min_{z \in \Re}\{\tilde{G}^*(z) = \tilde{G}_0^*(z) \wedge \left(\wedge_{i=1}^{p} \tilde{G}_i^*(z)\right)\} \tag{2.5}$$

which is equivalent to

$$\min_{z \in \Re} {}_{\alpha,\beta \in [0,1]} G^*(\overline{z}) = (-\alpha) \wedge \left(\wedge_{i=0}^{p} G_i^*(\overline{z})\right). \tag{2.6}$$

Therefore, for all $\alpha, \beta \in [0,1], G^*(\bar{z})$ is a convex set. And from Definition 2.1.1, we know it is a fuzzy-valued convex function, such that the local minimum solution to $G^*(\bar{z})$ is its global minimum one, therefore, the theorem holds. □

Theorem 2.2.4 *Let $\tilde{A}_i^*(0 \leqslant i \leqslant p)$ be a CSM function. Then the strict fuzzy local minimum solution to (2.1) denotes its strict fuzzy global minimum one.* ◊

Proof Since $\tilde{G}^*$ stands for a quasi-convex function (see Theorem 2.1.2), it denotes a fuzzy convex one for $\tilde{A}_i^*(0 \leqslant i \leqslant p)$. Similarly, we can confirm that the strict fuzzy local minimum solution to $\tilde{G}^*$ represents its strict fuzzy global minimum one as in Theorem 2.2.3. □

Similarly, we can prove the following.

Theorem 2.2.5 *Let $\tilde{A}_i^*(0 \leqslant i \leqslant p)$ be a CSM function. Then the fuzzy local minimum solution to (2.1) means its fuzzy global minimum one.* ◊

Theorem 2.2.6 *Let (2.1) be a strongly fuzzy convex GP. Then any fuzzy local minimum solution to it means its unique fuzzy global minimum one.* ◊

Proof Because $\tilde{A}_0^*(\bar{x}')$ and $\tilde{F}(\bar{x}')$ denote strongly fuzzy convex functions and $\tilde{Y}^* = \tilde{A}_0^* \cap \tilde{F}$, for all of $x^1, x^2 \in X, x^1 \neq x^2$ and $\lambda \in (0,1)$,

$$\begin{aligned}\tilde{Y}^*(\ & \lambda x^1 + (1-\lambda)x^2, \beta) \\ &= \tilde{A}_0^*(\lambda x^1 + (1-\lambda)x^2, \beta) \wedge \tilde{F}(\lambda x^1 + (1-\lambda)x^2, \beta) \\ &> \big(\tilde{A}_0^*(x^1,\beta) \wedge \tilde{A}_0^*(x^2,\beta)\big) \wedge \big(\tilde{F}(x^1,\beta) \wedge \tilde{F}(x^2,\beta)\big) \\ &= \tilde{Y}^*(x^1,\beta) \wedge \tilde{Y}^*(x^2,\beta),\end{aligned}$$

i.e. for all $\beta \in [0,1], \tilde{Y}^*$ is a strongly fuzzy convex set. Therefore, $\tilde{Y}^*(x,\beta)$ denotes a strongly fuzzy convex function that is strongly quasi-convex.

According to Theorem 6.4 in the Reference [Avr76], $\tilde{Y}^*(\bar{x}')$ is made to reach a minimum on X at most one point $\bar{x}'$, so is $\tilde{Y}^*(x)$ on X at most one point x. Therefore x is a unique global minumun solution to (2.1). □

Theorem 2.2.7 *Let $\tilde{c}_{ik} > 0$, for CSM function $\tilde{A}_i^*(1 \leqslant k \leqslant J_i, 0 \leqslant i \leqslant p)$. Then*

$$\log \tilde{G}_i(z) = \log \sum_{k=1}^{J_i} \tilde{c}_{ik} \exp\{\sum_{l=1}^{m} \tilde{\gamma}_{ikl} z_l\}$$

with $\log \tilde{G}_i(z)$ *being a fuzzy convex function about $\tilde{A}_i^*$.* ◊

Proof Since $\tilde{A}_i^*(\bar{z}) = \log G_i(\bar{z})$, with

$$\begin{aligned}G_i(\bar{z}) &= G_i(z,\beta)/\tilde{B}_i^{-1}(\alpha) \\ &= \log \sum_{k=1}^{J_i} \tilde{c}_{ik}^{-1}(\beta) \exp\{\sum_{l=1}^{m} \tilde{\gamma}_{ikl}^{-1}(\beta) z_l\}/\tilde{B}_i^{-1}(\alpha)\end{aligned}$$

here $\bar{z} = (z, \alpha, \beta)$, but $\tilde{c}_{ik} > 0$, iff $\tilde{c}_{ik}^{-1}(\alpha, \beta) > 0 (0 \leq i \leq p)$ hold for all of $\alpha, \beta \in [0, 1]$. It is easy to prove $\tilde{A}_i^*(\bar{z})$ to be a convex function as in Theorem 1.2.3 [WY82]. Therefore, $\log \tilde{G}_i(z)$ denotes a fuzzy convex function with respect to $\tilde{A}_i^*$. □

2.3 Dual Programming of Fuzzy PGP

Lemma 2.3.1 *Let $\tilde{B}_k(1 \leqslant k \leqslant J)$ be a continuous and strictly nondecreasing (CSN) function. If $\tilde{v}_k > 0, w_k > 0(1 \leqslant k \leqslant J)$ and $\sum_{k=1}^{J} w_k = 1$, then*

$$\prod_{k=1}^{J} \tilde{v}_k^{w_k} \lesssim \sum_{k=1}^{J} w_k \tilde{v}_k \tag{2.7}$$

where the sign of equality holds iff $\tilde{v}_1 = \tilde{v}_2 = \cdots = \tilde{v}_J$. ◊

Proof $\log v_k(\bar{x}')$ represents a quasi-concave function since it is a strictly concave function on $X \subset \Re^m$. Therefore, for all $v_k(x^1, \beta), v_k(x^2, \beta) \in X$ with two points $x^1 \neq x^2$ and $\lambda, \beta \in (0, 1)$,

$$\begin{aligned}&\log\big(\lambda v_k(x^1, \beta) + (1 - \lambda) v_k(x^2, \beta)\big)\\ &\quad > \min\{\log v_k(x^1, \beta), \log v_2(x^2, \beta)\}.\end{aligned}$$

From the strict nondecrease of $\tilde{B}_k(\cdot)$,

$$\begin{aligned}&\tilde{B}_k\big(\log\big(\lambda v_k(x^1, \beta) + (1 - \lambda) v_k(x^2, \beta)\big)\big)\\ &\quad > \tilde{B}_k\{\min[\log v_k(x^1, \beta), \log v_k(x^2, \beta)]\}\\ &\quad = \min\{\tilde{B}_k\big(\log v_k(x^1, \beta), \tilde{B}_k(\log v_k(x^2, \beta)\big)\},\end{aligned}$$

hence, $\tilde{B}_k(\log v_k)$ is a strictly fuzzy quasi-concave function valued on [0,1]. By applying Theorem 2.1.1, we know $\log \tilde{v}_k$ denotes a strictly fuzzy concave function with respect to $\tilde{B}_k$. Again

$$\tilde{B}_k\big(\log \sum_{k=1}^{J} \bar{w}'_k v_k(\bar{x}') - \sum_{k=1}^{J} \bar{w}'_k \log v_k(\bar{x}')\big) > \alpha$$

such that

$$\exp\{\sum_{k=1}^{J} \bar{w}'_k v_k(\bar{x}')\} \leqslant \exp\{\log \sum_{k=1}^{J} \bar{w}'_k v_k(\bar{x}')\} / e^{\tilde{B}_k^{-1}(\alpha)},$$

i.e.

$$\prod_{k=1}^{J} \big(v_k(\bar{x}')\big)^{\bar{w}'_k} \leqslant \sum_{k=1}^{J} \bar{w}'_k v_k(\bar{x}').$$

Here $\bar{x}' = (x, \beta)$, $\bar{w}'_k = (w_k, \beta)$.

Therefore

$$\prod_{k=1}^{J} \tilde{v}_k^{w_k} \lesssim \sum_{k=1}^{J} w_k \tilde{v}_k$$

and the sign of equality holds iff $\tilde{v}_1 = \tilde{v}_2 = \cdots = \tilde{v}_J$, so that (2.7) is true. □

Obviously, the lemma is viewed as Theorem 1.4.1 when $\tilde{A}_k$ embodies a distinct set.

Lemma 2.3.2 *Let $\tilde{v}_k > 0, w_k \geqslant 0 (1 \leqslant k \leqslant J)$. If $\tilde{A}_k$ means a CSN function, then*

$$\left(\sum_{k=1}^{J} \tilde{v}_k\right)^{w_0} \gtrsim \prod_{k=1}^{J} \left(\frac{\tilde{v}_k}{w_k}\right)^{w_k} w_0^{w_0} \tag{2.8}$$

where the sign of equality holds iff

$$w_k \sum_{k=1}^{J} \tilde{v}_k = \tilde{v}_k w_0 \quad (1 \leqslant k \leqslant J_0); \tag{2.9}$$

here $w_0 = \sum_{k=1}^{J} w_k$, stipulating $\left(\frac{\tilde{v}_k}{w_k}\right)^{w_k} |_{w_k=0} = 1$. ◇

Proof 1. If $w_k > 0 (1 \leqslant k \leqslant J)$, the conclusion holds from Lemma 2.3.1.

2. If each $w_k = 0 (1 \leqslant k \leqslant J)$, then $w_0 = 0$, with (2.8) being an equality. It naturally follows that the conclusion holds.

3. If w_k are not all zero, we might as well let $w_k \neq 0 (1 \leqslant k \leqslant J'), w_k = 0 (J' + 1 \leqslant k \leqslant J)$, such that (2.9) does not hold. But from 1, we know

$$\begin{aligned}\left(\sum_{k=1}^{J} \tilde{v}_k\right)^{w_0} &\gtrsim \left(\sum_{k=1}^{J'} \tilde{v}_k\right)^{w_0} \\ &\gtrsim \prod_{k=1}^{J'} \left(\frac{\tilde{v}_k}{w_k}\right)^{w_k} w_0^{w_0} \\ &= \prod_{k=1}^{J} \left(\frac{\tilde{v}_k}{w_k}\right)^{w_k} w_0^{w_0}.\end{aligned}$$

Therefore, the conclusion is also true. □

Theorem 2.3.1 *Let $\tilde{A}_i = \tilde{B}_i \circ \tilde{\phi}_i$ be a CSM fuzzy-valued function, where $\tilde{\phi}_i = \tilde{c}_{ik} \circ \tilde{\gamma}_{ikl}$ $(1 \leqslant k \leqslant J_i, 0 \leqslant i \leqslant p, 1 \leqslant l \leqslant m)$. Then a dual form of (2.1) means*

$$\begin{aligned} &\widetilde{\max}\ \tilde{d}(w) \\ &\text{s.t. } w_{00} = \sum_{k=1}^{J_0} w_{0k} = 1 \\ &\quad \tilde{\Gamma}^{\mathrm{T}} w = 0 \\ &\quad w \geqslant 0 \end{aligned} \tag{2.10}$$

where

$$\tilde{\Gamma} = \begin{pmatrix} \tilde{\gamma}_{011}\cdots & \tilde{\gamma}_{01l}\cdots & \tilde{\gamma}_{01m} \\ \vdots & \vdots & \vdots \\ \tilde{\gamma}_{0J_01}\cdots & \tilde{\gamma}_{0J_0l}\cdots & \tilde{\gamma}_{0J_0n} \\ \cdots & \cdots & \cdots \\ \tilde{\gamma}_{p11}\cdots & \tilde{\gamma}_{p1l}\cdots & \tilde{\gamma}_{p1m} \\ \cdots & \cdots & \cdots \\ \tilde{\gamma}_{pJ_p1}\cdots & \tilde{\gamma}_{pJ_pl}\cdots & \tilde{\gamma}_{pJ_pm} \end{pmatrix}$$

denote structure of fuzzy exponent to apiece term of variable x_l corresponding to an objective function $\tilde{g}_0(x)$ and each constraint function $\tilde{g}_i(x)(1 \le i \le p)$, called fuzzy exponent matrices. It contains $J = J_0 + J_1 + \cdot + J_p$ row and m column, J is equal to the sum of apiece term in $\tilde{g}_i(x)(0 \le i \le p)$, and

$$\tilde{d}(w) = (\frac{\tilde{a}_{00}}{w_{00}})^{w_{00}} \prod_{i=0}^{p}\prod_{k=1}^{J_i}(\frac{\tilde{c}_{ik}}{\tilde{a}_i w_{ik}})^{w_{ik}} \prod_{i=1}^{p}(w_{i0})^{w_{i0}}$$

where $w = (w_{01}, \ldots, w_{0J_0}, \ldots, w_{p1}\ldots, w_{pJ_p})^T$ is a $J-$dimensional variable vector, $w_{i0} = \sum\limits_{k=1}^{J_0} w_{ik}$ is the sum of each set of dual variables corresponding to an objective function $\tilde{g}_0(x)(i = 0)$ or constraint function $\tilde{g}_i^*(x)(0 < i \leqslant p)$. ◊

Proof Suppose $\tilde{B}_i, \tilde{c}_{ik}$ and $\tilde{\gamma}_{ikl}$ $(1 \leqslant k \leqslant J_i, 0 \leqslant i \leqslant p, 1 \leqslant l \leqslant m)$ are shown in (1.14)(1.15) and (1.16) or (1.14)(1.17) and (1.18), and from Theorem 2.2.1, we know that (2.1) is equivalent to (2.3), while (2.3) is a classical PGP with parametric variable α, β, such that for $\tilde{A}_i$, its dual form denotes

$$\begin{aligned} \max\ & \left(\frac{-\alpha}{\bar{w}_{00}}\right)^{\bar{w}_{00}} \prod_{i=0}^{p}\prod_{k=1}^{J_i}\left(\frac{\tilde{c}_{ik}^{-1}(\beta)/\tilde{B}_i^{-1}(\alpha)}{\bar{w}_{ik}}\right)^{\bar{w}_{ik}} \prod_{i=1}^{p}\bar{w}_{i0}^{\bar{w}_{i0}} \\ \text{s.t.}\ & \bar{w}_{00} = \sum_{k=1}^{J_0}\bar{w}_{0k} = 1 \\ & \bar{\Gamma}^{\mathrm{T}}\bar{w} = 0, \alpha, \beta \in [0,1] \\ & \bar{w} \geqslant 0 \end{aligned} \tag{2.11}$$

where $\bar{w}_{ik} = w_{ik}(\alpha, \beta)$, and

$$\overline{\Gamma} = \begin{pmatrix} \tilde{\gamma}_{011}^{-1}(\beta)\cdots & \tilde{\gamma}_{01l}^{-1}(\beta)\cdots & \tilde{\gamma}_{01m}^{-1}(\beta) \\ \cdots & \cdots & \cdots \\ \tilde{\gamma}_{0J_01}^{-1}(\beta)\cdots & \tilde{\gamma}_{0J_0l}^{-1}(\beta)\cdots & \tilde{\gamma}_{0J_0m}^{-1}(\beta) \\ \cdots & \cdots & \cdots \\ \tilde{\gamma}_{p11}^{-1}(\beta)\cdots & \tilde{\gamma}_{p1l}^{-1}(\beta)\cdots & \tilde{\gamma}_{p1m}^{-1}(\beta) \\ \cdots & \cdots & \cdots \\ \tilde{\gamma}_{pJ_p1}^{-1}(\beta)\cdots & \tilde{\gamma}_{pJ_pl}^{-1}(\beta)\cdots & \tilde{\gamma}_{pJ_pm}^{-1}(\beta) \end{pmatrix}$$

denote an exponent matrix with respect to β, where $\bar{w} = (\bar{w}_{01}, \ldots, \bar{w}_{0J_0}, \ldots, \bar{w}_{p1}, \ldots, \bar{w}_{pJ_p})^{\mathrm{T}}$ is a J–dimensional variable vector with respect to α, β, $\bar{w}_{00}$ is a dual variables corresponding to an objective coefficient 1 and $\bar{w}_{i0}$ is the sum of each dual variables corresponding to the functions $g_i^*(x)(0 \leqslant i \leqslant p)$ with respect to α, β.

Now, (2.11) means

$$\max\left(a_{00}/\bar{w}_{00}\right)^{\bar{w}_{00}} \prod_{i=1}^{p}\prod_{k=1}^{J_i}\left(c_{ik}/a_i\bar{w}_{ik}\right)^{\bar{w}_{ik}} \prod_{i=1}^{p} \bar{w}_{i0}^{\bar{w}_{i0}}$$

$$\begin{aligned} \text{s.t.}\quad & \overline{w}_{00} = \sum_{k=1}^{J_0} \bar{w}_{0k} = 1 \\ & \Gamma^{\mathrm{T}}\bar{w} = 0, \alpha, \beta \in [0,1] \\ & a_{00} = \tilde{B}_{00}^{-1}(\alpha) \\ & a_i = \tilde{B}_i^{-1}(\alpha), c_{ik} = \tilde{c}_{ik}^{-1}(\beta)(0 \leqslant i \leqslant p) \\ & \bar{w} \geqslant 0 \end{aligned}$$

where

$$\Gamma = \begin{pmatrix} \gamma_{011}\cdots & \gamma_{01l}\cdots & \gamma_{01m} \\ \cdots & \cdots & \cdots \\ \gamma_{0J_01}\cdots & \gamma_{0J_0l}\cdots & \gamma_{0J_0m} \\ \cdots & \cdots & \cdots \\ \gamma_{p11}\cdots & \gamma_{p1l}\cdots & \gamma_{p1m} \\ \cdots & \cdots & \cdots \\ \gamma_{pJ_p1}\cdots & \gamma_{pJ_pl}\cdots & \gamma_{pJ_pm} \end{pmatrix},$$

$\gamma_{ikl} = \tilde{\gamma}_{ikl}^{-1}(\beta)(1 \leqslant k \leqslant J_i, 0 \leqslant i \leqslant p, 1 \leqslant l \leqslant m)$.

Since $a_i = \tilde{B}_i^{-1}(\alpha)$, $a_i \leqslant \tilde{B}_i^{-1}(\alpha)$ can be induced, (2.11) is equivalent to (2.10), i.e, let $\alpha = 1 - \beta, \tilde{B}_i(\alpha) \geqslant 1 - \beta$.

Especially, we know

$$\tilde{B}_{00}(a_{00}) \geqslant 1 - \beta.$$

And

$$c_{ik} = \tilde{c}_{ik}^{-1}(\beta), \gamma_{ikl} = \tilde{\gamma}_{ikl}^{-1}(\beta)$$

iff

$$\tilde{c}_{ik}(c_{ik}) \geqslant 1 - \alpha, \tilde{\gamma}_{ikl}(\gamma_{ikl}) \geqslant 1 - \alpha.$$

□

Corollary 2.3.1 *Let $\tilde{B}_i : X \to [0,1](0 \leqslant i \leqslant p)$ be a CSM fuzzy-valued function. Then the model is*

$$(\tilde{P}_1) \qquad \begin{aligned} &\widetilde{\min}\ g_0(x) = \sum_{k=1}^{J_0} c_{0k} \prod_{l=1}^{m} x_l^{\gamma_{0kl}} \\ &\text{s.t. } g_i(x) = \sum_{k=1}^{J_i} c_{ik} \prod_{l=1}^{m} x_l^{\gamma_{ikl}} \lesssim 1 (1 \leqslant i \leqslant p) \\ &\quad x > 0 \end{aligned}$$

and $(\tilde{P}_1)$ dual form is

$$\widetilde{\max}(\tilde{a}_{00}/w_{00})^{w_{00}} \prod_{i=0}^{p} \prod_{k=1}^{J_i} (c_{ik}/\tilde{a}_i w_{ik})^{w_{ik}} \prod_{i=1}^{p} (w_{i0})^{w_{i0}}$$

$$\begin{aligned} \text{s.t. } & w_{00} = \sum_{k=1}^{J_0} w_{0k} = 1 \\ & \Gamma^{\mathrm{T}} w = 0 \\ & w \geqslant 0 \end{aligned}$$

where Γ stands for an exponent matrix (γ_{ikl}). ◊

Corollary 2.3.2 *Let $\tilde{c}_{ik}, \tilde{\gamma}_{ikl} : \Re \to [0,1](0 \leqslant i \leqslant p)$ be a CSM membership function. Then a model with fuzzy coefficients as well as fuzzy exponents is*

$$(\tilde{P}_2) \qquad \begin{aligned} &\widetilde{\min} \sum_{k=1}^{J_0} \tilde{c}_{0k} \prod_{l=1}^{m} x_l^{\tilde{\gamma}_{0kl}} \\ &\text{s.t. } \sum_{k=1}^{J_i} \tilde{c}_{ik} \prod_{l=1}^{m} x_l^{\tilde{\gamma}_{ikl}} \leqslant 1\ (1 \leqslant i \leqslant p) \\ &\quad \tilde{c}_{ik} \in [c_{ik}^-, c_{ik}^+], \tilde{\gamma}_{ikl} \in [\gamma_{ikl}^-, \gamma_{ikl}^+]\ (0 \leqslant i \leqslant p) \\ &\quad x > 0 \end{aligned}$$

whose dual form is

$$\begin{aligned}
&\widetilde{\max} \prod_{i=0}^{p} \prod_{k=1}^{J_i} \left(\tilde{c}_{ik}/w_{ik}\right)^{w_{ik}} \prod_{i=1}^{p} (w_{i0})^{w_{i0}} \\
&\text{s.t. } w_{00} = 1 \\
&\qquad \tilde{\Gamma}^{\mathrm{T}} w = 0 \\
&\qquad w \geqslant 0
\end{aligned}$$

where $\tilde{\Gamma}$ *means a fuzzy exponent matrix* $(\tilde{\gamma}_{ikl})$. ◊

Theorem 2.3.2 *Let* $\tilde{A}_i^*(0 \leqslant i \leqslant p)$ *be a CSM function. Then (2.1) contains the same fuzzy optimum value as its dual programming problem (2.10).* ◊

Proof The theorem holds because (2.1) can always be changed into a monomial fuzzy PGP, which can be changed into a fuzzy linear programming (LP) problem having an identical fuzzy optimum value with its dual problem [Cao87a,b][Cao89a][Cao95d] [WY82]. □

Theorem 2.3.3 *If a logarithm is taken for* $\tilde{d}(w)$ *in (2.10), then*

$$\begin{aligned}
&\widetilde{\max} \ \log \tilde{d}(w) \\
&\text{s.t. } w_{00} = \sum_{k=1}^{J_0} w_{0k} = 1 \\
&\qquad \Gamma^{\mathrm{T}} w = 0 \\
&\qquad w \geqslant 0
\end{aligned}$$

means a fuzzy concave programming, hence any fuzzy local maximal solution is its fuzzy global one. ◊

Proof A constrained set is written as $\Re$ meaning a linear equality, hence it is fuzzy concave. We prove only that

$$\begin{aligned}
\log \tilde{d}(w) &= w_{00}(\log \tilde{a}_0 - \log w_{00}) \\
&+ \sum_{i=1}^{p} \sum_{k=1}^{J_i} w_{ik}(\log \tilde{c}_{ik} - \log \tilde{a}_i - \log w_{ik}) + \sum_{i=1}^{p} w_{i0} \log w_{i0}
\end{aligned}$$

is a fuzzy concave function of w, i.e., the Hessian matrix $\overline{H}$ of $\log \tilde{d}(w)$ denotes a negative semi-definite form and, for all $\alpha, \beta \in [0,1]$, we find a partial derivative of $\log d(\bar{w})$ according to Definition 2.1.6 with respect to w:

$$\overline{H} = \begin{pmatrix} \overline{H}_0 & & & \\ & \overline{H}_1 & & \\ & & \ddots & \\ & & & \overline{H}_p \end{pmatrix},$$

where $\bar{w} = (w, \alpha, \beta)$, with

$$\overline{H}_0 = \begin{pmatrix} -\bar{w}_{01}^{-1} & & & \\ & -\bar{w}_{02}^{-1} & & \\ & & \ddots & \\ & & & -\bar{w}_{0J_0} \end{pmatrix},$$

$$\overline{H}_i = \begin{pmatrix} \bar{w}_{i0}^{-1} - \bar{w}_{i1}^{-1} & \bar{w}_{i0}^{-1} & \cdots & \bar{w}_{i0}^{-1} \\ \bar{w}_{i0}^{-1} & \bar{w}_{i0}^{-1} - \bar{w}_{i2}^{-1} & \cdots & \bar{w}_{i0}^{-1} \\ \vdots & \vdots & \ddots & \vdots \\ \bar{w}_{i0}^{-1} & \bar{w}_{i0}^{-1} & \cdots & \bar{w}_{i0}^{-1} - \bar{w}_{iJ_i} \end{pmatrix} (i > 0),$$

while

$$\begin{vmatrix} \bar{w}_{ik_1}^{-1} - \bar{w}_{i0}^{-1} & \bar{w}_{i0}^{-1} & \cdots & \bar{w}_{i0}^{-1} \\ \bar{w}_{i0}^{-1} & \bar{w}_{ik_2}^{-1} - \bar{w}_{i0}^{-1} & \cdots & \bar{w}_{i0}^{-1} \\ \vdots & \vdots & \ddots & \vdots \\ \bar{w}_{i0}^{-1} & \bar{w}_{i0}^{-1} & \cdots & \bar{w}_{ik_N}^{-1} - \bar{w}_{i0} \end{vmatrix}$$

$$= \Big(\prod_{j=1}^{N} \bar{w}_{ik_j}\Big)^{-1} - \bar{w}_{i0}^{-1} \sum_{j=1}^{N} \bar{w}_{ik_j} \Big(\prod_{j=1}^{N} \bar{w}_{ik_j}\Big)^{-1} \geqslant 0 (1 \leqslant i \leqslant p)$$

such that we know $\overline{H}_0$ is negative semidefinite form, and $-\overline{H}_i(1 \leqslant i \leqslant p)$ is positive semidefinite from $\bar{w} = w(\alpha, \beta)$, so $\overline{H}_i$ is negative semidefinite form. Therefore, the theorem holds. □

Theorem 2.3.4 *If the problem (2.10) is strongly fuzzy concave, then any fuzzy local maximum solution to (2.10) is its unique fuzzy global maximum solution.* ◊

Proof Since (2.10) is equivalent to

$$\max\{\tilde{D}^*(\bar{w}) = \tilde{d}(\bar{w}) \wedge \tilde{s}(\bar{w})\}$$

where $\tilde{s}(\bar{w}) = \wedge_{l=0}^{m+1} \tilde{s}_l(\bar{w})$, $\tilde{d}(\bar{w})$ represents an optimal point set corresponding to an objective function in (2.10) and $\tilde{s}(\bar{w})$ a feasible solution set corresponding to constraint functions in (2.10). $\tilde{d}(\bar{w}), \tilde{s}_0(\bar{w}), \ldots, \tilde{s}_{m+1}(\bar{w})$ is a strongly quasi-concave function (from Theorem 2.1.2) and, for all of $\lambda \in [0,1]$ and $\alpha, \beta \in [0,1], \bar{w}^1, \bar{w}^2, \bar{w}^1 \neq \bar{w}^2$, we have

$$\tilde{d}\big(\lambda\bar{w}^1 + (1-\lambda)\bar{w}^2\big) > \tilde{d}(\bar{w}^1) \wedge \tilde{d}(\bar{w}^2),$$
$$\tilde{s}_l\big(\lambda\bar{w}^1 + (1-\lambda)\bar{w}^2\big) > \tilde{s}_l(\bar{w}^1) \wedge \tilde{s}_l(\bar{w}^2) \quad (0 \leqslant l \leqslant m+1).$$

Hence

$$\begin{aligned}
&\tilde{D}^*\big(\lambda\bar{w}^1 + (1-\lambda)\bar{w}^2\big) \\
&\quad = \tilde{d}\big(\lambda\bar{w}^{-1} + (1-\lambda)\bar{w}^2\big) \wedge \tilde{s}\big(\lambda\bar{w}^1 + (1-\lambda)\bar{w}^2\big) \\
&\quad \geqslant \min\{\tilde{d}(\bar{w}^1), \tilde{d}(\bar{w}^2); \tilde{s}_0(\bar{w}^1), \tilde{s}_0(\bar{w}^2); \ldots; \tilde{s}_{m+1}(\bar{w}^1), \tilde{s}_{m+1}(\bar{w}^2)\} \\
&\quad = \min\{\tilde{d}(\bar{w}^1), \tilde{s}_0(\bar{w}^1), \ldots, \tilde{s}_{m+1}(\bar{w}^1); \tilde{d}(\bar{w}^2), \tilde{s}_0(\bar{w}^2), \\
&\qquad \ldots, \tilde{s}_{m+1}(\bar{w}^2)\} \\
&\quad = \min\{\tilde{d}(\bar{w}^1) \wedge \tilde{s}(\bar{w}^1), \tilde{d}(\bar{w}^2) \wedge \tilde{s}(\bar{w}^2)\} \\
&\quad = \tilde{D}^*(\bar{w}^1) \wedge \tilde{D}^*(\bar{w}^2)
\end{aligned}$$

such that $\tilde{D}^*(\bar{w})$ is a strongly quasi-concave function on $X \subset \Re^m$, where $\bar{w}^1 = (w^1, \alpha, \beta), \bar{w}^2 = (w^2, \alpha, \beta)$. Therefore, its local maximum solution is its unique global maximum one from Reference [Avr76]; the theorem holds from an arbitrary choice of α, β on $[0, 1]$. □

Theorem 2.3.5 *Let $\tilde{d}(w)$ be a strictly concave function. If a fuzzy maximum solution set is nonempty in (2.10), any of its fuzzy maximum solutions is unique.* ◊

Proof The theorem can be easily proved by noticing only that constraint conditions in (2.10) are fuzzy concave functions. □

2.4 Dual Theorem of Fuzzy PGP

Definition 2.4.1 If $x > 0$, $g_i(x) \leqslant 1$ is extended into

$$v\big(\tilde{g}_i(x) \lesssim 1\big) = \sup_{x,y|x \geqslant y} \min\{\mu_1(x), \mu_{\tilde{g}_i}(y)\} \ (1 \leqslant i \leqslant p).$$

If $v\big(\tilde{g}_i(x) \lesssim 1\big) = 1$, then (2.1) is called fuzzy compatible, and if $v\big(\tilde{g}_i(x) < 1\big) = 1$, then (2.1) is called fuzzy super-consistent by an extension principle, $v(\cdot)$ representing a degree of possibility of "$\cdot$".

Lemma 2.4.1 *Let x represent any fuzzy feasible solutions to (2.1) and w any feasible solutions to (2.10). If $\tilde{A}_i = \tilde{B}_i \circ \tilde{\phi}_i (0 \leqslant i \leqslant p)$ is a CSN function, then*

$$\tilde{g}_0(x) \gtrsim \tilde{g}_0(x) \prod_{i=1}^{p} \big(\tilde{g}_i(x)\big)^{w_{i0}} \gtrsim \tilde{d}(w) \tag{2.12}$$

and $\tilde{g}_0(x) = \tilde{d}(w)$ iff

$$w_{ik} = \begin{cases} \tilde{v}_{0k}/\tilde{g}_0(x) & (i = 0; 1 \leqslant k \leqslant J_0) \\ w_{i0}\tilde{v}_{ik}(x) & (i \neq 0; 1 \leqslant k \leqslant J_i) \end{cases} \tag{2.13}$$

holds; x and w represent fuzzy optimal solutions to (2.1) and optimal solutions to (2.10) with respect to $\tilde{A}_i$. ◊

Proof $\tilde{A}_i^{-1}$ exists if $\tilde{A}_i (0 \leqslant i \leqslant p)$ is a CSN function. From the proof of Theorem 2.2.1, we know (2.1) $\Longleftrightarrow$ (2.3), but from the proof of Theorem 2.3.1, we know the dual form of (2.3) is (2.11), with (2.11)$\Longleftrightarrow$(2.10). If x and w are fuzzy feasible solutions to (2.1) and feasible ones to (2.10), with $\bar{x} = (x, \alpha, \beta)$ and $\bar{w} = (w, \alpha, \beta)$ for $\alpha, \beta \in [0, 1]$ necessitating any feasible parametric solution to (2.3) and (2.11), respectively, then $\tilde{g}_i(x) \lesssim 1 \Longleftrightarrow g_i(\bar{x}) \leqslant 1 (1 \leqslant i \leqslant p)$. Hence, for all $\alpha, \beta \in [0, 1]$, by applying Lemma 1.5.3 in [WY82] and (2.3) and (2.11), we have

$$-\alpha \geqslant (-\alpha) \prod_{i=1}^{p} \big(g_i(\bar{x})\big)^{\bar{w}_{i0}} \geqslant d(\bar{w})$$

with $-\alpha = d(\bar{w})$ iff

$$\bar{w}_{ik} = \begin{cases} v_{00}(\bar{x})/(-\alpha) & (i = 0; k = 0, \alpha \neq 0) \\ \bar{w}_{i0}v_{ik}(\bar{x}) & (1 \leqslant k \leqslant J_i; 0 \leqslant i \leqslant p) \end{cases} \tag{2.14}$$

holds. Therefore, $\bar{x}$ and $\bar{w}$ represent optimal parametric solutions to (2.3) and (2.10), respectively.

Problem (2.1) re-done to (2.3), (2.10) can be induced from (2.11). From Theorem 2.1.2 and 2.1.3 and Lemma 2.3.2,

$$\tilde{B}_0^{-1}(-\alpha) \geqslant \tilde{B}_0^{-1}\Big[-\alpha \prod_{i=1}^{p} \big(g_i(\bar{x}')\big)^{\tilde{B}_i^{-1}(\bar{w}'_{00})}\Big] \geqslant \tilde{B}_0^{-1}\big(d(\bar{w}')\big)$$

with $\tilde{B}_0^{-1}(-\alpha) = \tilde{B}_0^{-1}\big(d(\bar{w}')\big)$ iff

$$\tilde{B}_i^{-1}(\bar{w}'_{ik}) = \begin{cases} \tilde{B}_0\big(v_{0k}(\bar{x}')\big)/\tilde{B}_0^{-1}(-\alpha), & (i=0; 1 \leqslant k \leqslant J_0, \tilde{B}_0^{-1}(-\alpha) \neq 0) \\ \tilde{B}_i^{-1}(\bar{w}'_{i0})\tilde{B}_i\big(v_{ik}(\bar{x}')\big), & (i \neq 0; 1 \leqslant k \leqslant J_i), \end{cases}$$

which means that (2.12) holds and $\tilde{g}_0(x) = \tilde{d}(w)$ iff (2.13) holds.

Hence x and w are a fuzzy optimal solution to (2.1) and an optimal one to (2.10), respectively. □

Corollary 2.4.1 *Let w^* be a fuzzy optimal solution to (2.10). If there exist CSN functions $\tilde{A}_i^* = \tilde{B}_i \circ \tilde{\phi}_i (0 \leqslant i \leqslant p)$, then a fuzzy feasible solution to (2.1) satisfying*

$$\sum_{l=1}^{m} \tilde{\gamma}_{ikl} \log x_l = \begin{cases} \log\big(w_{0k}^* \tilde{d}(w^*)/\tilde{c}_{0k}\big), & (i=0; 1 \leqslant k \leqslant J_0, \tilde{c}_{0k} \neq 0) \\ \log\big(w_{ik}^*/(w_{i0}^* \tilde{c}_{ik})\big), & (i \neq 0; w_{i0}^* \tilde{c}_{ik} \neq 0, 1 \leqslant k \leqslant J_i) \end{cases} \quad (2.15)$$

denotes a fuzzy optimal solution to (2.1), where $\tilde{\phi}_i = \tilde{c}_{ik} \circ \tilde{\gamma}_{ikl} (1 \leqslant k \leqslant J_i, 0 \leqslant i \leqslant p, 1 \leqslant l \leqslant m)$. ◇

Proof From the given condition, there must exist $\tilde{A}_i^* = \tilde{B}_i \circ \tilde{\phi}_i = \tilde{B}_i \circ \tilde{c}_{ik} \circ \tilde{\gamma}_{ikl} (1 \leqslant k \leqslant J_i, 0 \leqslant i \leqslant p, 1 \leqslant l \leqslant m)$, such that (2.15) is equivalent to

$$\sum_{l=1}^{m} \tilde{\gamma}_{ikl}^{-1}(\beta) \log \bar{x}_l = \begin{cases} -\log\big(\bar{w}_{00}^* \alpha\big), & (i=0, k=0, \alpha \neq 0) \\ \log\big(\bar{w}_{ik}^*/(\bar{w}_{i0}^* \tilde{c}_{ik}^{-1}(\beta))\big), & (1 \leqslant k \leqslant J_i; w_{i0}^* \tilde{c}_{ik}^{-1}(\beta) \neq 0, 0 \leqslant i \leqslant p). \end{cases} \quad (2.16)$$

Formula (2.14) can be obtained by taking exponents on both sides of (2.16), so $-\alpha = d(\bar{w}^*)$ holds. Again, $\bar{x}$ is an optimal parameter solution to (2.3), which means (2.13) holds, with $\tilde{g}_0(x^*) = \tilde{d}(w^*)$. And x^* denotes a fuzzy optimal solution to (2.1) on the basis of Lemma 2.4.1. □

Theorem 2.4.1 (the first fuzzy dual theorem) *Given that fuzzy PGP (2.1) denotes fuzzy super-consistence with a fuzzy optimal solution x^* existing, there must exist a Lagrange multiplier $\mu^* = (\mu_1^*, \mu_2^*, \ldots, \mu_p^*)^T \geqslant 0$ and a CSN* *function* $\tilde{A}_i^*(0 \leqslant i \leqslant p)$, *such that*

$$\nabla \tilde{g}_0(x^*) + \sum_{i=1}^{p} \mu_i^* \nabla \tilde{g}_i(x^*) = 0,$$
$$\mu_i^*\left(\tilde{g}_i(x^*) - 1\right) = 0 \quad (1 \leqslant i \leqslant p),$$

while w^* *defined by*

$$w_{ik}^* = \begin{cases} \tilde{v}_{0k}(x^*)/\tilde{g}_0(x^*), & (i = 0; 1 \leqslant k \leqslant J_0) \\ \mu_i^* \tilde{v}_{ik}(x^*)/\tilde{g}_0(x^*), & (i \neq 0; 1 \leqslant k \leqslant J_i) \end{cases}$$

denotes a dual optimal solution to (2.10), with

$$\tilde{g}_0(x^*) = \tilde{d}(w^*).$$

◊

Proof Because (2.1) is fuzzy super-consistent, programming problem (2.5) denotes a fuzzy convex programming satisfying a fuzzy Slater condition: $\tilde{G}_i(z) < 1$, which is equivalent to saying that parameter program (2.6) is convex, satisfying the parameter Slater condition: $G_i(\overline{z}) < 1$ (from Theorem 2.2.1 and 2.2.3). But (2.3) is convex, meeting the parameter Slater condition: $g_i(\overline{x}) < 1$, hence (2.1) fuzzy super-consistency means (2.6) super-consistency, so does (2.3). Again there exists a fuzzy optimal solution x^* to (2.1) iff (2.3) has a parameter optimal solution $\overline{x}^* = (x_1^*, \ldots, x_p^*; \alpha, \beta)^{\mathrm{T}}$ with components being positive, such that $\forall \alpha, \beta \in [0, 1]$; from Theorem 1.6.3 in [WY82], there must exist a Lagrange multiplier $\mu^* = (\mu_0^*, \mu_1^*, \ldots, \mu_p^*)^{\mathrm{T}} \geqslant 0$, such that

$$\nabla(-\alpha) + \sum_{i=0}^{p} \mu_i^* \nabla g_i(\overline{x}^*) = 0,$$
$$\mu_i^*\left(g_i(\overline{x}^*) - b_i\right) = 0 \quad (0 \leqslant i \leqslant p),$$

where $b_0 = z_0, b_i = 1(1 \leqslant i \leqslant p)$. And $\bar{w}^*$ defined by

$$\bar{w}_{ik}^* = \begin{cases} \tilde{v}_{0k}(\overline{x}^*)\bar{w}_{00}^*, & (i = 0; k = 0) \\ \mu_i^* v_{ik}(\overline{x}^*)/(-\alpha), & (1 \leqslant k \leqslant J_i, 1 \leqslant i \leqslant p, \alpha \neq 0) \end{cases}$$

is an optimal parameter solution to (2.11) with $-\alpha = d(\bar{w}^*)$, i.e., suppose $\mu_0^* = 1$, then

$$\nabla \tilde{B}_0\left(g_0(x^*, \beta)\right) + \sum_{i=1}^{p} \mu_i^* \nabla \tilde{B}_i\left(g_i(x^*, \beta)\right) = 0,$$
$$\mu_i^*\left(\tilde{B}_i(g_i(x^*, \beta) - b_i)\right) = 0 \quad (0 \leqslant i \leqslant p)$$

and

$$\tilde{B}_i^{-1}\big(w_{ik}^*(\beta)\big) = \begin{cases} \tilde{B}_0\big(v_{0k}(x^*,\beta)\big)\tilde{B}_0^{-1}\big(w_{00}^*(\beta)\big), (i=0, k=0) \\ \mu_i^*\tilde{B}_i\big(v_{ik}(x^*,\beta)\big)/\tilde{B}_0^{-1}(-\alpha), (1 \leqslant i \leqslant p, 1 \leqslant k \leqslant J_i; \tilde{B}_0^{-1}(-\alpha) \neq 0). \end{cases}$$

Therefore, the theorem is true. □

Theorem 2.4.2 (the second fuzzy dual theorem) *Let fuzzy PGP (2.1) be fuzzy compatible and (2.10) contain a fuzzy feasible solution with components being positive. Then there must exist a CSM function $\tilde{A}_i^*(0 \leqslant i \leqslant p)$, such that problem (2.1) contains a fuzzy optimal solution.* ◊

Proof Because (2.1)$\Longleftrightarrow$ a convex programming (2.6) with a parameter, (2.3) is consistent due to (2.1) fuzzy consistency $\Longleftrightarrow$(2.6) consistency. Again, because (2.10) $\Longleftrightarrow$ (2.11), there exists a fuzzy feasible solution with components being positive in (2.10) $\Longleftrightarrow$ there exists a parameter feasible solution with components being positive in (2.11). Therefore, for an arbitrary parameter $\alpha, \beta \in [0,1]$, we know there is an optimal parameter solution to (2.3) if Theorem 1.8.1 in [WY82] is applied in formula (2.3). And also there exists a fuzzy optimal solution to (2.1) by means of equivalence of (2.1) and (2.3) and an arbitrariness of α, β in interval [0,1], so the theorem holds. □

2.5 Primal Algorithm for Fuzzy PGP

2.5.1 Introduction

Fuzzy PGP has proceeded further in its algorithmic development since it was originated at the IFSA Congress held in Tokyo in 1987. In [Cao87a,b][Cao89a] [Cao93a][Cao01a,e], direct and dual algorithms were evoked on fuzzy valued sets. In this section, three direct algorithms are advanced in fuzzy PGP including: 1. a primal direct algorithm; 2. a direct-repeated algorithm with added constraints; 3. a direct-repeated algorithm deleting constraints, whose effectiveness is confirmed by numerical examples. The three algorithms are most effective in solving problems of difficult degree (termed as DD henceforth) $D > 0$ in fuzzy PGP and multi-objective fuzzy PGP as well.

2.5.2 Primal Direct Algorithms

Consider fuzzy PGP (2.1).

Lemma 2.5.1 *If constraint $\tilde{C}$ denotes a strictly convex fuzzy set on $\Re^m$, i.e., $\forall \lambda \in (0,1)$, if there exist $x^1, x^2 \in \Re^m$, and $x^1 \neq x^2$, we have*

$$\tilde{C}\big(\lambda x^1 + (1-\lambda)x^2\big) > \tilde{C}(x^1) \wedge \tilde{C}(x^2),$$

then $G(\alpha) \triangleq \sup_{x \in C_\alpha} |G_0(x)|$ is called continuance, where $\alpha \in [0,1]$. ◊

Lemma 2.5.2 *If it is differentiable, then $G(x)$ must continue on $X \subset \Re^m$.* ◊

Since $\min G_0(x)$ can be taken as $\max[-G_0(x)]$ for solution purposes, only max state is to be considered. It can be confirmed that there exists a unique fixed point $\overline{\alpha} = G(\overline{\alpha})$ if $G_0(x)$ continues.

As fuzzy PGP (2.1) is a fuzzy convex programming problem (otherwise, it can be changed into a convex programming problem), (2.2) is also convex. $\tilde{M} \triangleq \sup_{x \in A_\alpha} \tilde{Y}^*(x)$ continues with a unique fixed point α existing. Because of $\tilde{M} : M \to [0,1]$, finding fuzzy PGP is concluded into obtaining a fixed point, such that a fuzzy optimal solution is determined in programming problem (2.1).

Algorithm I Steps for calculation:

1^0 Let $k = 1$ and select α_1 by means of formula $1 - kh = \alpha_1$.

2^0 Calculate

$$\tilde{M}^{(k)} = \sup_{x \in A_\alpha} |G_0(\bar{x}')| = \sup_{x \in A_{\alpha_k}} |\tilde{B}_0(\sum_{k=1}^{J_0} \tilde{c}_{0k}^{-1}(\beta) \prod_{l=1}^{m} x_l^{\tilde{\gamma}_{0kl}^{-1}(\beta)})|.$$

3^0 Calculate $\varepsilon_k = \alpha_k - \tilde{M}^{(k)}$. If $\varepsilon_k > \varepsilon$, back to step 2^0, otherwise directly to step 5^0.

4^0 Select $r_k \in [0,1]$ properly. If $\alpha_{k+1} = \alpha_k - r_k\varepsilon_k$ and $k = k+1$, we turn back to step 2^0.

5^0 Find x^* for $\alpha = \alpha_k$, such that $\sup_{x\in A_{\alpha_k}} |G_0(x)| = \tilde{M}^{(k)}(x^*)$.

Interpretation:

In step 1^0, the proper selection pattern of α_1 is to take $\alpha_1 \in [0.9, 1)$ and to take $\alpha_1 \in [0.75, 0.9)$ for $\tilde{g}_0(x) \uparrow$ and for $\tilde{g}_0(x) \downarrow$, respectively. And the determination of k value assures that it shall be smaller when a constraint constant $b(>0)$ is very large. Otherwise, k value is proper to be larger (for example $k = 20$).

In Step 4^0, r_k is taken by the bisection method, i.e., take $r_k = 0.5$. If $\varepsilon_k \gg \varepsilon_i (k < i)$, then $r_k = 1$. If $\varepsilon_1 \uparrow \varepsilon_2$ changes little, we properly take $r_k = 0.618$. If $\varepsilon_1 \ll \varepsilon_2$, we properly take $r_k = 0.382$, otherwise, we might obtain the conclusion of $\tilde{M}^{(k)} > 1, \alpha_k > 1$ but $\varepsilon_k > \varepsilon$ [Cao87a,b].

Here 0.618 and 0.382 are proportionality factors of the Fibonacci method.

Example 2.3 Find a fuzzy GP

$$\begin{aligned}
&\max\ (x_1^5 x_2^{-2})\\
&\text{s.t.}\ x_1^2 \lesssim 20\\
&\qquad x_2^2 \lesssim 3, x_1 > 0\\
&\qquad x_1^2 x_2^3 \lesssim 30, x_2 > 0
\end{aligned}$$

Solution. Take a membership function like

$$\tilde{A}_1(d_1) = \tilde{A}_2(d_2) = \tilde{A}_3(d_3) = \begin{cases} 1 - 20\log d, & 0 \leqslant d < 1.051 \\ 0, & d \geqslant 1.051. \end{cases}$$

For $d_i = 1.051$, $\tilde{M}^{(k)} = 0.000551317 x_1^5 x_2^{-2}$ and $(x_1, x_2) \doteq (4.702, 1.126)$.

$\forall \alpha_1 = 0.8$, $\varepsilon = 0.001$ and $r_k = 1$, we can get an optimal level $\alpha = \alpha_3 = 0.972$ and an optimal solution $x_1^* \doteq 4.531$, $x_2^* \doteq 1.135$ and optimal value $g_0(x^*) = 1482.405$, through three steps of calculations.

Besides, an fuzzy PGP problem (2.1) can be turned into $(\tilde{P}_\varepsilon)$ because a multi-term fuzzy posynomial is approached by a fuzzy monomial through a fuzzy geometric inequality. $(\tilde{P}_\varepsilon)$ can be solved by methods of fuzzy LP [Cao87a,b][Cao93a] since it is equivalent to a fuzzy LP whose variables are subject to lower and upper bounds. We have many solution methods to fuzzy LP, what we obtain for a fuzzy LP approximates an optimal level and a solution, which have something in common with those obtained by the direct algorithm, but they are not as brief and accurate as the direct algorithm in solving a fuzzy GP problem.

You might try turning Example 2.3 into LP for solution and what you gain is the same as the result mentioned above.

2.5.3 Direct-repeated Algorithm with Added Constraints

Let $\tilde{g}_0(x) = \tilde{x}_0$, then change (2.1) into

$$\begin{aligned} &\widetilde{\min}\ \tilde{x}_0 \\ &\text{s.t. } \tilde{g}_0(x) \lesssim \tilde{x}_0 \\ &\quad\ \tilde{g}_i(x) \lesssim 1\ (1 \leqslant i \leqslant p) \\ &\quad\ \tilde{x}_0, x > 0 \end{aligned}$$

which, with only one more variable in it than in (2.1), still belongs to a fuzzy PGP and is obviously equivalent to (2.1). We might as well let the objective function in (2.1) be

$$\tilde{g}_0(x) = \tilde{x}_1$$

Practically, we can estimate the extent where solutions exist. Therefore, we consider a fuzzy PGP with a variable subject to lower and upper bounds

$$\begin{aligned} &\widetilde{\min}\ \tilde{x}_1 \\ &\text{s.t. } \tilde{g}_i(x) \lesssim 1\ (1 \leqslant i \leqslant p) \\ &\quad\ 0 < x_L \leqslant x \leqslant x_U, \end{aligned} \tag{2.17}$$

here, x_L and x_U represent two positive vectors.

$\forall y > 0$, suppose

$$\varepsilon_{ik} = \tilde{v}_{ik}(y)/\tilde{g}_i(y)\ (1 \leqslant k \leqslant J_i;\ 0 \leqslant i \leqslant p)$$

where

$$\begin{aligned} \tilde{v}_{ik}(y) &= \tilde{c}_{ik} \prod_{l=1}^{m} y_{ik}^{\tilde{\gamma}_{ikl}}, \\ \tilde{g}_i(y) &= \sum_{k=1}^{J_i} \tilde{v}_{ik}(y) = \sum_{k=1}^{J_i} \tilde{c}_{ik} \prod_{l=1}^{m} y_{ik}^{\tilde{\gamma}_{ikl}}. \end{aligned}$$

Obviously,

$$\begin{aligned} \sum_{k=1}^{J_i} \varepsilon_{ik} &= \sum_{k=1}^{J_i} \left[\tilde{c}_{ik} \prod_{l=1}^{m} y_{ik}^{\tilde{\gamma}_{ikl}} \Big/ \sum_{k=1}^{J_i} \tilde{c}_{ik} \prod_{l=1}^{m} y_{ik}^{\tilde{\gamma}_{ikl}} \right] \\ &= \sum_{k=1}^{J} \tilde{c}_{ik} \prod_{l=1}^{m} y_{ik}^{\tilde{\gamma}_{ikl}} \Big/ \sum_{k=1}^{J_i} \tilde{c}_{ik} \prod_{l=1}^{m} y_{ik}^{\tilde{\gamma}_{ikl}} = 1. \end{aligned}$$

We suppose

$$\tilde{\tilde{g}}_i(x, y) = \prod_{k=1}^{J_i} (\tilde{v}_{ik}(x)/\varepsilon_{ik})^{\varepsilon_{ik}} = \tilde{\tilde{c}}_i \prod_{k=1}^{J_i} x_l^{\tilde{\tilde{\gamma}}_{ikl}}$$

where

$$\tilde{\tilde{c}}_i = \prod_{k=1}^{J_i} \left(\frac{\tilde{c}_{ik}}{\varepsilon_{ik}}\right)^{\varepsilon_{ik}}, \tilde{\tilde{\gamma}}_{il} = \sum_{k=1}^{J_i} \tilde{\gamma}_{il}\varepsilon_{ik} \tag{2.18}$$

and while

$$\tilde{g}_i(x) = \sum_{k=1}^{J_i} \tilde{c}_{ik} \prod_{l=1}^{m} x_l^{\tilde{\gamma}_{ikl}}.$$

Theorem 2.5.1 *Let $\tilde{A}_i = \tilde{B}_i \circ \tilde{\phi}_i (1 \leqslant i \leqslant p)$ be a CSN function. If $\tilde{v}_{ik} > 0, \varepsilon_{ik} > 0$ $(1 \leqslant k \leqslant J_i;\ 0 \leqslant i \leqslant p)$ with $\sum_{k=1}^{J_i} \varepsilon_{ik} = 1$, then*

$$\tilde{\tilde{g}}_i(x, y) \lesssim \tilde{g}_i(x). \tag{2.19}$$

◊

Proof On the presumption of the theorem and, by Lemma 2.3.1, we have

$$\prod_{k=1}^{J_i} (\tilde{v}_{ik}(x))^{\varepsilon_{ik}} \lesssim \sum_{k=1}^{J_i} \varepsilon_{ik}\tilde{v}_{ik}(x). \tag{2.20}$$

Applying (2.20), we have

$$\begin{aligned}
\tilde{\tilde{g}}_i(x, y) &= \prod_{k=1}^{J_i} \left(\frac{\tilde{c}_{ik}}{\varepsilon_{ik}} \prod_{l=1}^{m} x_l^{\tilde{\gamma}_{ikl}}\right)^{\varepsilon_{ik}} \\
&\lesssim \sum_{k=1}^{J_i} \varepsilon_{ik} \left(\frac{\tilde{c}_{ik}}{\varepsilon_{ik}} \prod_{l=1}^{m} x_l^{\tilde{\gamma}_{ikl}}\right) \\
&= \sum_{k=1}^{J_i} \tilde{c}_{ik} \prod_{l=1}^{m} x_l^{\tilde{\gamma}_{ikl}} \\
&= \tilde{g}_i(x).
\end{aligned}$$

Therefore,

$$\tilde{\tilde{g}}_i(x, y) \lesssim \tilde{g}_i(x).$$

Moreover

$$\tilde{\tilde{g}}_i(y, y) = \tilde{g}_i(y).$$

Now, we change (2.17) into a monomial fuzzy PGP for solution by means of (2.19), such that we can obtain a fuzzy optimal solution in (2.1) without difficulty.

Let $\tilde{F}^0 = \{x | \tilde{\tilde{g}}_i(x, x_0) \lesssim 1 (0 \leqslant i \leqslant p), x_L \leqslant x \leqslant x_U\}$ for arbitrariness $x_0 > 0$; when $k = 1$ in $\tilde{F}^{k-1}$, we consider

$$\begin{aligned}
&\widetilde{\min}\ \tilde{x}_1 \\
&\text{s.t. } x \in \tilde{F}^{k-1}.
\end{aligned} \tag{2.21}$$

□

Theorem 2.5.2 *Suppose $\tilde{A}_i (0 \leqslant i \leqslant p)$ to be a CSM function. If there exists a fuzzy minimal solution x^k to (2.21) for a certain k, such that*

$$\tilde{g}_{i(k)}(x^k) = \max_{1 \leqslant i \leqslant k} \tilde{g}_i(x^k) \lesssim 1, \tag{2.22}$$

then i) x^k represents a fuzzy optimal solution of (2.17). ii) Otherwise, if (2.22) is no longer true for infinite k, then any limit point in a point range $\{x^k\}$ is a fuzzy optimal point of (2.17). ◊

Proof i) If $\tilde{F} = \{x | \tilde{g}_i(x) \lesssim 1 (1 \leqslant i \leqslant p, 0 < x_L \leqslant x \leqslant x_U)\}$ are feasible solution sets to (2.17), we suppose

$$\tilde{C}^k = \{x | \tilde{\bar{g}}_i(x, x^k) \lesssim 1, 0 < x_L \leqslant x \leqslant x_U\},$$
$$\tilde{F}^k = \tilde{F}^{k-1} \bigcap \tilde{C}^k \triangleq \min\{\tilde{F}^{k-1}(x), \tilde{C}^k(x)\}.$$

And again,

$$\tilde{F} = \bigcup_{\alpha \in [0,1]} \alpha F_\alpha, \ \tilde{F}^k = \bigcup_{\alpha \in [0,1]} \alpha F^k_\alpha, \ \tilde{C}^k = \bigcup_{\alpha \in [0,1]} \alpha C^k_\alpha.$$

Because Formula (2.19) holds, there exists $\tilde{F} \subset \tilde{C}^k$ as for $\forall k$ and $\alpha \in [0,1]$. Obviously, again $\tilde{F} \subset \tilde{F}^0$, therefore, it is not difficult to prove $\tilde{F} \subset \tilde{F}^k (k = 0, 1, \cdots)$ by mathematical induction.

When Formula (2.22) holds, there exists $x^k \in \tilde{F}$, while x^k is a fuzzy optimal point of $\tilde{x}_1$ on $\tilde{F}^{k-1}$, so is on $\tilde{F}$, such that i) holds.

ii) Suppose that $\{x^k\}$ is an infinite point range, x^k is bounded, then $\forall \varepsilon > 0, \exists : N > 0$, when $n > N$, we have $d(x^k, x^*) < \varepsilon$, such that there exists a limit point x^*.

Again, $\tilde{x}_1^{k_n} = \min_{x \in \tilde{F}^{k-1}} \tilde{x}_1 \lesssim \min_{x \in \tilde{F}} \tilde{x}_1$. Let $x^{k_n} \to x^* (n \to \infty)$, then

$$\tilde{x}_1^* \lesssim \min_{x \in \tilde{F}} \tilde{x}_1.$$

Therefore, we prove only $x^* \in \tilde{F}$.

Assume oppositely

$$\tilde{\bar{g}}_{i(k_n)}(x^{k_n}, x^{k_n}) = \tilde{g}_{i(k_n)}(x^{k_n}) \gtrsim 1,$$

hence, $x^{k_n} \notin \tilde{C}^{k_n}$.

In fact, this condition is equivalent to the truth of

$$\tilde{B}_i[\bar{g}_{i(k_n)}(x^{k_n}, x^{k_n}, \beta) - \bar{g}_{i(k_n)}(x^{k_n}, \beta)] = 0$$

while

$$\tilde{B}_i[\bar{g}_{i(k_n)}(x^{k_n}, \beta) - 1] > 0$$
$$\Longrightarrow \tilde{B}_i[\bar{g}_{i(k_n)}(x^{k_n}, x^{k_n}, \beta)] = \tilde{B}_i[g_{i(k_n)}(x^{k_n}, \beta)] > 1$$

for $\forall\beta \in [0,1]$. And $\forall\alpha \in [0,1]$, we have $x^{k_n} \notin C_\alpha^{k_n}$.

Again, for $k_n + 1 > k_n$,

$$x^{k_n+1} \in F_\alpha^{k_{n+1}-1} \subset F_\alpha^{k_{n+1}-2} \subset \cdots \subset F_\alpha^{k_n} \subset C_\alpha^{k_n}.$$

Therefore

$$d(x^{k_n}, x^{k_n+1}) > d(x^{k_n}, C_\alpha^{k_n}) \to 0(x^{k_n} \to x^*). \tag{2.23}$$

Now, assume $x^* \notin F_\alpha^c$, i.e.

$$\tilde{g}_i(x^*) \gtrsim 1(1 \leqslant i \leqslant p).$$

According to continuity, if we take $\varepsilon = 1/2(\tilde{g}_i(x^*) - 1)$, there exists $k > 0$, such that

$$\bar{g}_{i(k_n)}(x^{k_n}, x^{k_n}, \beta) = g_{i(k_n)}(x^{k_n}, \beta) \geqslant g_i(x^{k_n}, \beta) > 1 + \varepsilon \tag{2.24}$$

for $k_n > k$.

But $\bar{g}_{i(k_n)}(x, y, \beta)(1 \leqslant i \leqslant p)$ is a function continuing on nonempty compact region $\bar{X} = \{(x,y)|x_L \leqslant x \leqslant x_U, x_L \leqslant y \leqslant x_U\}$, so it is uniformly continuous. Then, $\forall\varepsilon > 0, \exists : \delta = \delta(\varepsilon) > 0$, such that $\forall(x^1, y^1), (x^2, y^2)$, for

$$d[(x^1, y^1), (x^2, y^2)] < \delta$$

we have

$$d[g_i(x^1, y^1, \beta), g_i(x^2, y^2, \beta)] < \varepsilon.$$

But $x \in C_\alpha^{k_n}, \tilde{B}_i[g_{i(k_n)}(x, x^{k_n}, \beta) - 1] \leqslant 0$ from (2.24), we have

$$d[\bar{g}_{i(k_n)}(x, x^{k_n}, \beta), \bar{g}_{i(k_n)}(x^{k_n}, x^{k_n}, \beta)] > \varepsilon.$$

Then, $\forall\bar{x}' \in C_\alpha^{k_n}$, we have $d(x, x^{k_n}) \geqslant \delta$, which contradicts (2.23), so $x^* \in \tilde{F}$.

□

Therefore, we conclude the direct-repeated algorithm with added constraints of a fuzzy PGP (2.1) as follows:

Algorithm II

1^0 Change (2.1) into a variable limited by lower and upper bounds for problem (2.17).

2^0 $\forall x^0 > 0$, let $\tilde{F}^0 = \{x|\tilde{\bar{g}}_i(x, x^0) \lesssim 1(1 \leqslant i \leqslant p), x_L \leqslant x \leqslant x_U\}$. From (2.19), we know a fuzzy feasible solution set is $\tilde{F}^1 \subset \tilde{F}^0$ in (2.17) and take $k = 1$.

3^0 Find (2.17). Solution to (2.17) can be got only by finding its equivalent program (2.21). Apparently, constraint functions of (2.21) are fuzzy monomials, which is called a monomial fuzzy PGP of variable subject to lower and upper bounds, and is equivalent to fuzzy LP of variable subject to lower and upper bounds from [Cao87a,b].

There are many solution methods to (2.21), such as

a. A fuzzy dual method.

b. A fuzzy direct method.

We obtained a fuzzy optimal solution to (2.17) from fuzzy LP. If no fuzzy optimal solution exists in (2.21), nor does it in (2.17). Stop, otherwise we turn to 4^0 for help.

4^0 If (2.21) has a fuzzy optimal solution, one of the methods in 3^0 can be used to get a solution and the fuzzy optimal one is denoted by x^k.

5^0 If $\tilde{g}_{i(k)}(x^k) = \max_{1\leqslant i\leqslant k} \tilde{g}_i(x^k) \lesssim 1$, then x^k is the fuzzy optimal solution to (2.17). It ends. Otherwise we turn to 6^0 for help.

6^0 Let

$$\tilde{C}^k = \{x|\tilde{\bar{g}}_{i(k)}(x, x^k) \lesssim 1, 0 < x_L \leqslant x \leqslant x_U\},$$
$$\tilde{F}^k = \tilde{F}^{k-1} \bigcap \tilde{C}^k \triangleq \min\{\tilde{F}^{k-1}(x), \tilde{C}^k(x)\}.$$

Use k as $k+1$ and then back to 3^0.

Note 2.2 We choose a membership function as

$$\tilde{A}(x) = \frac{\tilde{g}_0(x) - M}{M - m}$$

where m, M are infimum and supremum of $\tilde{g}_0(x)$, respectively.

The two methods from 3^0 are as follows:

a. Fuzzy dual method [Cao87a,b][Cao93a] with its calculation found in Example 2.4.

b. Fuzzy direct method with its calculation below

i) Let $k = 1$, from $1 - hd = \alpha_1$, we properly choose α_1 for (2.1).

ii) Calculate $\tilde{M}^{(k)} = \sup_{x\in C_\alpha} |\tilde{A}_0(g_0(x))|$.

iii) Calculate $\varepsilon_k = \alpha_k - |\tilde{M}^{(k)}|$, if $\varepsilon_k > \varepsilon$, then back to ii). Otherwise to v).

iv) Properly choose $r_k \in [0, 1]$, suppose $\alpha_{k+1} = \alpha_k - r_k\varepsilon_k$, and $k = k+1$, we turn back to ii).

v) For $\bar{\alpha} = \alpha_k$, we find x^*, such that $\sup_{x\in C_{\bar{\alpha}}} |\tilde{A}_0(g_0(x))| = \tilde{M}^{(k)}(x^*)$.

Solution by changing program (2.1) into a fuzzy LP. Take the following for example.

Let $z_l = \log x_l$, then (2.1) is equivalent to a fuzzy LP [Cao87a,b] with variable subject to lower and upper bounds

$$\begin{aligned} &\widetilde{\min}\ \tilde{z}_1 \\ &\text{s.t. } \tilde{G}_i(z) \lesssim 0 (1 \leqslant i \leqslant p) \\ &\quad z_L \leqslant z \leqslant z_U \end{aligned} \tag{2.25}$$

where $\tilde{G}_i(z) = \sum_{l=1}^{m} \tilde{\tilde{\gamma}}_{ik} z_l + \log \tilde{\tilde{c}}_{ik}$, z_L and z_U represent two positive vectors.

By means of membership function $\tilde{A}_i = \frac{\tilde{G}_i(z)-M}{M-m}$, (2.25) can be obtained from Zimmermann algorithm [Zim78] or the algorithms advanced in References [Cao87a,b][Cao93a] (i.e., *b*. mentioned above, in which we only substitute $\tilde{z}_1$ for $\tilde{g}_0(x)$).

The above algorithms can be flexibly chosen according to practical states. Generally, when DD of the problem $D = 0$ or $D = 1$, it is proper to use the dual algorithm for solution, and to use the other one for $D > 1$.

2.5.4 Direct-Repeated Algorithm Deleting Constraints

With previous constraints stored, we shall increase our calculation since every repeat gains a new constraint, so that we advance an improved algorithm deleting some added constraints.

This algorithm depend on follow convergence theorem:

Theorem 2.5.3 *If (2.22) holds for a CSM function $B_i(\cdot)$, then i) x^k is a fuzzy optimal solution to (2.17); ii) Otherwise, $\{x^k\}$ is an infinite point range with any fuzzy limit point of it, which means a fuzzy optimal solution to (2.17).* ◊

Proof i) We can let $\tilde{F} \subset \tilde{F}^{k-1}$ because of $\tilde{F} \subset \tilde{F}^0$. Since x^k is a fuzzy optimal point in $\tilde{x}_1$ on $\tilde{F}^{k-1}$, then it holds $\forall x \in \tilde{F}^{k-1}$.

Especially, $\forall x \in \tilde{F}$, we have

$$\tilde{x}_1^k \lesssim \tilde{x}_1.$$

So, we substitute lower bound of $\tilde{x}_1$ for $\tilde{x}_1^k$, i.e., exchanging x_L^{k-1} for x_L^k with $\tilde{F}^c$ and $\tilde{C}^l$ unchangeable. Since $\tilde{F} \subset \tilde{F}^0$, with $\tilde{F} \subset \tilde{C}^l$, then $\forall k \geqslant 1$, there must exist $\tilde{F} \subset \tilde{F}^k$. Therefore, when (2.22) holds, i) is true.

ii) Let $\{x^k\}$ be an infinite point range. Only $x^* \in \tilde{F}$ needs proving if any fuzzy limit point x^* is proved to be a fuzzy optimal point of (2.17).

Given $x^{k_n} \to x^*(n \to \infty)$, let

$$\delta_2(x^{k_n}) = \inf\{d(x^{k_n}, x) | \tilde{x}_1 \gtrsim \tilde{x}_1^{k_n} + \delta_1(x^{k_n}), x \in \tilde{F}^0\} \tag{2.26}$$

where

$$\delta_1(x^l) = \tfrac{1}{H}\{\tilde{g}_{i(l)}(x^l) - 1\}. \tag{2.27}$$

H represents a given positive number. We write

$$\begin{aligned} \mathcal{I}_k &= \{l | \tilde{x}_1^k \lesssim \tilde{x}_1^l - \delta_1(x^l), l \in \mathcal{I}_{k-1}\} \bigcup \{k\}, \\ \tilde{F}^k &= \tilde{F}^0 \bigcap\nolimits_{l \in \mathcal{I}_k} C^l. \end{aligned} \tag{2.28}$$

If $k_n \in \mathcal{I}_{k_n+1} - 1$, then $\forall \alpha \in [0, 1]$, because

$$x^{k_n+1} \in F_\alpha^{k_{n+1}-1} \subset C_\alpha^{k_n}, x^{k_n} \notin C_\alpha^{k_n},$$

we get that

$$d(x^{k_n}, x^{k_n+1}) \geqslant d(x^{k_n}, C_\alpha^{k_n}). \tag{2.29}$$

If $k_n \notin \mathcal{I}_{k_n+1} - 1$, there must exist $k_n \leqslant l < k_{n+1} - 1$ because of $k_n \in \mathcal{I}_{k_n}$, such that $k_n \in \mathcal{I}_l$, but $k_n \notin \mathcal{I}_{l+1}$.

Again $x^{k_n+1} \in F_\alpha^0$, and

$$\tilde{x}_1^{k_n+1} \gtrsim \tilde{x}_1^{k_{n+1}-1} \gtrsim \cdots \gtrsim \tilde{x}_1^{l+1} \gtrsim \tilde{x}_1^{k_n} + \delta_1(x^{k_n})$$

such that from the definition of $\delta_2(x^{k_n})$, i.e., from (2.26), we can see that

$$d(x^{k_n}, x^{k_n+1}) \geqslant \delta_2(x^{k_n}) \text{ (if } k_n \notin \mathcal{I}_{k_n+1} - 1). \tag{2.30}$$

Let $x^{k_n} \to x^*(n \to \infty)$, then $d(x^{k_n}, x^{k_n+1}) \to 0$. From (2.29) and (2.30), we know at least either

$$d(x^{k_n}, C_\alpha^{k_n}) \to 0 \quad \text{or} \quad \delta_2(x^{k_n}) \to 0, \alpha \in [0,1]$$

holds.

If $d(x^{k_n}, C_\alpha^{k_n}) \to 0$ for any $\alpha \in [0,1]$ holds. Repeating some proofs from Theorem 2.5.2, we know $x^* \in \tilde{F}$.

Given $\delta_2(x^{k_n}) \to 0$, since a set

$$\{x | \tilde{x}_1 \gtrsim \tilde{x}_1^{k_n} + \delta_1(x^{k_n}), x \in \tilde{F}^0\}$$

is compact, there exists $\hat{x}^{k_n}$, such that $d(x^{k_n}, \hat{x}^{k_n}) = \delta_2(x^{k_n})$ with

$$0 \leqslant \delta_1(x^{k_n}) \lesssim \hat{x}_1^{k_n} - \tilde{x}_1^{k_n} \lesssim d(x^{k_n}, \hat{x}^{k_n}) \to 0.$$

From (2.27), we know that a monotonic and continuous function $\tilde{B}_i(\cdot)$ exists, such that

$$\max_{1 \leqslant i \leqslant p} \tilde{B}_i[g_i(x^{k_n}, \beta) - 1] \to 0.$$

Finally, from the continuation of $g_i(x, \beta)$, we can see

$$\tilde{B}_i[g_i(x^*, \beta) - 1] \leqslant 0(1 \leqslant i \leqslant p),$$

i.e. $\tilde{g}_i(x^*) \lesssim 1$; also $x^* \in F_\alpha$ holds for any $\alpha \in [0,1]$, such that $ii)$ holds. □

Therefore, the steps for the direct-repeated algorithm to (2.1) by deleting constraints are concluded as follows:

Direct algorithm III

Write x_L as x_L^0, with $\mathcal{I}_0 = \phi$, then

1^0 Change (2.1) into problem (2.17) with variable subject to lower and upper bounds.

2^0 Suppose

$$\tilde{F}^0 = \{x | \tilde{\bar{g}}_i(x, x^0) \lesssim 1 (1 \leqslant i \leqslant p), x_L^0 \leqslant x \leqslant x_U\}$$

$\forall x^0 > 0$ and we take $k = 1$.

3^0 Solve (2.20) in the way given in Note 2.2 and let x^k be its fuzzy optimal solution.

4^0 If (2.21) holds, x^k is an optimal solution to (2.17), now calculation ends. Otherwise, we turn to 5^0 for help.

5^0 Redefine

$$\tilde{F}^0 = \{x | \tilde{\bar{g}}_i(x, x^0) \lesssim 1 (1 \leqslant i \leqslant p), x_L^k \leqslant x \leqslant x_U\};$$

here, x_L^k is obtained by changing the first component of x_L^{k-1} into x_1^k, and C^k is determined as it is done at the 6th step in Algorithm II, and let $\delta_1(x^l)$ be like (2.27) and (2.28). Again suppose k to be $k+1$ before we turn to 3^0.

2.5.5 Numerical Example

Example 2.4 Find:

$$\begin{aligned} &\widetilde{\min}\ 4x_2^{-1} \\ &\text{s.t.}\ \frac{x_2}{1 + x_1/2} \lesssim 1 \\ &x_1, x_2 > 0 \end{aligned} \tag{2.31}$$

Solution. With the method in References [Cao93a][Cao98a,b], the problem is concluded as solutions to three determined GP as follows:

1^0 The determined programming corresponding to (2.31) is

$$\begin{aligned} &\min\ 4x_2^{-1} \\ &\text{s.t.}\ \frac{x_2}{1 + x_1/2} \leqslant 1 \\ &x_1, x_2 > 0 \end{aligned}$$

Take

$$\begin{aligned} &x^0 = (x_1, x_2) = (2, 1.5), \\ &\varepsilon_1 = \frac{1}{1 + x_1/2}|_{x^0} = \frac{1}{2} = 0.5,\ \varepsilon_2 = \frac{x_1/2}{1 + x_1/2}|_{x^0} = \frac{1}{2}, \\ &1 + x_1/2 \geqslant \left(\frac{1}{\varepsilon_1}\right)^{\varepsilon_1} \left(\frac{x_1/2}{\varepsilon_2}\right)^{\varepsilon_2} = \sqrt{2} x_1^{\frac{1}{2}}. \end{aligned}$$

The optimal solution is denoted by $(x_1, x_2) = (2, 2)^{\mathrm{T}}$ and the optimal value is $g_0 = 2$.

2^0 Take $d_1 = -0.6, d_2 = 1$, the determined programming corresponding to (2.31) is

$$\begin{aligned} &\min\{4x_2^{-1} \leqslant 0.4\} \\ &\text{s.t. } \frac{x_2}{1+x_1/2} \leqslant 2 \\ &\quad x_1, x_2 > 0 \end{aligned}$$

$$\overset{x_1 \leqslant 4x_2^{-1}-0.4}{\Longrightarrow} \quad \begin{aligned} &\min x_1 \\ &\text{s.t. } 4x_1^{-1}x_2^{-1} \leqslant 0.4 \\ &\quad \frac{1}{\sqrt{2}}x_1^{-\frac{1}{2}}x_2 \leqslant 2 \\ &\quad x_1, x_2 > 0 \end{aligned}$$

$$\Longleftrightarrow$$

$$\begin{aligned} &\max\left(\frac{1}{w_0}\right)^{w_0}\left(\frac{\frac{4}{0.4}}{w_1}\right)^{w_1}\left(\frac{1}{2\sqrt{2}w_2}\right)^{w_2} w_1^{w_1} w_2^{w_2} \\ &\text{s.t. } w_0 = 1 \\ &\quad w_0 - w_1 - \frac{1}{2}w_2 = 0 \\ &\quad -w_1 + w_2 = 0 \\ &\quad w_1 \geqslant 0, w_2 \geqslant 0 \end{aligned}$$

$$\Longrightarrow \quad w_0 = 1, w_1 = w_2 = \frac{2}{3}$$

$$\begin{aligned} D &= \left(\frac{4}{0.4x}\right)^{\frac{2}{3}}\left(\frac{1}{2\sqrt{2}x}\right)^{\frac{2}{3}} \\ &= (10)^{\frac{2}{3}} \times 0.5 \approx 2.32 \end{aligned}$$

$$\Longrightarrow \quad x_1 = 2.32, 10x_1^{-1}x_2^{-1} = 1, \frac{1}{2\sqrt{2}}x_1^{-1/2}x_2 = 1$$

$$\Longrightarrow \quad \begin{cases} x_1 \approx 2.32, \\ x_2 \approx 4.31, \end{cases} \quad g_0' = 2.32$$

3^0 We are given

$$\begin{aligned} &g_0 + d_0 = g_0' = 2.32 \\ &\Longrightarrow d_0 = 0.32 \end{aligned}$$

and

$$d_1 = -0.6, d_2 = 1.$$

Suppose d_0 and d_1 to be flexible indexes of object and constraint and we find a crisp programming corresponding to (2.31)

$$\begin{aligned}
&\min(-\alpha)\\
&\text{s.t. } \frac{-x_1}{-2.32+0.32\alpha} \leqslant 1\\
&\qquad \frac{4x_1^{-1}x_2^{-1}}{1-0.6\alpha} \leqslant 1\\
&\qquad \frac{x_2}{(1+x_1/2)(1+\alpha)} \leqslant 1, \alpha \in [0,1]\\
&\qquad x_1, x_2 > 0
\end{aligned}$$

$$\begin{aligned}
&\Longleftrightarrow\\
&\max\left(\frac{-\alpha}{w_0}\right)^{w_0}\left(\frac{\frac{1}{w_1}}{2.32-0.32\alpha}\right)^{w_1}\left(\frac{\frac{4}{w_2}}{1-0.6\alpha}\right)^{w_2}\\
&\qquad \left(\frac{\frac{1}{w_3}}{\sqrt{2}(1+\alpha)}\right)^{w_3} w_1^{w_1}w_2^{w_2}w_3^{w_3}\\
&\text{s.t. } w_0 = w_1 = 1\\
&\qquad w_1 - w_2 - \frac{1}{2}w_3 = 0\\
&\qquad -w_2 + w_3 = 0\\
&\qquad \alpha \in [0,1]\\
&\qquad w_1 \geqslant 0, w_2 \geqslant 0, w_3 \geqslant 0\\
\Longrightarrow \quad & w_0 = w_1 = 1, w_2 = w_3 = \frac{2}{3}
\end{aligned}$$

From

$$\begin{aligned}
d(\overline{w}) &= (-\alpha)\left(\frac{1}{2.32-0.32\alpha}\right)\left(\frac{4\times\frac{3}{2}}{1-0.6\alpha}\right)^{2/3}\\
&\qquad \left(\frac{3}{2\sqrt{2}(1+\alpha)}\right)^{\frac{2}{3}}\left(\frac{2}{3}\right)^{2/3}\left(\frac{2}{3}\right)^{2/3}\\
&= -\alpha\\
\Longrightarrow &\left(\frac{2\sqrt{2}}{(1-0.6\alpha)(1+\alpha)}\right)^{2/3} = 2.32-0.32\alpha
\end{aligned}$$

by searching methods, we determine

$$\begin{aligned} &\alpha \approx 0.7786, \\ &D(w^*) = d(\overline{w})(2.32 - 0.32\alpha)/(-\alpha) \\ &\qquad\; = 2.0714, \\ \Longrightarrow\ &x_1 \approx 2.0714, x_2 \approx 3.62 \end{aligned}$$

such that $g_0(x) = x_1 = 2.0714$.

2.5.6 Conclusion

The three algorithms mentioned in this section can be programmed directly to calculate the primal fuzzy PGP (2.1) and multi-objective GP. Of course, a fuzzy direct algorithm and a dual one advanced in [Cao87a,b][Cao93a] can also be used to find their solutions. We have the flexibility to choose an algorithm for an application according to its operation and speed.

2.6 Dual Algorithm for Fuzzy PGP

2.6.1 Introduction

On the basis of a dual theory on a fuzzy PGP, a dual algorithm is proposed with two states considered according to the DD $D = 0$ and $D > 0$.

Definition 2.6.1 We call

$$D = J - m - 1 = \sum_{i=0}^{p} J_i - m - 1$$

a DD of a fuzzy PGP $(\tilde{P})$ and its dual program $(\tilde{D})$, where J denotes the terms for the primal programming (resp. numbers of variables in a dual programming), $m + 1$ denotes numbers of variables in the primal programming (resp. numbers of a dual programming equation group).

2.6.2 DD $D = 0$

If a DD $D = 0$, and $J = m + 1$ is a rank of coefficient matrix in a dual program $(\tilde{D})$, then there exists a unique solution w^* in

$$w_{00} = 1, \quad \Gamma^{\mathrm{T}} w = 0.$$

If w^* fails to meet a nonnegative condition, then there exists no feasible solution to $(\tilde{D})$, such that there must exist no constraint minimal solution to the fuzzy PGP $(\tilde{P})$, even no infimum in a positive constraint $M_{\tilde{P}}$[Cao96b]. If $w^* \geqslant 0$, then w^* is a fuzzy optimal solution to $(\tilde{D})$ because it is a unique fuzzy feasible solution to $(\tilde{D})$. Therefore, if $(\tilde{P})$ has a constraint minimal solution x^*, then the solution x^* must be found out through (2.13) or (2.15), otherwise, any fuzzy feasible solution to (2.13) or (2.15) must be also a fuzzy optimal solution to $(\tilde{P})$. At this moment, we can find out the primal fuzzy PGP $(\tilde{P})$ without difficulty, i.e, we call $D = 0$.

According to a dual theorem in Section 2.4 in Chapter 2, we can deduce a dual algorithm to the fuzzy PGP. Now steps of the dual algorithm for solution to (2.1) are given as follows:

Step 1. Fuzzify (2.1) into (2.3) and consider the equivalence form (2.4) of (2.3).

Step 2. Write out a dual form (2.11) of (2.4).

Step 3. Find a feasible solution $\bar{w}$ to (2.11), determine an optimal level α, β, and obtain a dual parameter optimal solution by means of $d(\bar{w}) = -\alpha$, where

$$d(\bar{w}) = \max \left(\frac{-\alpha}{\bar{w}_{00}}\right)^{\bar{w}_{00}} \prod_{i=0}^{p} \prod_{k=1}^{J_i} \left(\frac{\tilde{c}_{ik}^{-1}(\beta)/\tilde{B}_i^{-1}(\alpha)}{\bar{w}_{ik}}\right)^{\bar{w}_{ik}} \prod_{i=1}^{p} \bar{w}_{i0}^{\bar{w}_{i0}}.$$

Step 4. Find a fuzzy optimal solution x^* to the primal problem (2.1) by means of (2.13) and (2.15).

Example 2.5 Find the fuzzy PGP:

$$\begin{aligned}&\widetilde{\min}(40x_1^{\tilde{\gamma}_1}x_2^{-1}x_3^{-1}+40x_2x_3)\\ &\text{s.t.}\quad 2x_1x_3+x_1x_2^{\tilde{\gamma}_2}\lesssim 2.4\\ &\qquad x_1,x_2,x_3>0\end{aligned}$$

where $\tilde{\gamma}_1\in[-\frac{3}{2},-\frac{1}{2}],\tilde{\gamma}_2\in[\frac{1}{2},\frac{3}{2}]$.

Solution 1^0 Take a membership function as (1.14), (1.15) and (1.16) (for $n=1$), and we turn the problem into (2.11), where $z_0=60,d_0=5,d_1=2$ are taken before we obtain

$$\begin{aligned}&\max\alpha=\min(-\alpha)\\ &\text{s.t.}\ (-40x_1^{\beta-\frac{3}{2}}x_2^{-1}x_3^{-1}-40x_2x_3)/(-64+5\alpha)\leqslant 1\\ &\quad(2x_1x_3+x_1x_2^{\frac{1}{2}+\beta})/(2.4+2\alpha)\leqslant 1,\alpha,\beta\in[0,1]\\ &\quad x_1,x_2,x_3>0\end{aligned}$$

2^0 Write out its dual form

$$\begin{aligned}\max\quad&(\frac{-\alpha}{w_{00}})^{w_{00}}(\frac{40/w_{01}}{64-5\alpha})^{w_{01}}(\frac{40/w_{02}}{64-5\alpha})^{w_{02}}(\frac{2/w_{11}}{2.4+2\alpha})^{w_{11}}\\ &(\frac{1/w_{12}}{2.4+2\alpha})^{w_{12}}(w_{01}+w_{02})^{w_{01}+w_{02}}(w_{11}+w_{12})^{w_{11}+w_{12}}\\ \text{s.t.}\quad&w_{00}=1\\ &w_{01}+w_{02}=w_{00}\\ &(\beta-\frac{3}{2})w_{01}+w_{11}+w_{12}=0\\ &-w_{01}+w_{02}+(\frac{1}{2}+\beta)w_{12}=0\\ &-w_{01}+w_{02}+w_{11}=0,\alpha,\beta\in[0,1]\\ &w\geqslant 0\end{aligned}$$

3^0 For $\beta\in[0,1]$, a feasible solution to the problem is

$$\begin{aligned}w_{00}&=1,\\ w_{01}&=(\frac{1}{4}+\frac{\beta}{2})\frac{\frac{1}{2}\beta-\beta}{(\frac{1}{2}+\beta)^2+2}+\frac{1}{2},\\ w_{02}&=-(\frac{1}{4}+\frac{\beta}{2})\frac{\frac{3}{2}-\beta}{(\frac{1}{2}+\beta)^2+2},\\ w_{11}&=(\tfrac{1}{2}+\beta)\frac{\frac{3}{2}-\beta}{(\frac{1}{2}+\beta)^2+2},\\ w_{12}&=\frac{\frac{3}{2}-\beta}{(\frac{1}{2}+\beta)^2+2}.\end{aligned}$$

Substitute them for an objective function, and we can determine α by $d(\bar{w}) = -\alpha$. If we take $\beta = \frac{1}{2}$, we can get a feasible solution to the problem $w^0 = (1, \frac{2}{3}, \frac{1}{3}, \frac{1}{3}, \frac{1}{3})^{\mathrm{T}}$, with its corresponding objective function

$$d(\bar{w}) = (-\alpha)\left(\frac{60}{64-5\alpha}\right)^{\frac{2}{3}}\left(\frac{120}{64-5\alpha}\right)^{\frac{1}{3}}\left(\frac{6}{2.4+2\alpha}\right)^{\frac{1}{3}}\left(\frac{3}{2.4+\alpha}\right)^{\frac{1}{3}}\left(\frac{2}{3}\right)^{\frac{2}{3}}.$$

Let $d(\bar{w}) = -\alpha$. Then

$$120(2)^{\frac{1}{3}}/(64-5\alpha)(2.4+2\alpha)^{\frac{2}{3}} = 1$$

such that $\alpha = 0.8$ is obtained. For $\alpha = 0.8$, from $d(w^*) = d(\bar{w})(64 - 5\alpha)/(-\alpha)$, we have

$$d(w^*) = 60(1)^{\frac{2}{3}}(2)^{\frac{1}{3}}\left(\frac{3}{2}\right)^{\frac{1}{3}}\left(\frac{3}{4}\right)^{\frac{1}{3}}\left(\frac{2}{3}\right)^{\frac{2}{3}} = 60.$$

4^0 According to (2.14),

$$\begin{cases} (-40x_1^{-1}x_2^{-1}x_3^{-1})/(-60) = w_{01}/(w_{01}+w_{02}) \\ (-40x_2x_3)(-60) = w_{02}/(w_{01}+w_{02}) \\ 2x_1x_3/4 = w_{11}/(w_{11}+w_{12}) \\ x_1x_2/4 = w_{12}/(w_{11}+w_{12}) \end{cases}$$

Find the above simultaneous equations; we obtain an optimal solution $x_1^* = 2, x_2^* = 1, x_3^* = \frac{1}{2}$ for $\alpha = 0.8$ and $\beta = \frac{1}{2}$, and an optimal value is denoted by $g_0(x^*) = 60$.

An optimal valued set of the primal problem is obtained $\forall \alpha, \beta \in [0, 1]$, so it is easy to determine their functional expression of optimal fuzzy solutions by means of figures.

2.6.3 DD $D > 0$

As for a CSM function $\tilde{A}_i = \tilde{B}_i \circ \tilde{\phi}_i$, where $\phi = \tilde{c}_{ik} \circ \tilde{\gamma}_{ikl}(1 \leqslant k \leqslant J_i, 0 \leqslant i \leqslant p, 1 \leqslant l \leqslant m)$, when we have a DD $D > 0$, $(\tilde{P})$ has the same fuzzy optimum value as a dual program $(\tilde{D})$.

In fact, since

$$(\tilde{P}) \qquad \Longleftrightarrow$$

$$\begin{aligned} &\max \alpha = \min(-\alpha) \\ &\text{s.t. } G_i(\bar{x}) = \sum_{k=1}^{J_i} \frac{\tilde{c}_{ik}^{-1}(\beta)}{\tilde{B}_i^{-1}(\alpha)} \prod_{l=1}^{m} x_l^{\tilde{\gamma}_{ikl}^{-1}(\beta)} \leqslant 1 \\ &\alpha, \beta \in [0,1] (0 \leqslant i \leqslant p) \\ &x > 0 \end{aligned} \tag{2.32}$$

while

$$
\begin{aligned}
&(2.32) \Longleftrightarrow \\
&\max\left(\frac{-\alpha}{w_{00}}\right)^{w_{00}} \prod_{i=0}^{p} \prod_{k=1}^{J_i} \left(\frac{\tilde{c}_{ik}^{-1}(\beta)/\tilde{B}_i^{-1}(\alpha)}{w_{ik}}\right)^{w_{ik}} \prod_{i=1}^{p} w_{i0}^{w_{i0}} \\
&\text{s.t. } w_{00} = 1, w_{00} = \sum_{k=1}^{J_0} w_{0k} \\
&\quad \Gamma^{\mathrm{T}} w = 0, \alpha, \beta \in [0,1] \\
&\quad w \geqslant 0
\end{aligned} \tag{2.33}
$$

but (2.33)$\Longleftrightarrow$ $(\tilde{D})$. Obviously, $(\tilde{P})$ denotes the extension of (2.1), then the conclusion holds according to Theorem 2.3.2.

Meanwhile, it is easy to see that their DD are the same in problems around nonfuzzification.

In most cases, it is simple to find a dual problem and easy to win an analytic solution. Therefore, we turn the primal problem into a dual one for solution, generally. But for a DD $D > 0$, an equation group $w_{00} = 1, \Gamma^{\mathrm{T}} w = 0$ contains infinite groups of solutions, such that a dual program $(\tilde{D})$ also keeps infinite groups of feasible solutions, that is the reason why it is not easy to find an optimal solution from them. Generally speaking, the larger a DD is, the more difficulties we have in the solution. Now we shall exhibit steps for dual algorithms in the fuzzy PGP $(\tilde{P})$ in the case of a DD $D > 0$ and unknown states in a flexible index:

Step 1 Change $(\tilde{P})$ into two deterministic PGP

$$
\begin{aligned}
&\min\ g_0(\bar{x}') \\
&\text{s.t. } g_i(\bar{x}') \leqslant 1, \beta \in [0,1] (1 \leqslant i \leqslant p) \\
&\quad x > 0
\end{aligned}
$$

and

$$
\begin{aligned}
&\min\ g_0(\bar{x}') \\
&\text{s.t. } g_i(\bar{x}') \leqslant 1 + d_i, \beta \in [0,1] (1 \leqslant i \leqslant p) \\
&\quad x > 0
\end{aligned}
$$

where $\bar{x}' = (x, \beta)$.

Step 2 Write out their dual form

$$
\begin{aligned}
&\max d(\bar{w}) = \left(\frac{-\alpha}{\bar{w}_{00}}\right)^{\bar{w}_{00}} \prod_{i=0}^{p} \prod_{k=1}^{J_i} \left(\frac{\tilde{c}_{ik}^{-1}(\beta)/(1+d_i)}{\bar{w}_{ik}}\right)^{\bar{w}_{ik}} \prod_{i=1}^{p} \bar{w}_{i0}^{\bar{w}_{i0}} \\
&\text{s.t. } \bar{w}_{00} = 1 \\
&\quad \Gamma^{\mathrm{T}} \bar{w} = 0, \alpha, \beta \in [0,1] \\
&\quad \bar{w} \geqslant 0
\end{aligned} \tag{2.34}
$$

where d_i is selected as 0 or d_i.

Step 3 Find (2.32) with $d_i = 0$ or d_i, respectively. We can find a feasible solution by means of optimizing methods such as a brief gradient, a gradient projection, feasible direction or variable elimination.

Especially for $D = 1$, search methods (resp. methods of 0.618, twice or three times in interpolation) are applied into obtaining its optimal solution, such that we get two optimal values g_i^* and g_i^{*1}.

Step 4 Find an ordinary PGP with parameters α, β:

$$\begin{aligned} &\max \alpha = \min(-\alpha) \\ &\text{s.t. } g_0(\bar{x}') \leqslant z_0 + d_0\alpha \\ &\quad g_i(\bar{x}') \leqslant 1 + d_i\alpha \\ &\quad \alpha, \beta \in [0,1](1 \leqslant i \leqslant p) \\ &\quad x > 0 \end{aligned}$$

where $z_0 = g_0^{*1}$ (we had better suppose $g_0^{*1} < g_0^*$), $d_0 = g_0^{*1} - g_0^*$ and its dual programming denotes (2.11). Of Step 3 one is used to solve (2.11) before obtaining optimal parameter solutions, accordingly, we determine value of α, β by $d(\bar{w}) = -\alpha$.

Step 5 According to formula

$$\bar{w}_{ik} = \begin{cases} v_{00}(\bar{x})/(-\alpha), & (i = 0, k = 0, \alpha \neq 0) \\ \bar{w}_{i0}v_{ik}(\bar{x}), & (0 \leqslant i \leqslant p, 1 \leqslant k \leqslant J_i) \end{cases}$$

we find $\bar{x}$, i.e, it is optimal solution to (2.4) with respect to parameters α, β.

Step 6 As $\forall \alpha, \beta \in [0,1]$, a fuzzy optimal solution x to $(\tilde{P})$ is compounded by $\bar{x}$.

In application, we only go ahead to Step 5. Some practical examples are to be shown in Chapter 8.

2.6.4 Conclusion

Note 2.3

1. The steps for solution to DD $D > 0$ are suitable to the state of $D = 0$. But we have a difference in Step 3: for $D = 0$, a unique solution w^* can be found only by a direct solution to a set of simultaneous equations in Step 3, but for $D > 0$, a solution w^*, which is not a unique solution, can be found by all optimized methods dependent on each situation.

2. We can determine $(\tilde{P})$ by using membership functions in (1.15), (1.16) or (1.17), (1.18), whichever is used depends on the case itself.

3. The solution to the fuzzy PGP are effective by direct or dual algorithm, whichever method is used depends on itself. Take Example 2.3 for example, we can get an optimal solution by a direct algorithm and no feasible solution by a dual one.

Chapter 3

FUZZY STRONGLY DUAL RESULTS FOR FUZZY PGP

In the previous chapter, we discussed the relationship between fuzzy PGP and its dual programming built on a fuzzy weak dual theorem. With a generalized weak dual theorem, we present in this chapter a fuzzy strongly dual theory on fuzzy PGP.

In Section 1, we introduce a fuzzy dual problem. In Section 2, we define fuzzy sub-consistency and pose a converse theorem. In Sections 3 and 4, we relate fuzzy monomial PGP and its equivalence and renew to testify a dual theorem on fuzzy PGP, respectively. And lastly, we discuss classification and properties of corresponding classes in fuzzy PGP.

3.1 Basic Theorem

Consider a more general fuzzy PGP than the fuzzy one in References[Cao89] [Cao93a]

$$(\tilde{P})\qquad \begin{array}{ll} \widetilde{\inf} & \tilde{g}_0(x) \\ \text{s.t.} & \tilde{g}_i(x) <_* 1 \ (1 \leqslant i \leqslant p) \\ & x > 0, \end{array} \tag{3.1}$$

where the sign "$<_*$" denotes "$\leqslant$" or "$\lesssim$" and its dual form stands for

$$(\tilde{D})\qquad \begin{array}{ll} \widetilde{\sup} & \tilde{d}(w) \\ \text{s.t.} & w_{00} = 1, \\ & \sum\limits_{k=1}^{J_0} w_{0k} = w_{00} \\ & \tilde{\Gamma}^{\mathrm{T}} w = 0; \alpha, \beta \in [0,1] \\ & w \geqslant 0. \end{array} \tag{3.2}$$

Here

$$\tilde{d}(w) = (\tilde{a}_0/w_{00})^{w_{00}} \prod_{k=1}^{J_0} (\tilde{c}_{0k}/\tilde{a}_{0k}w_{0k})^{w_{0k}} \prod_{i=1}^{p}\prod_{k=1}^{J_i} \left(\frac{\tilde{c}_{ik}}{\tilde{a}_{ik}w_{ik}}\right)^{w_{ik}} (w_{i0})^{w_{i0}}$$

and $\widetilde{\min}$ and $\widetilde{\max}$ in the References [Cao89a][Cao93a] are replaced by $\widetilde{\inf}$ and $\widetilde{\sup}$, respectively. The former is said to obtain a fuzzy infimum, and the latter a fuzzy supremum. When $(\tilde{P})$ and $(\tilde{D})$ are fuzzy consistent, $M_{\tilde{P}}$ and $M_{\tilde{D}}$ denote a fuzzy constraint infimum of $(\tilde{P})$ and a fuzzy constraint supremum of $(\tilde{D})$, respectively. Obviously, $M_{\tilde{P}}$ is a finite fuzzy number.

Theorem 3.1.1 *Let a fuzzy-valued function $\tilde{B}_i$ and membership functions $\tilde{c}_{ik}, \tilde{\gamma}_{ikl}$ $(1 \leqslant k \leqslant J_i, 0 \leqslant i \leqslant p, 1 \leqslant l \leqslant m)$ be CSM. Then $(\tilde{P})$ is equivalent to*

$$\begin{aligned}
(\bar{P}) \quad & \inf\ (-\alpha) \\
& \text{s.t. } G_0(\overline{x}) = \sum_{k=1}^{J_0} \frac{\tilde{c}_{0k}^{-1}(\beta)}{\tilde{B}_0^{-1}(\alpha)} \prod_{l=1}^{m} x_l^{\tilde{\gamma}_{0kl}^{-1}(\beta)} \leqslant 1 \\
& G_i(\overline{x}) = \sum_{k=1}^{J_i} \frac{\tilde{c}_{ik}^{-1}(\beta)}{\tilde{B}_i^{-1}(\alpha)} \prod_{l=1}^{m} x_l^{\tilde{\gamma}_{ikl}^{-1}(\beta)} \leqslant 1 \\
& \alpha, \beta \in [0,1] (1 \leqslant i \leqslant p) \\
& x > 0
\end{aligned} \tag{3.3}$$

where $\overline{x} = (x, \alpha, \beta)$, and the dual form of (3.3) is

$$\begin{aligned}
(\bar{D}) \qquad & \sup d(\bar{w}) = \sup\left(\frac{-\alpha}{\bar{w}_{00}}\right)^{\bar{w}_{00}} \prod_{i=0}^{p}\prod_{k=1}^{J_i}\left(\frac{\tilde{c}_{ik}^{-1}(\beta)}{\tilde{B}_i^{-1}(\alpha)\bar{w}_{ik}}\right)^{\bar{w}_{ik}} \prod_{i=1}^{p} (\bar{w}_{i0})^{\bar{w}_{i0}} \\
& \text{s.t. } \quad \bar{w}_{00} = 1 \\
& \bar{\Gamma}^{\mathrm{T}}\bar{w} = 0, \alpha, \beta \in [0,1] \\
& \bar{w} \geqslant 0
\end{aligned}$$

where $\bar{\Gamma} = (\tilde{\gamma}_{ikl}^{-1}(\beta))$, $\bar{w} = w(\alpha, \beta)$. ◊

Proof The theorem can be proved as Theorem 3.1 and 4.1 in the Reference [Cao93a]. □

Theorem 3.1.2 *Let $\tilde{A}_i (0 \leqslant i \leqslant p)$ be a CSM function. When fuzzy PGP $(\tilde{P})$ and its dual form $(\tilde{D})$ are all consistent, then*

$$0 < M_{\tilde{D}} \lesssim M_{\tilde{P}} < \infty. \tag{3.4}$$

◊

Proof Because $(\tilde{D})$ fuzzy consistence is equivalent to $(\bar{D})$ consistence for $\tilde{A}_i (0 \leqslant i \leqslant p)$, $M_{\bar{D}} > 0$, i.e. $M_{\tilde{D}} > 0$. Again, since $(\tilde{P})$ fuzzy consistence is equivalent to $(\bar{P})$ consistence for $\tilde{A}_i$, $M_{\bar{P}} < \infty$, then $M_{\tilde{P}} < \infty$.

Hence, we know (3.4) holds from (2.12). □

Theorem 3.1.3 *Let $\tilde{A}_i(0 \leqslant i \leqslant p)$ be a CSM function. If x is a fuzzy feasible solution to $(\tilde{P})$ and if $i(0 \leqslant i \leqslant p)$ exists, such that $\tilde{g}_i(x) < 1$, then for any feasible solution w of $(\tilde{D})$ and i above, we have*

$$w_{i0} \leqslant \log(\tilde{g}_0(x)/\tilde{d}(w))/(-\log \tilde{g}_i(x)). \tag{3.5}$$

◊

Proof From (2.12), we know

$$\tilde{g}_i(x)(\tilde{g}_i(x))^{w_{i0}} \geqslant \tilde{d}(w)$$

holds and we can get (3.5) by taking logarithms from both sides of the above formula. □

Theorem 3.1.4 *Let $\tilde{A}_i(x) = \tilde{B}_i \circ \tilde{\phi}_i$ $(0 \leqslant i \leqslant p)$ be a CSM function for $\Re \to [0,1]$. If $(\tilde{P})$ is super-consistent and $M_{\tilde{P}} > 0$, $(\tilde{D})$ must be consistent, with optimal solution w^*, and*

$$M_{\tilde{D}} = \tilde{d}(w^*) = M_{\tilde{P}}. \tag{3.6}$$

◊

Proof If $M_{\tilde{P}}$ is a fuzzy constraint minimum of $(\tilde{P})$, the theorem is true from Theorem 2.4.1. Now we can suppose that there is no constraint minimum in $(\tilde{P})$.

As for $h > 0$, we consider an auxiliary PGP of problem $(\tilde{P})$

$$(\tilde{P}_h) \qquad \begin{array}{l} \inf\ \tilde{g}_0(x) \\ \text{s.t. } \tilde{g}_i(x) \lesssim 1 \ (1 \leqslant i \leqslant p+2m) \\ \quad x > 0 \end{array} \tag{3.7}$$

where

$$\tilde{g}_{p+l}(x) \cong h x_l, \quad \tilde{g}_{p+m+l}(x) \cong h x_l^{-1} \ (1 \leqslant l \leqslant m). \tag{3.8}$$

Therefore, $(\tilde{P})$ is equivalent to $(\overline{P})$ from Theorem 3.1.1, such that (3.7) and (3.8) are equivalent to

$$(\overline{P}_h) \qquad \begin{array}{l} \inf(-\alpha) \\ \text{s.t. } g_0(\bar{x}) \leqslant 1 \\ \quad G_i(\bar{x}) \leqslant 1 \ (1 \leqslant i \leqslant p+2m) \\ \quad \alpha, \beta \in [0,1] \\ \quad \bar{x} > 0 \end{array} \tag{3.9}$$

where

$$\begin{aligned}
G_{p+l}(\bar{x}) &= \sum_{k=1}^{J_{p+l}} \tilde{c}^{-1}_{(p+l)k}(\beta)\tilde{B}^{-1}_{(p+l)k}(\alpha)\prod_{l=1}^{m} x_l^{\tilde{\gamma}^{-1}_{(p+l)kl}(\beta)} \\
&= hx_l, \\
G_{p+m+l}(\bar{x}) &= \sum_{k=1}^{J_{p+m+l}} \tilde{c}^{-1}_{(p+m+l)k}(\beta)\tilde{B}^{-1}_{(p+m+l)k}(\alpha)\prod_{l=1}^{m} x_l^{\tilde{\gamma}^{-1}_{(p+m+l)kl}(\beta)} \\
&= hx_l^{-1}(1 \leqslant l \leqslant m).
\end{aligned}$$

Here,

$$\tilde{g}_{p+l}(x)\tilde{=}hx_l \iff \begin{cases} \tilde{g}_{p+l}(x) \lesssim hx_l \\ -\tilde{g}_{p+l}(x) \gtrsim -hx_l \end{cases} \iff \begin{cases} G_{p+l}(\bar{x}) \leqslant hx_l \\ -G_{p+l}(\bar{x}) \geqslant -hx_l \end{cases}$$

i.e., it is equivalent to $G_{p+l}(\bar{x}) = hx_l^{-1}(1 \leqslant l \leqslant m)$.

Similarly, we know $G_{p+m+l}(\bar{x}) = hx_l^{-1}$ holds.

$(\tilde{P})$ fuzzy super-consistence is equivalent to $(\bar{P})$ super-consistence, i.e., it is equivalent to $(\bar{P}_h)$ super-consistence when h is sufficiently small, such that it is equivalent to the $(\tilde{P}_h)$ fuzzy super-consistence.

Again, from Theorem 3.1.1, the dual form of $(\bar{P})$ is $(\bar{D})$, while that of $(\bar{P}_h)$ is

$$\begin{aligned}
(\bar{D}_h) \qquad & \sup\ d_k(\bar{w},\bar{u},\bar{v}) = d(\bar{w})h^{\sum_{l=1}^{m}(\bar{u}_l+\bar{v}_l)} \\
& \text{s.t. } \bar{w}_{00} = 1 \\
& \quad \bar{\Gamma}^{\mathrm{T}}\bar{w} + \bar{u} - \bar{v} = 0, \alpha, \beta \in [0,1] \\
& \quad \bar{w} \geqslant 0
\end{aligned}$$

where $\bar{u}_l = u(\bar{x}')$ and $\bar{v}_l = v(\bar{x}')$ are corresponding dual variables of $G_{p+l}(\bar{x})$ and $G_{p+m+l}(\bar{x})$, respectively.

Because a parametric feasible solution set of $(\bar{P}_h)$ is a compact set, $(\bar{P}_h)$ has a constraint minimum point x^h for sufficiently small h, such that we know from Theorem 2.4.1, the dual form $(\bar{D}_h)$ of $(\bar{P}_h)$ has an optimal parametric solution $(\bar{w}^h, \bar{u}^h, \bar{v}^h)$, and

$$-\alpha^h = d_h(\bar{w}^h, \bar{u}^h, \bar{v}^h) \leqslant d(\bar{w}^h). \tag{3.10}$$

Let $h \to 0$; obviously, $-\alpha^h \to -M_{\bar{P}}$. It is easy to prove that $\bar{w}^h$ has a constraint-condition limit point $\bar{w}^*$ satisfying $(\bar{D})$ in a similar way to Theorem 2.1.1 in Reference [WY82] and by Theorem 3.1.2 in this section. Hence, on the one hand, from (3.10), we have

$$M_{\overline{P}} \leqslant d(\bar{w}^*) \leqslant M_{\overline{D}}.$$

On the other hand, from proof of Theorem 3.1.2, we have $M_{\bar{D}} \leqslant M_{\bar{P}}$. Therefore,

$$M_{\bar{D}} = d(\bar{w}^*) = M_{\bar{P}} \tag{3.11}$$

holds.

Since there must be a limit point $\bar{w}^*$ in $\bar{w}^h$, obviously, $\bar{w}^*_{00} = 1, \bar{w}^* \geqslant 0$, and from

$$\bar{\Gamma}^{\mathrm{T}}\bar{w}^h + \bar{u}^h - \bar{v}^h = 0$$

and $\bar{u}^h \to 0, \bar{v}^h \to 0$, we know $\Gamma^{\mathrm{T}}\bar{w}^* = 0$, hence $\bar{w}^*$ is truly a parametric feasible solution to $(\bar{D})$.

Again, according to the equivalence of $(\tilde{P})$ and $(\bar{P}_h)$, $(\tilde{D})$ and $(\bar{D}_h)$, and to the equivalent transitivity between them, $\bar{w}^*$ is equivalent to w^*, such that

$$(3.11) \Longleftrightarrow (3.6).$$

Therefore, the theorem is true. □

3.2 Fuzzy Subconsistency and Converse Theorem

3.2.1 Fuzzy Subconsistency

Definition 3.2.1 If the fuzzy PGP

$$(\tilde{P}^\theta) \qquad \begin{aligned} &\widetilde{\inf}\ \theta\tilde{g}_0(x) \\ &\text{s.t. } \theta\tilde{g}_i(x) \lesssim 1\ (1 \leqslant i \leqslant p) \\ &\quad x > 0 \end{aligned}$$

presents an identically fuzzy consistence to all $\theta \in (0,1)$ and a CSM function $\tilde{A}_i (0 \leqslant i \leqslant p)$, then we call $(\tilde{P})$ a fuzzy sub-consistency or conditional groups $\tilde{g}_i(x) \lesssim 1\ (1 \leqslant i \leqslant p)$ a fuzzy sub-consistency.

For $\theta < 1$, we can prove the truth of $\theta g_i(x,\beta) < g_i(x,\beta)$ in a similar way to the Reference [WY82], such that, $\theta\tilde{g}_i(x) < \tilde{g}_i(x)$, so the fuzzy consistency of $(\tilde{P})$ contains its fuzzy sub-consistency. In the converse case, it is no longer valid.

Example 3.1

$$\begin{aligned} &\widetilde{\inf}\ x \\ &\text{s.t. } 1+x \lesssim 1 \\ &\quad x > 0 \end{aligned}$$

Obviously, this is not a fuzzy consistence, but the problem is fuzzy sub-consistent because

$$\begin{aligned} &\widetilde{\inf}\ x \\ &\text{s.t. } \theta(1+x) \lesssim 1 \\ &\quad x > 0 \end{aligned}$$

always means a consistence for $\theta \in (0,1)$.

Therefore, for $\forall\theta \in (0,1)$, the fuzzy sub-consistence of $(\tilde{P})$ contains the fuzzy super-consistence of $(\tilde{P}^\theta)$.

Proposition 3.2.1 *Let $\tilde{A}_i (0 \leqslant i \leqslant p)$ be a CSM function and $(\tilde{P})$ be fuzzy sub-consistent. Then $(\tilde{P}^\theta)$ is fuzzy consistent. If $M_{\tilde{P}^\theta}$ stands for the fuzzy constraint infimum of $(\tilde{P}^\theta)$, then $M_{\tilde{P}^\theta}$ is a fuzzy nondecreasing function of θ, with limit*

$$\mu_{\tilde{P}} = \lim_{\theta \to 1^-} M_{\tilde{P}^\theta}$$

existing (or probably $+\infty$). ◊

Proof From an assumption of the proposition, problem $(\tilde{P})$ is equivalent to $(\overline{P})$. Again, $(\tilde{P})$ is fuzzy sub-consistent, i.e. it is equivalent that $(\overline{P})$ is sub-consistent, such that, for certain $\alpha, \beta \in [0,1]$, $(\overline{P})$ becomes $(\overline{P})_\alpha$ which represents an ordinary PGP and a sub-consistence. Therefore, we have three points as follows:

1^0 $(\overline{P})_\alpha$ is consistent.

2^0 $M_{(\overline{P}^\theta)_\alpha}$, the constraint infimum of $(\overline{P}^\theta)_\alpha$, is a nondecreasing function of θ.

3^0 $\mu_{\bar{P}} = \lim_{\theta\to 1^-} M_{(\overline{P}^\theta)_\alpha}$ exists (maybe it is equal to $+\infty$) [WY82].

Because α, β is arbitrary in $[0, 1]$, the proposition holds. □

Definition 3.2.2 $\mu_{\tilde{P}}$ is defined as a fuzzy subinfimum of the fuzzy PGP $(\tilde{P})$.

Theorem 3.2.1 *Suppose $\tilde{A}_i (1 \leqslant i \leqslant p)$ to be a CSM function. Then if $(\tilde{P})$ is fuzzy sub-consistent and $\mu_{\tilde{P}} > 0$, $(\tilde{D})$ must be fuzzy consistent, with*

$$M_{\tilde{D}} = \mu_{\tilde{P}} \tag{3.12}$$

i.e., $M_{\tilde{D}}$ and $\mu_{\tilde{P}}$ are either equal to $+\infty$ or two equally finite fuzzy numbers. ◊

Proof Since $(\tilde{P})$ is equivalent to $(\overline{P})$, and a dual form of $(\overline{P})$ stands for $(\overline{D})$, then $(\tilde{P})$ fuzzy sub-consistence and $\mu_{\tilde{P}} > 0$ are equivalent to $(\overline{P})$ sub-consistence and $\mu_{\overline{P}} > 0$. Therefore, under the assumption of sub-consistence of $(\overline{P})$ and $\mu_{\overline{P}} > 0$, it is not difficult to prove, as in the proof in Theorem 2.2.1 in Reference [WY82], $(\overline{D})$ must be consistent, with $M_{\overline{D}} = \mu_{\overline{P}}$, i.e., $M_{\overline{D}}$ and $\mu_{\overline{P}}$ are together equal to $+\infty$, or two equal finite parameters. Again, according to equivalence of $(\tilde{P})$ and $(\overline{P})$, the duality of $(\tilde{P})$ and $(\tilde{D})$, as well as $(\overline{P})$ and $(\overline{D})$, and their transitive equivalence where there exists $(\tilde{D})$ equivalent to $(\overline{D})$, we know the theorem holds. □

There must exist $M_{\tilde{D}} > 0$ if $(\tilde{D})$ is fuzzy consistent. Theorem 3.2.1 holds provided that $\mu_{\tilde{P}} > 0$ exists according to formula (3.12). The condition of $\mu_{\tilde{D}} > 0$ is necessary even if we only assure a fuzzy consistence of $(\tilde{D})$, otherwise, we fail to assure it.

Take Example 3.1 for example, its fuzzy subinfimum is equal to 0, correspondingly, fuzzy dual programming

$$\begin{aligned}
&\sup\Big(\frac{\tilde{a}_{00}}{w_{00}}\Big)^{w_{00}}\Big(\frac{1}{\tilde{a}_{01}w_{01}}\Big)^{w_{01}}\Big(\frac{1}{\tilde{a}_{11}w_{11}}\Big)^{w_{11}}\Big(\frac{1}{\tilde{a}_{12}w_{12}}\Big)^{w_{12}}\\
&\quad (w_{11}+w_{12})^{w_{11}+w_{12}}\\
&\text{s.t. } w_{00} = 1\\
&\quad w_{01} = w_{00}\\
&\quad w_{11} + w_{12} = 0\\
&\quad w \geqslant 0
\end{aligned}$$

is not fuzzy consistent.

3.2.2 $M_{\tilde{P}} = \mu_{\tilde{P}}$

As for consistent fuzzy PGP $(\tilde{P})$, due to fuzzy sub-consistence, there exists $\mu_{\tilde{P}} \leqslant M_{\tilde{P}}$ for its fuzzy constraint infimum $M_{\tilde{P}}$ and fuzzy subinfimum $\mu_{\tilde{P}}$; take for example

$$\begin{aligned} &\widetilde{\inf}\ \frac{1}{x} \\ &\text{s.t. } 1+x \lesssim 1 \\ &\qquad x > 0 \end{aligned}$$

$M_{\tilde{P}} = 1+\infty$, while $\mu_{\tilde{P}}$ is a finite number, i.e $\mu_{\tilde{P}} < M_{\tilde{P}}$. Under what condition does $\mu_{\tilde{P}} = M_{\tilde{P}}$? Here we have an account of it as follows:

Theorem 3.2.2 *Suppose that $\tilde{A}_i (0 \leqslant i \leqslant p)$ represents a CSN function. If fuzzy PGP $(\tilde{P})$ is fuzzy consistent, there must exist*

$$M_{\tilde{P}} = \mu_{\tilde{P}}.$$

◊

Proof From the theorem assumption that $(\tilde{P})$ is equivalent to $(\overline{P})$, $(\tilde{P})$ is changed into an ordinary PGP $(\overline{P})$ for certain $\alpha_0, \beta_0 \in [0,1]$. Therefore, $(\tilde{P})$ fuzzy consistence is equivalent to $(\overline{P})$ consistence, such that it is equivalent to $(\overline{P})_{\alpha_0\beta_0}$, for which there must exist $M_{(\overline{P})_{\alpha_0\beta_0}} = \mu_{(\bar{P})_{\alpha_0\beta_0}}$ by applying Theorem 2.3.5 in Reference [WY82]. From the arbitrariness of α_0, β_0 in $[0,1]$, we know $M_{\overline{P}} = \mu_{\overline{P}}$, such that $M_{\tilde{P}} = \mu_{\tilde{P}}$ holds. □

Theorem 3.2.3 *Let $\tilde{A}_i (0 \leqslant i \leqslant p)$ be a CSN function. If $(\tilde{P})$ is fuzzy consistent, with $M_{\tilde{P}} > 0$, so is $(\tilde{D})$. And fuzzy constraint supremum $M_{\tilde{D}} = M_{\tilde{P}}$ is a finite fuzzy number.* ◊

Proof The theorem holds by combining Theorem 3.2.1 with Theorem 3.2.2. □

3.2.3 Fuzzy Converse Theorem

From Theorem 3.2.1, we know that once $(\tilde{P})$ is fuzzy sub-consistent having fuzzy subinfimum $\mu_{\tilde{P}} > 0$, $(\tilde{D})$ must be fuzzy consistent having $M_{\tilde{D}} = \mu_{\tilde{P}}$, and vice versa.

Theorem 3.2.4 (fuzzy converse theorem) *Let $\tilde{A}_i (0 \leqslant i \leqslant p)$ be a CSN function. If a dual program $(\tilde{D})$ is fuzzy consistent and $M_{\tilde{D}} < \infty$, then $(\tilde{P})$ must be fuzzy sub-consistent, and*

$$\mu_{\tilde{P}} = M_{\tilde{D}}.$$

◊

Proof We only need to prove $(\tilde{P})$ a fuzzy sub-consistence. But in the assumption of the theorem, $(\tilde{P})$ is equivalent to an ordinary PGP $(\overline{P})$ with parameters. Therefore, $(\tilde{P})$ fuzzy consistence is evolved from sub-consistence of $(\overline{P})$, such that the problem means only confirming sub-consistence of $(\overline{P})$.

A proof by contradiction will be adopted for the following.

Conversely, suppose $(\overline{P})$ to be nonsubconsistent. Then there must exist p' suitable for $1 \leqslant p' \leqslant p$ and $\bar{\theta}$ suitable for $1 < \bar{\theta} < p$, such that

$$\bar{\theta} G_i(\overline{x}) \leqslant 1 \ (1 \leqslant i \leqslant p'), x > 0, \alpha, \beta \in [0,1]$$

contains no parameter feasible solution. But for all $\theta \in (0,1)$,

$$\theta G_i(\overline{x}) \leqslant 1 \ (1 \leqslant i \leqslant p' - 1), x > 0, \alpha, \beta \in [0,1] \tag{3.13}$$

contains a parameter feasible solution. So, for all θ in $[\bar{\theta}, 1)$ and any parameter feasible solution to (3.13), we have

$$\theta G_p(\overline{x}) > \theta/\bar{\theta} \geqslant 1, \alpha, \beta \in [0,1]. \tag{3.14}$$

Now we inspect a sub-consistent PGP with parameters

$$(\overline{P}') \qquad \begin{array}{l} \inf\ G_{p'}(\overline{x}) \\ \text{s.t. } G_i(\overline{x}) \leqslant 1, (1 \leqslant i \leqslant p' - 1) \\ \quad \alpha, \beta \in [0,1] \\ \quad x > 0 \end{array}$$

and its dual program

$$(\overline{D}') \qquad \begin{array}{l} \sup\ d'(\overline{\lambda}) = \prod\limits_{i=1}^{p'} \prod\limits_{k=1}^{J_i} \left(\dfrac{\tilde{c}_{ik}^{-1}(\beta)/\tilde{B}_i^{-1}(\alpha)}{\overline{\lambda}_{ik}}\right)^{\overline{\lambda}_{ik}} \prod\limits_{i=0}^{p'} \overline{\lambda}_{i0}^{\overline{\lambda}_{i0}} \\ \text{s.t. } \sum\limits_{i=1}^{p'} \sum\limits_{k=1}^{J_i} \tilde{\gamma}_{ikl}^{-1}(\beta)\overline{\lambda}_{ik} = 0, \alpha, \beta \in [0,1] (0 \leqslant l \leqslant m) \\ \quad \overline{\lambda}_{ik} \geqslant 0; \end{array}$$

here $\overline{\lambda} = (\overline{\lambda}_{01}, \ldots, \overline{\lambda}_{0J_0}, \overline{\lambda}_{11}, \ldots, \overline{\lambda}_{p'1}, \ldots, \overline{\lambda}_{p'J_{p'}})^{\mathrm{T}}, \overline{\lambda} = \lambda(\alpha, \beta)$ and $\overline{\lambda}_{ik} = \lambda_{ik}(\alpha, \beta)$.

Since (3.14) holds for all $\alpha, \beta \in [0,1]$, $\theta \in [\bar{\theta}, 1)$ and all feasible solutions x in (3.13), then $\mu_{\overline{P}'} \geqslant 1/\bar{\theta} > 1$. Again, from Theorem 3.2.1, we know $(\overline{D}')$ is consistent, with $M_{\overline{D}'} = \mu_{\overline{P}'} \geqslant 1/\bar{\theta} > 1$. Hence, a feasible solution $\overline{\lambda}$ of $(\overline{D}')$ exists, such that

$$\overline{d}'(\overline{\lambda}) > \frac{1 + \overline{\theta}}{2\overline{\theta}} > 1. \tag{3.15}$$

Now, let $\bar{w}$ be any feasible solution to $(\overline{D}')$. Given $\bar{w}' = (\bar{w}_{01}, \ldots, \bar{w}_{0J_0}, \bar{w}_{11}, \ldots, \bar{w}_{1J_1}, \ldots, \bar{w}_{p'1}, \ldots, \bar{w}_{p'J_{p'}})^{\mathrm{T}}$, from (3.15) and its continuity, for sufficiently large M, we have

$$\overline{d}'(\overline{\lambda} + \bar{w}'/M) \geqslant \frac{1 + \overline{\theta}}{2\overline{\theta}} > 1,$$

and if

$$\overline{\lambda}_{ik} = 0(i = 0, p' + 1, \ldots, p'; 1 \leqslant k \leqslant J_i) \text{ and } \overline{\lambda}_0 = 0, \qquad (3.16)$$

obviously the extended $\overline{\lambda} = (\overline{\lambda}_0, \overline{\lambda}_{01}, ..., \overline{\lambda}_{0J_0}, ..., \overline{\lambda}_{p1}, \ldots, \overline{\lambda}_{pJ_p})^{\mathrm{T}}, \bar{w} + M\overline{\lambda}$ is also a parametric feasible solution to $(\overline{D})$, and because of $d'(\overline{\lambda}) = d'(\overline{\lambda}/M)^{\mathrm{M}}$, we have

$$\frac{d(\bar{w} + M\overline{\lambda})}{d(\bar{w})} = \frac{d'(\bar{w}' + M\overline{\lambda})}{d'(\bar{w}')} = \frac{d'(\overline{\lambda} + \bar{w}'/M)^{\mathrm{M}}}{d'(\bar{w}')} > \frac{1}{\overline{d}'(\bar{w}')}(\frac{1 + \overline{\theta}}{2\overline{\theta}})^{\mathrm{M}}.$$

Then, $M_{\overline{D}} \geqslant d(\overline{w} + M\overline{\lambda}) \to \infty$ $(M \to \infty)$, which contradicts $M_{\overline{D}}$, a finite parameter. Semantically, $(\overline{D})$ must be sub-consistent and

$$\mu_{\overline{P}} = M_{\overline{D}}. \qquad (3.17)$$

Therefore, from the theorem assumption, there exists an inverse function $\tilde{A}_i^{-1}$ of $\tilde{A}_i$, such that $(\overline{D})$ is equivalent to $(\tilde{D})$. $(\overline{D})$ sub-consistence is equivalent to $(\tilde{D})$ fuzzy sub-consistence, and the truth of (3.17) means that of (3.12), so that the theorem is true. □

Corollary 3.2.1 *As for a CSM function $\tilde{A}_i(0 \leqslant i \leqslant p)$, if an optimum solution w^* exists in a fuzzy dual program $(\tilde{D})$, then $(\tilde{P})$ must be fuzzy sub-consistence, with*

$$\mu_{\tilde{P}} = \tilde{d}(w^*)$$

◊

Proof The corollary can be proved without difficulty in the same way as in Theorem 3.2.4. □

It follows that condition $M_{\tilde{D}} < \infty$ can not be omitted in Theorem 3.2.4, therefore, we have the following theorem:

Theorem 3.2.5 (fuzzy strong dual theorem) *Let $\tilde{A}_i(0 \leqslant i \leqslant p)$ be a CSN function. Then $(\tilde{P})$ is fuzzy sub-consistent and $0 < \mu_{\tilde{P}} < \infty$ iff $(\tilde{D})$ is fuzzy consistent, with $M_{\tilde{D}} < \infty$ and $M_{\tilde{D}} = \mu_{\tilde{P}}$.* ◊

Proof The theorem holds by combining Theorems 3.2.1 and 3.2.4. □

If the condition $M_{\tilde{D}} < \infty$ is removed from it, the conclusion in Theorem 3.2.4 generally does not hold, accordingly neither does Theorem 3.2.5.

Example 3.2 Fuzzy PGP is

$$\begin{aligned} &\widetilde{\inf}\ x \\ &\text{s.t. } \frac{1}{x} + x \lesssim 1 \\ &\quad x > 0 \end{aligned} \qquad (3.18)$$

Its corresponding parametric programming is

$$\begin{aligned} &\sup\ \alpha \\ &\text{s.t.}\ \tilde{B}_0(\alpha) \geqslant \alpha \\ &\quad\ \tilde{B}_1(\frac{1}{x} + x) \geqslant \alpha \\ &\quad\ \alpha \in [0,1], x > 0 \end{aligned} \tag{3.19}$$

while (3.19) can be turned into

$$\begin{aligned} &\sup\ \alpha = \inf(-\alpha) \\ &\text{s.t.}\ x \leqslant \tilde{B}_0^{-1}(\alpha) \\ &\quad\ \frac{1}{x} + x \leqslant \tilde{B}_1^{-1}(\alpha) \\ &\quad\ \alpha \in [0,1], x > 0 \end{aligned} \tag{3.20}$$

From $\theta(\frac{1}{x} + x) \leqslant \tilde{B}_1^{-1}(\alpha)$, and by proceeding to $x^2 - \frac{1}{\theta}\tilde{B}_1^{-1}(\alpha)x + 1 = 0$, we find $x = \frac{\tilde{B}_1^{-1}(\alpha)/\theta \pm \sqrt{(\tilde{B}_1^{-1}(\alpha))^2/\theta^2 - 4}}{2}$. Now we take only $\tilde{B}_i^{-1}(\alpha)(i = 0,1) \in [0,1]$, obviously (3.20) is not sub-consistent, such that (3.18) is not fuzzy sub-consistent, while a corresponding parametric dual problem of (3.20) is

$$\begin{aligned} &\sup(\frac{-\alpha}{\bar{w}_{00}})^{\bar{w}_{00}}(\frac{1}{\tilde{B}_0^{-1}(\alpha)\bar{w}_{01}})^{\bar{w}_{01}}(\frac{1}{\tilde{B}_1^{-1}(\alpha)\bar{w}_{11}})^{\bar{w}_{11}} \\ &\quad(\frac{1}{\tilde{B}_1^{-1}(\alpha)\bar{w}_{12}})^{\bar{w}_{12}}(\bar{w}_{11} + \bar{w}_{12})^{(\bar{w}_{11}+\bar{w}_{12})} \\ &\text{s.t.}\ \bar{w}_{00} = 1 \\ &\quad\ \bar{w}_{01} = \bar{w}_{00} \\ &\quad\ \bar{w}_{01} - \bar{w}_{11} + \bar{w}_{12} = 0 \\ &\quad\ \alpha \in [0,1], \bar{w} \geqslant 0 \end{aligned} \tag{3.21}$$

Obviously, it is consistent, and

$$\forall \bar{w}_{12} \geqslant 0, \bar{w}_{01} = 1, \bar{w}_{11} = 1 + \bar{w}_{12}, \bar{w}_{12}$$

are all of its dual feasible solutions; the reason is that a dual objective function is

$$\begin{aligned} d' &= \frac{-\alpha}{\tilde{B}_0^{-1}(\alpha)}(\frac{1/\tilde{B}_1^{-1}(\alpha)}{1+\bar{w}_{12}})^{1+\bar{w}_{12}}(\frac{1/\tilde{B}_1^{-1}(\alpha)}{\bar{w}_{12}})^{\bar{w}_{12}}(1+2\bar{w}_{12})^{1+2\bar{w}_{12}} \\ &= \frac{-\alpha}{\tilde{B}_0^{-1}(\alpha)}(\frac{1+2\bar{w}_{12}}{\tilde{B}_1^{-1}(\alpha)(1+\bar{w}_{12})})^{1+\bar{w}_{12}}(\frac{1+2\bar{w}_{12}}{\tilde{B}_1^{-1}(\alpha)\bar{w}_{12}})^{\bar{w}_{12}} \\ &\geqslant \frac{-\alpha}{\tilde{B}_0^{-1}(\alpha)}\frac{2^{\bar{w}_{12}}}{(\tilde{B}_1^{-1}(\alpha))^{1+2\bar{w}_{12}}}. \end{aligned}$$

Then $+\infty$ becomes the constraint supremum of parametric dual program (3.21). Especially, if we take $\alpha = 0$, then

$$d' \geqslant \frac{2^{\bar{w}_{12}}}{\tilde{B}_0^{-1}(0)\left(\tilde{B}_1^{-1}(0)\right)^{1+\bar{w}_{12}}} \geqslant 2^{\bar{w}_{12}} \to \infty(\bar{w}_{12} \to \infty);$$

the constraint infimum of the dual programming obviously is $+\infty$. And (3.21) is equivalent to (3.18), while corresponding fuzzy dual programming of (3.18)

$$\begin{aligned} \widetilde{\sup}\ & \left(\frac{\tilde{a}_{00}}{w_{00}}\right)^{w_{00}}\left(\frac{1}{\tilde{a}_{01}w_{01}}\right)^{w_{01}}\left(\frac{1}{\tilde{a}_{11}w_{11}}\right)^{w_{11}}\left(\frac{1}{\tilde{a}_{12}w_{12}}\right)^{w_{12}} \\ & (w_{11}+w_{12})^{w_{11}+w_{12}} \\ \text{s.t. }\ & w_{00} = 1 \\ & w_{01} = w_{00} \\ & w_{01} - w_{11} + w_{12} = 0 \\ & w \geqslant 0 \end{aligned} \tag{3.22}$$

is fuzzy consistent. The reason is that $+\infty$ becomes the fuzzy constraint supremum of (3.22).

3.3 Fuzzy Monomial PGP and Its Equivalence

Definition 3.3.1 Let $g_i(x) = c_i \prod_{l=1}^{m} x_l^{\gamma_{il}} (0 \leqslant i \leqslant p)$ be monomial. Then we call

$$\begin{aligned} &\widetilde{\min}\ g_0(x) \\ &\text{s.t. } g_i(x) \lesssim 1\ (1 \leqslant i \leqslant p) \\ &\quad\ x > 0 \end{aligned} \tag{3.23}$$

a monomial fuzzy PGP.

Theorem 3.3.1 *Suppose that the monomial fuzzy PGP is (3.23). For a CSM function $\tilde{A}_i(1 \leqslant i \leqslant p)$, (3.23) can always be changed into a fuzzy LP*

$$\begin{aligned} &\widetilde{\min} \sum_{l=1}^{m} \gamma_{0l} z_l \\ &\text{s.t. } \sum_{l=1}^{m} \gamma_{il} z_l + \log c_i \lesssim 0\ (1 \leqslant i \leqslant p) \end{aligned} \tag{3.24}$$

or

$$\begin{aligned} &\widetilde{\max} \sum_{l=1}^{m} (-\gamma_{0l} z_l) \\ &\text{s.t. } \sum_{l=1}^{m} \gamma_{il} z_l \lesssim -\log c_i\ (1 \leqslant i \leqslant p) \end{aligned} \tag{3.25}$$

◊

Proof Let $z_l = \log\ x_l (1 \leqslant l \leqslant m)$. We might as well take $c_0 = 1$, then

$$g_i(x) = c_i e^{\gamma_{il} \sum_{l=1}^{m} z_l}$$

such that (3.23) can be changed into

$$\begin{aligned} &\widetilde{\min}\ e^{\gamma_{0l} \sum_{l=1}^{m} z_l} \\ &\text{s.t. } c_i e^{\gamma_{il} \sum_{l=1}^{m} z_l} \lesssim 1 (1 \leqslant i \leqslant p) \end{aligned} \tag{3.26}$$

For a CSM fuzzy-valued function $\tilde{A}_i \in (0, 1)\ (0 \leqslant i \leqslant p)$, (3.26) is equivalent to

$$\begin{aligned} &\max \alpha = \min(-\alpha) \\ &\text{s.t. } \gamma_{0l} \sum_{l=1}^{m} z_l \leqslant \log \tilde{B}_0^{-1}(\alpha) \\ &\quad \gamma_{il} \sum_{l=1}^{m} z_l + \log c_i \leqslant \log \tilde{B}_i^{-1}(\alpha)(1 \leqslant i \leqslant p) \end{aligned} \tag{3.27}$$

Obviously, (3.24) can be changed into (3.27). Semantically, the theorem holds. □

Theorem 3.3.2 *The dual programming of (3.23) is*

$$\begin{aligned} &\widetilde{\max} \prod_{i=1}^{p} c_i^{w_i} \\ &\text{s.t. } w_0 = 1 \\ &\quad \sum_{i=1}^{p} \gamma_{il} w_i = 0 \; (1 \leqslant l \leqslant m) \\ &\quad w \geqslant 0 \end{aligned} \tag{3.28}$$

and (3.28) with

$$\begin{aligned} &\widetilde{\max} \sum_{i=1}^{p} (\log c_i) w_i \\ &\text{s.t. } \sum_{i=1}^{p} \gamma_{il} w_i = -\gamma_{0l} \; (1 \leqslant l \leqslant m) \\ &\quad w \geqslant 0 \end{aligned}$$

or

$$\begin{aligned} &\widetilde{\min} \sum_{i=1}^{p} (-\log c_i) w_i \\ &\text{s.t. } \sum_{i=1}^{p} \gamma_{il} w_i = -\gamma_{0l} \; (1 \leqslant l \leqslant m) \\ &\quad w \geqslant 0 \end{aligned} \tag{3.29}$$

keeps the same constraint conditions and fuzzy optimum points, while (3.25) and (3.29) are exactly a pair of fuzzy dual LP. ◊

Proof This theorem is a special case of Theorem 2.3.1 for $k = 1$. Now (3.28) holds. And it is easy to prove the truth of the theorem with the aid of Theorem 3.2.1. □

Therefore, we know, a fuzzy optimal solution to the monomial fuzzy PGP can be acquired by changing the latter into an equivalent fuzzy LP, while we can construct an infinite monomial fuzzy PGP for a non-monomial fuzzy PGP by applying a fuzzy geometric inequality.

Given a primal fuzzy PGP $(\tilde{P})$, a J−dimensional nonnegative vector satisfying $\sum_{k=1}^{J_i} \varepsilon_{ik} = 1 (0 \leqslant i \leqslant p)$:

$$\varepsilon = (\varepsilon_{01}, \ldots, \varepsilon_{0J_0}, \ldots, \varepsilon_{p1}, \ldots, \varepsilon_{pJ_p})^{\mathrm{T}} \geqslant 0;$$

from Theorem 2.5.1, we have

$$\bar{\tilde{g}}_i(x) = \prod_{k=1}^{J_i} \Big(\frac{\tilde{c}_{ik}}{\varepsilon_{ik}} \prod_{l=1}^{m} x_l^{\gamma_{ikl}}\Big)^{\varepsilon_{ikl}} = \bar{\tilde{c}}_i \prod_{l=1}^{m} x_l^{\bar{\gamma}_{ikl}} \lesssim \tilde{g}_i(x) \tag{3.30}$$

where

$$\tilde{\bar{c}}_i = \prod_{k=1}^{J_i} \left(\frac{c_{ik}}{\varepsilon_{ik}}\right)^{\varepsilon_{ik}}, \quad \overline{\gamma}_{il} = \sum_{k=1}^{J_i} \gamma_{ikl}\varepsilon_{ik}.$$

We can now change $(\tilde{P})$ into a monomial fuzzy PGP

$$(\tilde{P}_\varepsilon) \qquad \widetilde{\min}\, \tilde{\bar{g}}_0(x) \quad \text{s.t. } \tilde{\bar{g}}_i(x) \lesssim 1 \; (1 \leqslant i \leqslant p), \quad x > 0$$

Example 3.3 Find a monomial fuzzy PGP [Cao87a,b]

$$\begin{aligned} &\max x_1^5 x_2^{-2} \\ &\text{s.t. } x_1^2 \lesssim 20 \\ &\quad x_2^2 \lesssim 3 \\ &\quad x_1 x_2^3 \lesssim 30 \\ &\quad x_1 \geqslant 0, x_2 > 0 \end{aligned} \tag{3.31}$$

Solution 1^0 Let $z_l = \log x_l$ $(l = 1, 2)$. Then problem (3.31) can be turned into a fuzzy LP

$$\begin{aligned} &\max(5z_1 - 2z_2) \\ &\text{s.t. } z_1 \lesssim 1.5 \\ &\quad z_2 \lesssim 0.55 \\ &\quad z_1 + 3z_2 \lesssim 3.4 \end{aligned} \tag{3.32}$$

2^0 Because of a minor constant term in constraint condition $\tilde{C}$, we had better select

$$\tilde{C}(d_i) = \begin{cases} 1 - 20d_i, & 0 \leqslant d_i \leqslant 0.5 \\ 0, & d_i > 0.5, \end{cases}$$

then, $d_i = 0.05 \Rightarrow \sup\limits_{z \in C_{0.05}} \alpha z = 6.48 \Rightarrow \tilde{M} = 0.7716z_1 - 0.3086z_2$.

3^0 Find a fixed point $\overline{\alpha} = \sup\limits_{z \in C_{\overline{\alpha}}} \tilde{M}(z)$.

Since $\tilde{B}(G_0(z))$ increases slowly, for $\varepsilon = 0.001$, we would better select $\alpha_1 = 0.7, r_k = 0.618$ before we get the approximately fuzzy optimal solution to (3.32) through ten steps of calculations: $z_1 = 1.53136, z_2 = 0.628, \alpha_{10} = 0.98681$, and the approximately fuzzy optimal value: $G_0(z) = 6.4008$.

4^0 Substitute back.

A fuzzy satisfactory solution can be obtained to the primal problem (3.31) by means of substituting the fuzzy optimal solution for (2.14), i.e., $x_1 = 4.6245, x_2 = 1.8739$ and the fuzzy optimal value $g_0(x) = 602.33$.

This conforms with practice.

Note: If we select

$$\tilde{C}(d_i) = \begin{cases} 1 - 2d_i, & 0 \leqslant d_i \leqslant 0.5 \\ 0, & d_i > 0.5, \end{cases}$$

then $\tilde{M} = 0.5725z_1 - 0.229z_2$. Given $\alpha_0 = 0.75, \varepsilon = 0.01$, for $r_i = 0.5$, we obtain $\alpha_9 = 1.0053 > 1, \tilde{M}^{(8)} = 1.0142 > 1$, through eight steps, but now it is inconvenient for calculation. For $r_i = 0.618$, we have $\tilde{M}^{(9)} = 1.00257 > 1, \alpha_{10} > 1$, through nine steps. Although the solution to (3.32) can be calculated as: $z_1 = 1.9599, z_2 = 0.5217, \varepsilon_9 < \varepsilon, G_0 = 8.7561$, by means of back substitution, the objective function value for (3.31) is $g_0(x) = 6349.3$, which differs greatly from practical numbers.

Therefore, when a direct algorithm of fuzzy LP [Asa82] is used to solve a fuzzy GP which can be linearized, we shall make the following improvements.

1^0 Let $\tilde{M}^{(k)} = \sup_{z \in c_{\alpha_k}} |B(G_0(z))|$, such that a solution can be found when objective functions of linearized GP problem are negative values.

2^0 The choice of parameter in $1 - kd = \alpha_1$ is not arbitrary. Generally, when $G_0(z)$ strictly and monotonicity increases, we would better select $\alpha_1 \in [0.8, 0.9]$; when $G_0(z)$ does not increase strictly, we would better select $\alpha_1 \in [0.7, 0.75]$. Meanwhile, if constant b is very large, larger, smaller and very small in constraint $\tilde{C}, k$ values are taken as 0.02, 0.2, 2 and 20, respectively. Otherwise, if one of the cases above does not hold, through finite steps, we obtain: $\tilde{M}^{(k)} > 1, \alpha_k > 1$, but $\varepsilon_k > \varepsilon$, which contradicts $\tilde{M}^{(k)} \in [0,1], \varepsilon_k \in [0,1]$.

3^0 Choose r_k. It is proper to try calculating $r_k = 0.5$ (i.e., the middle point in the interval [0,1]) by use of the bisection method before dealing with the following according to the concrete situation. If $\varepsilon_1 \gg \varepsilon_2$, it is proper to select $r_k = 0.5$, but if $\varepsilon_1 \searrow \varepsilon_2$ changes a little, it is proper to select $r_k \in [0.618, 1]$ instead of $r_k = 0.5$. If $\varepsilon_1 \ll \varepsilon_2$, it is proper to select $r_k \in [0.382, 0.4]$, otherwise, we can gain the conclusion of $\tilde{M}^{(k)} > 1, \alpha_k > 1$, while $\varepsilon_k > \varepsilon$ [Cao87a,b].

Lastly, it is pointed out that, if the constraint in (3.23) looks like

$$\begin{cases} x_1^{\alpha_1} x_2^{\alpha_2} \lesssim k_1 \\ x_1^{\beta_1} x_2^{\beta_2} \lesssim k_2 \end{cases}$$

where α_i, β_i, k_i $(i = 1, 2)$ are all real constants, then the constraint of (3.23) can be changed into

$$\begin{cases} \alpha_1 z_1 + \alpha_2 z_2 \lesssim \log k_1 \\ \beta_1 z_1 + \beta_2 z_2 \lesssim \log k_2 \end{cases}$$

by supposing $x_i = e^{\log z_i} (i = 1, 2)$. (3.23) can be converted into a fuzzy LP with variables subject to lower and upper bounds for solution.

Readers can probe further into some nature concerning the work.

3.4 Another Proof of A Fuzzy PGP Dual Theorem

In Section 2.4 of Chapter 2, we demonstrated a fuzzy Kuhn-Tucker type theorem for a fuzzy convex programming, and proved the dual theorem of a fuzzy PGP. In this section, we shall give another proof of a fuzzy dual theorem [Cao92b,c].

Lemma 3.4.1 *For the CSN function $\tilde{A}_i (0 \leqslant i \leqslant p)$, with $\tilde{g}_0(x^*) = \tilde{d}(w^*)$, x^* must be fuzzy optimal solutions to $(\tilde{P})$ and w^* must be an optimal solution to $(\tilde{D})$, and for any feasible solution w in $(\tilde{D})$, we have*

$$\tilde{g}_0(x^*) \prod_{i=0}^{p} \prod_{k=1}^{J_i} (w_{ik}^*)^{w_{ik}^*} = \prod_{i=1}^{p} (w_{i0}^*)^{w_{i0}^*} \prod_{i=0}^{p} \prod_{k=1}^{J_i} \tilde{c}_{ik}^{w_{ik}^*}. \tag{3.33}$$

◊

Proof For any fuzzy feasible solution x to $(\tilde{P})$, from Lemma 4.1 in Reference [Cao93a], we have

$$\tilde{g}_0(x) \gtrsim \tilde{d}(w^*) = \tilde{g}_0(x^*),$$

therefore x^* must be a fuzzy optimal solution to $(\tilde{P})$.

In a similar way, we can prove that x^* and w^* must satisfy

$$w_{ik}^* = \begin{cases} \tilde{v}_{0k}(x^*)/\tilde{g}_0(x^*), & (i = 0; 1 \leqslant k \leqslant J_0) \\ w_{i0}^* \tilde{v}_{ik}(x^*), & (i \neq 0; 1 \leqslant k \leqslant J_i), \end{cases}$$

hence

$$(\tilde{g}_0(x^*) w_{0k}^*)^{w_{0k}^*} = (\tilde{v}_{0k}(x^*))^{w_{0k}^*}$$
$$(w_{ik}^*)^{w_{ik}^*} = (w_{i0}^* \tilde{v}_{ik}(x^*))^{w_{ik}^*} (i \neq 0).$$

Thus, from $w_{00}^* = 1$,

$$\tilde{g}_0(x^*) \prod_{i=0}^{p} \prod_{k=1}^{J_i} (w_{ik}^*)^{w_{ik}^*} = \prod_{i=1}^{p} (w_{i0}^*)^{w_{i0}^*} \prod_{i=0}^{p} \prod_{k=1}^{J_i} (\tilde{v}_{ik}(x^*))^{w_{ik}^*}$$

$$= \prod_{i=1}^{p} (w_{i0}^*)^{w_{i0}^*} \prod_{i=0}^{p} \prod_{k=1}^{J_i} \tilde{c}_{ik}^{w_{ik}^*} \prod_{l=1}^{m} x_l^{\sum\limits_{i=0}^{p} \sum\limits_{k=1}^{J_i} \tilde{\gamma}_{ikl} w_{ik}^*}$$

and from the orthogonal condition,

$$\tilde{\Gamma}^{\mathrm{T}} w^* = \sum_{i=0}^{p} \sum_{k=1}^{J_i} \tilde{\gamma}_{ikl} w_{ik}^* = 0 \; (1 \leqslant l \leqslant m).$$

Therefore, equation (3.33) is shown to be true. □

Theorem 3.4.1 *Let $\tilde{A}_i = \tilde{B}_i \circ \tilde{\phi}_i$ be a CSN function, where $\tilde{\phi}_i = \tilde{c}_{ik} \circ \tilde{\gamma}_{ikl} (1 \leqslant k \leqslant J_i, 0 \leqslant i \leqslant p, 1 \leqslant l \leqslant m)$. Suppose that x denotes a fuzzy* *feasible solution to $(\tilde{P})$, and so does x to a monomial fuzzy PGP $(\tilde{P}_\varepsilon)$. If $(\tilde{P})$ is fuzzy consistent, so is $(\tilde{P}_\varepsilon)$. Again $M_{\tilde{P}_\varepsilon} \lesssim M_{\tilde{P}}$, then, the dual form of $(\tilde{P}_\varepsilon)$ is*

$$
\begin{aligned}
(\tilde{D}_\varepsilon) \qquad & \max \ \left(\frac{\tilde{a}'_0}{w'_0}\right)^{w'_0} \prod_{i=0}^{p} \left(\frac{\tilde{\tilde{c}}_i \tilde{a}_i}{w_i}\right)^{w_i} \\
& \text{s.t. } w'_0 = 1 \\
& \quad \tilde{\tilde{\gamma}}_{01} w_0 = w'_0 \\
& \quad \sum_{i=1}^{p} \tilde{\tilde{\gamma}}_{il} \bar{w}_i = 0, (1 \leqslant l \leqslant m) \\
& \quad w \geqslant 0
\end{aligned}
$$

where $M_{\tilde{P}}$ and $M_{\tilde{P}_\varepsilon}$ denote the constraint infimum of $(\tilde{P})$ and of $(\tilde{P}_\varepsilon)$, respectively. $\tilde{\tilde{\gamma}}_{il}$ is an exponent of $\tilde{g}_i(x)(0 \leqslant i \leqslant p)$. $w = (w'_0, w_0, w_1, \ldots, w_p)^T$ is a $(p+2)$–dimensional nonnegative dual variable vector. ◊

Proof If we substitute $\tilde{\tilde{g}}_i(x)$ for $\tilde{g}_i(x)(0 \leqslant i \leqslant p)$, then we turn $(\tilde{P})$ into $(\tilde{P}_\varepsilon)$. Owing to

$$
\left(\sum_{k=1}^{J_i} \tilde{v}_k(x)\right)^{w_0} \gtrsim \prod_{k=1}^{J_i} (\tilde{v}_k(x)/w_k)^{w_k} w_0^{w_0}
$$

with $w_k/w_0 = \varepsilon_{ik}$, the ε_{ik} is viewed as a weight vector. From

$$
\begin{aligned}
& \prod_{k=1}^{J_i} (\tilde{v}_{ik}(x)/w_k)^{w_k/w_0} \lesssim \sum_{k=1}^{J_i} \tilde{v}_{ik}(x) \\
\Longrightarrow \quad & \prod_{k=1}^{J_i} (\tilde{v}_{ik}(x)/\varepsilon_{ik} w_0)^{\varepsilon_{ik}} = \prod_{k=1}^{J_i} (\tilde{v}_{ik}(x)/\varepsilon_{ik})^{\varepsilon_{ik}} \\
& = \tilde{\tilde{c}}_i \prod_{l=1}^{m} x_l^{\tilde{\tilde{\gamma}}_{il}} \overset{(3.30)}{\lesssim} \sum_{k=1}^{J_i} \tilde{v}_{ik}(x).
\end{aligned}
$$

Obviously, $\tilde{\tilde{g}}_i(x) \lesssim \tilde{g}_i(x)$ is deduced, such that a fuzzy feasible solution x to $(\tilde{P})$ is that to $(\tilde{P}_\varepsilon)$.

By introducing new variables α and β, $\alpha, \beta \in [0, 1]$, similar to References [Cao93a][Ver84] and by means of (1.14) in Section 1.6.3 in Chapter 1, we have

$$
\begin{aligned}
& \tilde{\phi}_i(\tilde{\tilde{c}}_i) \geqslant 1 - \alpha \\
\Longrightarrow \quad & \tilde{\tilde{c}}_i \geqslant \tilde{\phi}_i^{-1}(1 - \alpha).
\end{aligned}
$$

Writing $1 - \alpha = \beta$, and substituting $\tilde{\tilde{c}}_i^{-1}$ for $\tilde{\phi}_i^{-1}$, we arrive at $\tilde{\tilde{c}}_i \geqslant \tilde{\tilde{c}}_i^{-1}(\beta)$.

Similarly, by induction we can show that $\tilde{\tilde{\gamma}}_{0l} \geqslant \tilde{\tilde{\gamma}}_{0l}^{-1}(\beta)$.

We determine a fuzzy function inequality by applying (1.14) and (1.15)(1.16) in Section 1.6.3 in Chapter 1, then

$$\begin{aligned}
&\tilde{\bar{g}}_i(x) = \tilde{\bar{c}}_i \prod_{l=1}^{m} x_l^{\tilde{\bar{\gamma}}_{il}} \lesssim 1 \\
\Longleftrightarrow\ & \tilde{\bar{c}}_i^{-1}(\beta) \prod_{l=1}^{m} x_l^{\tilde{\bar{\gamma}}_{il}^{-1}(\beta)} \tilde{\bar{c}}_i^{-1}(\beta) \prod_{l=1}^{m} x_l^{\tilde{\bar{\gamma}}_{il}^{-1}(\beta)} \lesssim 1 \\
\Longrightarrow\ & \tilde{A}_i(\tilde{\bar{c}}_i^{-1}(\beta) \prod_{l=1}^{m} x_l^{\tilde{\bar{\gamma}}_{il}^{-1}(\beta)}) \geqslant \alpha \ \ (0 \leqslant i \leqslant p).
\end{aligned}$$

Therefore

$$\begin{aligned}
(\tilde{P}_\varepsilon) \quad & \widetilde{\min}\ \tilde{\bar{c}}_0 \prod_{l=1}^{m} x_l^{\tilde{\bar{\gamma}}_{0l}} \\
& \text{s.t. } \tilde{\bar{c}}_i \prod_{l=1}^{m} x_l^{\tilde{\bar{\gamma}}_{il}} \lesssim 1 (1 \leqslant i \leqslant p) \\
& x > 0
\end{aligned}$$

$$\Longleftrightarrow$$

$$\begin{aligned}
& \max \alpha = \min(-\alpha) \\
& \text{s.t. } \tilde{A}_0(\tilde{\bar{c}}_0^{-1}(\beta) \prod_{l=1}^{m} x_l^{\tilde{\bar{\gamma}}_{0l}^{-1}(\beta)}) \geqslant \alpha \\
& \tilde{A}_i(\tilde{\bar{c}}_i^{-1}(\beta) \prod_{l=1}^{m} x_l^{\tilde{\bar{\gamma}}_{il}^{-1}(\beta)}) \geqslant \alpha \ \ (1 \leqslant i \leqslant p) \\
& \alpha, \beta \in [0, 1] \\
& x > 0
\end{aligned}$$

$$\Longleftrightarrow$$

$$\begin{aligned}
(\bar{P}_\varepsilon) \quad & \min(-\alpha) \\
& \text{s.t. } \tilde{\bar{c}}_i^{-1}(\beta) / \bar{B}_i^{-1}(\alpha) \prod_{l=1}^{m} x_l^{\tilde{\bar{\gamma}}_{il}^{-1}(\beta)} \leqslant 1 \ \ (0 \leqslant i \leqslant p) \\
& \alpha, \beta \in [0, 1] \\
& x > 0
\end{aligned}$$

$$\Longleftrightarrow$$

$$(\bar{D}_\varepsilon) \qquad \max\ (-\alpha/\bar{w}_0)^{\bar{w}_0} \prod_{i=1}^{p} (\tilde{\bar{c}}_i^{-1}(\beta)/\tilde{B}_i^{-1}(\alpha)\bar{w}_i)^{\bar{w}_i}$$

$$\text{s.t. } \bar{w}_0' = 1$$

$$\tilde{\bar{\gamma}}_{01}^{-1}(\beta)\bar{w}_0 = \bar{w}_0'$$

$$\sum_{i=0}^{p} \tilde{\bar{\gamma}}_{il}^{-1}(\beta)\bar{w}_i = 0, \alpha, \beta \in [0,1] \ \ (1 \leqslant l \leqslant m)$$

$$\bar{w} \geqslant 0$$

where $\tilde{B}_i^{-1}, \tilde{\bar{c}}_i^{-1}$ and $\tilde{\bar{\gamma}}_{il}^{-1}$ $(0 \leqslant i \leqslant p, 1 \leqslant l \leqslant m)$ are inverse functions with respect to $\tilde{B}_i$, $\tilde{\bar{c}}_i$ and $\tilde{\bar{\gamma}}_{il}$, respectively. $\bar{w} = (\bar{w}_0', \bar{w}_0, \bar{w}_1, \cdots, \bar{w}_p)^{\mathrm{T}}$ is a $(p+2)$-dimension dual parameter variable vector, here $\bar{w}_i = (w_i, \alpha, \beta)$.

So it is easy to prove that $(\bar{D}_\varepsilon) \Longleftrightarrow (\tilde{D}_\varepsilon)$ in a similar way as Theorem 2.3.1, such that $(\tilde{P}_\varepsilon) \Longleftrightarrow (\tilde{D}_\varepsilon)$. Therefore, the theorem holds. □

Theorem 3.4.2 *Let $\tilde{A}_i = \tilde{B}_i \circ \tilde{\phi}_i$ be a CSM function, where $\tilde{\phi}_i = \tilde{c}_{ik} \circ \tilde{\gamma}_{ikl}(1 \leqslant k \leqslant J_i, 0 \leqslant i \leqslant p, 1 \leqslant l \leqslant m)$. If $(\tilde{P}_\varepsilon)$ is fuzzily consistent, with $M_{\tilde{P}_\varepsilon} > 0$, then there must exist a dual feasible solution w', such that we have a weight vector $\varepsilon > 0$,*

$$\varepsilon_{ik} w_{i0}' = w_{ik}' \quad (1 \leqslant k \leqslant J_i; 0 \leqslant i \leqslant p) \tag{3.34}$$

and $M_{\tilde{P}_\varepsilon} = \tilde{d}(w')$.

Conversely, if (3.34) holds at a dual feasible solution w', there must exist $M_{\tilde{P}_\varepsilon} > 0$. ◊

Proof We know that

$$(\tilde{P}_\varepsilon) \Longleftrightarrow (\overline{P}_\varepsilon)$$

in which

$$(\overline{P}_\varepsilon) \overset{z_l = \log x_l}{\Longleftrightarrow} \quad \max(-\alpha)$$

$$\text{s.t. } \sum_{l=1}^{m} \tilde{\bar{\gamma}}_{il}^{-1}(\beta) z_l \leqslant \log(\tilde{B}_i^{-1}(\alpha)/\tilde{\bar{c}}_i^{-1}(\beta)) (0 \leqslant i \leqslant p)$$

$$\alpha, \beta \in [0,1].$$

Given $M_{\overline{P}_\varepsilon} > 0$, and according to the proof in Theorem 1.7.2 [WY82], we know that LP $(\overline{P}_\varepsilon)$ corresponding to $(\tilde{P}_\varepsilon)$ possesses the constraint infimum $\log M_{\overline{P}_\varepsilon}$. Hence, from the dual theory in an ordinary LP, an LP corresponding to $(\tilde{\bar{D}}_\varepsilon)$

$$\begin{aligned}
&\max\ \{-\bar{w}'_0\log\alpha + \sum_{i=0}^{p}\bar{w}_i'\log(\tilde{\bar{c}}_i^{-1}(\beta)/\tilde{B}_i^{-1}(\alpha))\}\\
&\text{s.t. } \bar{w}_0' = 1\\
&\quad \tilde{\bar{\gamma}}_{0l}^{-1}(\beta)\bar{w}_0' = 1\\
&\quad \sum_{i=0}^{p}\tilde{\bar{\gamma}}_{il}^{-1}(\beta)\bar{w}_i' = 0, \alpha, \beta \in [0,1]\ (1 \leqslant l \leqslant m)\\
&\quad w' \geqslant 0
\end{aligned} \tag{3.35}$$

contains maximum value $\log M_{\tilde{P}_\varepsilon}$, such that there exists a fuzzy maximum value $M_{\tilde{P}_\varepsilon}$ in $(\tilde{\bar{D}}_\varepsilon)$. Now, let $\overline{w}^{*'}$ be an optimal solution to $(\overline{D}_\varepsilon)$ and to (3.35), then

$$M_{\overline{P}_\varepsilon} = (-\alpha/\varepsilon_{0k})^{\varepsilon_{0k}\bar{w}_0^{*'}} \prod_{i=0}^{p}\prod_{k=1}^{J_i}(\tilde{\bar{c}}_{ik}^{-1}(\beta)\tilde{B}_i^{-1}(\alpha)/\varepsilon_{ik})^{\varepsilon_{ik}\bar{w}_i^{*'}}. \tag{3.36}$$

Let

$$\bar{w}_{ik}' = \varepsilon_{ik}\bar{w}_i^{*'}, (1 \leqslant k \leqslant J_i; 0 \leqslant i \leqslant p).$$

Because $\sum_{k=1}^{J_i}\varepsilon_{ik} = 1(0 \leqslant i \leqslant p)$, we have

$$\bar{w}_{i0}' = \sum_{k=1}^{J_i}\bar{w}_{ik}' = \sum_{k=1}^{J_i}\varepsilon_{ik}\bar{w}_i^{*'} = \bar{w}_i^{*'} \tag{3.37}$$

such that (3.34) holds. Again, according to (3.36), we arrive at

$$\begin{aligned}
M_{\tilde{P}_\varepsilon} &= (-\alpha\bar{w}_{00}'/\bar{w}_{0k}')^{\bar{w}_{0k}'}\prod_{i=0}^{p}\prod_{k=1}^{J_i}(\tilde{\bar{c}}_{ik}^{-1}(\beta)\tilde{B}_i^{-1}(\alpha)\bar{w}_{i0}'/\bar{w}_{ik}')^{\bar{w}_{ik}'}\\
&= d(\bar{w}').
\end{aligned} \tag{3.38}$$

Hence, $M_{\tilde{P}_\varepsilon} = d(\bar{w}')$, such that $M_{\tilde{P}_\varepsilon} = \tilde{d}(w')$ holds.

Since $\bar{w}^{*'}$ is a dual feasible solution to $(\overline{D}_\varepsilon)$, we have from (2.18) and (3.37) that

$$\begin{aligned}
\sum_{i=0}^{p}\sum_{k=1}^{J_i}\tilde{\gamma}_{ikl}^{-1}(\beta)\bar{w}_{ik}' &= \sum_{i=0}^{p}\sum_{k=1}^{J_i}\tilde{\gamma}_{ikl}^{-1}(\beta)\varepsilon_{ik}\bar{w}_i^{*'}\\
&= \sum_{i=0}^{p}\tilde{\gamma}_{il}^{-1}(\beta)\bar{w}_i^{*'} = 0.
\end{aligned}$$

Again, $\bar{w}^{*'} \geqslant 0, \varepsilon \geqslant 0$, then $\bar{w}' \geqslant 0$, with $\bar{w}_{00}' = \bar{w}_0^{*'} = 1$. So, $\bar{w}'$ is indeed a feasible solution to $(\overline{D}_\varepsilon)$, such that w' is a fuzzy feasible solution to (2.10).

Conversely, let $\bar{w}'$ denote a dual feasible solution satisfying (3.37) and define $\bar{w}_i^{*'} = \bar{w}'_{i0}(0 \leqslant i \leqslant p)$. Then $\bar{w}_0^{*'} = 1$ and $\bar{w}_0^{*'} \geqslant 0$ satisfying

$$\sum_{i=0}^{p} \bar{\gamma}_{il}^{-1}(\beta)\bar{w}_i^{*'} = \sum_{i=0}^{p}\sum_{k=1}^{J_i} \tilde{\gamma}_{ikl}^{-1}(\beta)\varepsilon_{ik}\bar{w}_i^{*'}$$
$$= \sum_{i=0}^{p}\sum_{k=1}^{J_i} \tilde{\gamma}_{ikl}^{-1}(\beta)\bar{w}'_{ik}\ (1 \leqslant l \leqslant m).$$

Accordingly, from Theorem 3.1.2, $\bar{w}^{*'}$ is a feasible solution to $(\overline{D}_\varepsilon)$ and $M_{\tilde{P}_\varepsilon} > 0$.

Since $(\tilde{P}_\varepsilon)$ consistence $\Longleftrightarrow(\overline{P}_\varepsilon)$ consistence $\Longleftrightarrow(\overline{D}_\varepsilon)$ consistence, then

$$\begin{aligned}
&(3.37) \Longleftrightarrow \tilde{A}_i(\varepsilon_{ik}\bar{w}_i^{*'}) = \tilde{A}_i(\varepsilon_{ik}\bar{w}_i^{*}) = \tilde{A}_i(\bar{w}'_{ik}),\\
&\qquad\qquad \text{i.e., } \varepsilon_{ik}w'_{i0} = w'_{ik},\\
&(3.38) \Longleftrightarrow \tilde{A}_i(M_{\overline{P}_\varepsilon)}) = \tilde{A}_i(d(\bar{w}')),\\
&\qquad\qquad \text{i.e., } M_{\tilde{P}\varepsilon} = \tilde{d}(w')\ (1 \leqslant k \leqslant J_i; 0 \leqslant i \leqslant p),\\
&M_{\overline{P}_\varepsilon} > 0 \Longleftrightarrow M_{\tilde{P}_\varepsilon} > 0.
\end{aligned}$$

Hence, the theorem holds. □

Theorem 3.4.3 *Let $\tilde{A}_i = \tilde{B} \circ \tilde{\phi}_i$ be a CSM function, where $\tilde{\phi}_i = \tilde{c}_{ik} \circ \tilde{\gamma}_{ikl}(1 \leqslant k \leqslant J_i, 0 \leqslant i \leqslant p, 1 \leqslant l \leqslant m)$. If fuzzy PGP $(\tilde{P})$ is fuzzy supper consistent having a minimal point x^*, then x^* is a minimal point of a monomial fuzzy PGP $(\tilde{P}_\varepsilon)$ corresponding to the weight*

$$\varepsilon_{ik} = \tilde{v}_{ik}(x^*)/\tilde{g}_i(x^*)(1 \leqslant k \leqslant J_i; 0 \leqslant i \leqslant p)$$

and

$$M_{\tilde{P}_\varepsilon} = M_{\tilde{P}}.$$

◊

Proof From the condition of the theorem, $(\tilde{P})$ is equivalent to

$$\begin{aligned}
&\min(-\alpha)\\
&\text{s.t. } \sum_{k=1}^{J_i} \tilde{c}_{ik}^{-1}(\beta)/\tilde{B}_i^{-1}(\alpha)\prod_{l=1}^{m} x_l^{\tilde{\gamma}_{ikl}^{-1}(\beta)} \leqslant 1\ (0 \leqslant i \leqslant p)\\
&\quad\ x > 0, \alpha, \beta \in [0,1]
\end{aligned}$$

Obviously, it is super-consistent with a minimal point $\overline{x}^*$ existing. For a certain determined α, β, by applying Theorem 1.7.3 in Reference [WY82], we know $\overline{x}^*$ is a minimal point at a monomial posynomial parameter GP $(\overline{P}\varepsilon)$ corresponding to the weight

$$\varepsilon'_{ik} = v_{ik}(\overline{x}^*)/g_i(\overline{x}^*)\ (1 \leqslant k \leqslant J_i; 0 \leqslant i \leqslant p). \tag{3.39}$$

and $M_{\tilde{P}_\varepsilon} = M_{\tilde{P}}$, such that

$$\varepsilon_{ik} = \tilde{v}_{ik}(x^*)/\tilde{g}_i(x^*),$$

and

$$M_{\tilde{P}_\varepsilon} = M_{\tilde{P}}$$

from arbitrary α, β in [0,1].

Therefore, this theorem holds. □

Theorem 3.4.4 *Let the fuzzy PGP ($\tilde{P}$) be fuzzy super-consistent with a fuzzy optimal solution x^*. Then there must exist a Lagrange multiplier $\mu^* = (\mu_1^*, \mu_2^*, \dots, \mu_p^*)^T \geqslant 0$ and a CSN function $\tilde{A}_i$, such that*

$$\nabla \tilde{g}_0(x^*) + \sum_{i=1}^{p} \mu_i^* \nabla \tilde{g}_i(x^*) = 0,$$

$$\mu_i^*(\tilde{g}_i(x^*) - 1) = 0 \ (1 \leqslant i \leqslant p),$$

whereas w^* defined by

$$w_{ik}^* = \begin{cases} \tilde{v}_{0k}(x^*)/\tilde{g}_0(x^*) & (i = 0, 1 \leqslant k \leqslant J_0) \\ \mu_i^* \tilde{v}_{ik}(x^*)/\tilde{g}_0(x^*) & (i \neq 0; 1 \leqslant k \leqslant J_i) \end{cases} \tag{3.40}$$

is a dual optimal solution to ($\tilde{D}$) with

$$\tilde{g}_0(x^*) = \tilde{d}(w^*).$$

◊

Proof From (3.39) and Theorem 3.4.3, x^* is the minimal point corresponding to ($\tilde{P}_\varepsilon$) with minimum $\tilde{\bar{g}}_0(x^*) = \tilde{g}_0(x^*) \gtrsim 0$, but from Theorem 3.4.2, we know there exists w^* satisfying (3.34), such that

$$\tilde{d}(w^*) = \tilde{\bar{g}}_0(x^*) = \tilde{g}_0(x^*).$$

From Lemma 3.4.1, w^* is a fuzzy optimal solution to ($\tilde{D}$).

And then, from (3.34) and (3.39), we arrive at

$$w_{ik}^* = \varepsilon_{ik} w_{i0}^* = \tilde{v}_{ik}(x^*) w_{i0}^* / \tilde{g}_i(x^*).$$

Since $w_{00}^* = 1$, then

$$w_{0k}^* = \tilde{v}_{0k}(x^*)/\tilde{g}_0(x^*) \text{ for } i = 0;$$

and

$$\begin{aligned} w_{i0}^* &= \sum_{k=1}^{J_i} w_{ik}^* = \mu_i^* \sum_{k=1}^{J_i} \tilde{v}_{ik}(x^*)/\tilde{g}_0(x^*) \\ &= \mu_i^* \tilde{g}_i(x^*)/\tilde{g}_0(x^*) \\ &\overset{\text{from(3.39)}}{=} \mu_i^*/\tilde{g}_0(x^*) \text{ for } i \neq 0, \end{aligned}$$

therefore,

$$w_{ik}^* = \tilde{v}_{ik}(x^*)\mu_i^*/\tilde{g}_i(x^*)\tilde{g}_0(x^*) = \mu_i^*\tilde{v}_{ik}(x^*)/\tilde{g}_0(x^*).$$

In conclusion, we can get (3.40), so that the theorem holds. □

3.5 Fuzzy PGP Classification and Its Corresponding Class Properties

In this section we research the classification of the fuzzy PGP and three types of programming properties including a canonical one, a totally degenerate one and a degenerate one of nontotal degeneration.[Cao95e]

3.5.1 Classification of Fuzzy PGP

Let the formula description of fuzzy PGP (3.1) be

$$(\tilde{P}) \qquad \widetilde{\inf}\, \tilde{g}_0(x)$$
$$\text{s.t. } \tilde{g}_i(x) \lesssim 1 (1 \leqslant i \leqslant p)$$
$$x > 0$$

where $x = (x_1, x_2, \ldots, x_m)^{\mathrm{T}}$ is a variable vector, $\tilde{g}_i(x) = \sum\limits_{k=1}^{J_i} \tilde{c}_{ik} \prod\limits_{l=1}^{m} x_l^{\gamma_{ikl}}$ is a fuzzy posynomial for x and its dual fuzzy programming serves as

$$(\tilde{D}) \qquad \sup \tilde{d}(w) = \prod_{i=1}^{p} \prod_{k=1}^{J_i} \left(\frac{\tilde{c}_{ik}}{w_{ik}}\right)^{w_{ik}} \prod_{l=1}^{m} w_{i0}^{w_{i0}}$$
$$\text{s.t. } w_{00} = 1$$
$$\Gamma^{\mathrm{T}} w = 0$$
$$w \geqslant 0$$

where $w = (w_{01}, \ldots, w_{0J_0}, \ldots, w_{p1}, \ldots, w_{pJ_p})^{\mathrm{T}}$ is a $J-$dimensional variable vector, $w_{i0} = \sum\limits_{k=1}^{J_i} w_{ik} = w_{i1} + w_{i2} + \cdots + w_{iJ_i} (0 \leqslant i \leqslant p)$ is exactly the sum of each dual variable corresponding to an objective function $\tilde{g}_0(x)$ (if $i = 0$) or constraint functions $\tilde{g}_i(x)$ (if $i \neq 0$), and Γ is an exponent matrix.

Consider the constraint condition of $(\tilde{D})$

$$\Gamma^{\mathrm{T}} w = 0, w \geqslant 0 \tag{3.41}$$

and classify $(\tilde{P})$ according to a solution to (3.41).

Definition 3.5.1 Let $w = (w_{01}, \ldots, w_{0J_0}, \ldots, w_{p1}, \ldots, w_{pJ_p})^{\mathrm{T}}$, denoted by $\mathcal{I} = \{(i,k) |$ (3.41) with a solution w satisfying $w_{ik} > 0\}$. Then $\mathcal{I}$ is called an unreduced subscript set.

Lemma 3.5.1 *There must exist a solution $\bar{w}$ in group (3.41) satisfying*

$$\bar{w}_{ik} > 0, \forall (i,k) \in \mathcal{I}. \tag{3.42}$$

Again, for any solution w and any $(i,k) \notin \mathcal{I}$ in (3.41), there exists

$$w_{ik} = 0, \forall (i,k) \notin \mathcal{I}.$$

◊

Proof The lemma holds because the sum of any two solutions to (3.41) is still a solution to (3.41). □

Definition 3.5.2 If an unreduced set

$$\mathcal{I} = \{(i,k)|1 \leqslant k \leqslant J_i, 0 \leqslant i \leqslant p\}$$

i.e., $\mathcal{I}$ includes all subscript pairs (i,k), then the primal fuzzy PGP $(\tilde{P})$ and dual program $(\tilde{D})$ are said to be canonical types. Otherwise, $(\tilde{P})$ and $(\tilde{D})$ degenerate. More specifically, if $\mathcal{I}$ fails to contain any $(0,k)(1 \leqslant k \leqslant J_0)$, $(\tilde{P})$ and $(\tilde{D})$ are called totally degenerate types.

Definition 3.5.3 Let $\tilde{g}_i^*(x) = \sum\limits_{\substack{k=1\\(i,k)\in\mathcal{I}}}^{J_i} \tilde{c}_{ik} \prod\limits_{l=1}^{m} x_l^{\gamma_{ikl}} (0 \leqslant i \leqslant p)$, calling

$$(\tilde{P}^*) \qquad \begin{aligned} &\inf\ \tilde{g}_0^*(x) \\ &\text{s.t. } \tilde{g}_i^*(x) \lesssim 1 (1 \leqslant i \leqslant p) \\ &\qquad x > 0 \end{aligned}$$

and

$$(\tilde{D}^*) \qquad \begin{aligned} &\widetilde{\sup}\ \tilde{d}(w^*) = \prod_{i=0}^{p} \prod_{k=1}^{J_i} \left(\frac{\tilde{c}_{ik}}{w_{ik}^*}\right)^{w_{ik}^*} \prod_{i=1}^{p} (w_{ik}^*)^{w_{ik}^*} \\ &\text{s.t. } w_{00}^* = 1 \\ &\qquad \Gamma^{*\mathrm{T}} w^* = 0 \\ &\qquad w^* \geqslant 0 \end{aligned}$$

a reduced programming of two fuzzy PGP induced from $(\tilde{P})$ and $(\tilde{D})$, where $w_{i0}^* = \sum\limits_{k=0}^{J_i} w_{ik}^* (0 \leqslant i \leqslant p)$. Here $\tilde{g}_i^*(x)$ represents a polynomial resulting from $\tilde{g}_i(x)$ where terms failing to belong to $\mathcal{I}$ are cut out; w^* represents a vector coming from w, where the subscripts failing to belong to components of $\mathcal{I}$ are deleted; $\tilde{\Gamma}^*$ represents a fuzzy exponent matrix coming from Γ, where the subscripts failing to belong to a row of $\mathcal{I}$ are struck out.

Definition 3.5.4 If all of the corresponding $(i,k)(1 \leqslant k \leqslant J_i)$ do not belong to $\mathcal{I}$ for a certain $i(\neq 0)$, then $\tilde{g}_i^*(x) = 0$. But in the fuzzy reduced programming, a corresponding fuzzy constraint condition $\tilde{g}_i^*(x) \lesssim 1$ does not appear again; $(\tilde{P}^*)$ now is called a degenerate fuzzy GP of nontotal degeneration.

Theorem 3.5.1 *For monotonic continuous function $\tilde{A}_i(0 \leqslant i \leqslant p)$, $(\tilde{P})$ and $(\tilde{D})$ is canonical iff there exists a fuzzy feasible solution $\bar{w} > 0$ to $(\tilde{D})$. $(\tilde{P})$ and $(\tilde{D})$ is total degeneration iff $(\tilde{D})$ is not fuzzy consistent, and degenerate* *programming of nontotal degeneration in $(\tilde{P})$ and $(\tilde{D})$ iff $(\tilde{D})$ is fuzzy consistent without all positive fuzzy feasible solutions.* ◊

Proof The theorem holds from Lemma 3.5.1 and Definition 3.5.2. □

3.5.2 Three Type Properties of Fuzzy PGP

The fuzzy PGP can be classified into three types including: A) a total degeneration; B) a canonical one and C) a degenerate one of nontotal degeneration by means of Definition 3.5.2 and 3.5.3.

Now, we discuss their properties as follows:

A) Total Degeneration.

Such a programming is very simple in its properties, which are described as the following theorems.

Theorem 3.5.2 *Let $\tilde{A}$ be a monotonic and continuous function. If $(\tilde{P})$ and $(\tilde{D})$ are total-degenerate fuzzy PGP, then*

1^0 *$(\tilde{D})$ is not fuzzy consistent.*

2^0 *If $(\tilde{P})$ is fuzzy consistent, then there must exist $M_{\tilde{P}} = 0$.*

3^0 *If $(\tilde{P})$ is fuzzy sub-consistent, there must exist $\mu_{\tilde{P}} = 0$.*

4^0 *There exists no fuzzy optimal solution to $(\tilde{P})$.* ◊

Proof Next indicate proofs for 2^0 and 3^0 since 1^0 is included in Theorem 3.5.1.

Given that $(\tilde{P})$ is fuzzy sub-consistent, with $\mu_{\tilde{P}} > 0$, $(\tilde{D})$ has fuzzy consistency from Theorem 3.2.1, which contradicts 1^0, such that $\mu_{\tilde{P}} = 0$. Hence 3^0 holds. Again, let $(\tilde{P})$ be fuzzy consistent. Then $(\tilde{P})$ is fuzzy sub-consistent, too. Again from Theorem 3.2.2, we know $M_{\tilde{P}} = \mu_{\tilde{P}} = 0$, so 2^0 holds.

4^0 holds by conclusions of $1^0, 2^0, 3^0$ and in combination with Theorem 3.2.5. □

B) Canonical Type.

A further result can be obtained for such a programming.

Theorem 3.5.3 *Let $\tilde{A}_i (0 \leqslant i \leqslant p)$ be a monotonic and continuous function. If $(\tilde{P})$ and $(\tilde{D})$ mean canonical types of the fuzzy PGP, then*

1^0 *There exists a fuzzy feasible solution $\tilde{w} > 0$ to $(\tilde{D})$.*

2^0 *A necessary and sufficient condition for fuzzy consistence of $(\tilde{P})$ is $M_{\tilde{D}} < \infty$.*

3^0 *If $M_{\tilde{D}} < \infty$, then $(\tilde{P})$ contains a fuzzy optimal solution x^* and $M_{\tilde{P}} = \tilde{g}_0(x^*) = M_{\tilde{D}}$.*

4^0 *$(\tilde{P})$ contains a fuzzy optimal solution if and only if $M_{\tilde{D}} < \infty$.*

5^0 $(\tilde{P})$ *contains a fuzzy optimal solution if and only if* $(\tilde{P})$ *is fuzzy consistent.*

$\Diamond$

Proof Only 2^0 and 3^0 need proving with 1^0 encompassed in Theorem 3.5.1.

2^0 $(\tilde{P})$ is fuzzy consistent, so is $(\tilde{D})$, such that $M_{\tilde{D}} > 0, M_{\tilde{P}} < \infty$. From Theorem 3.1.2, $0 < M_{\tilde{P}} < \infty$ exists and from (3.6) in Theorem 3.1.4, $M_{\tilde{D}} = M_{\tilde{P}} < \infty$.

3^0 From the assumption and Theorem 3.2.4, we know that $(\tilde{P})$ is fuzzy sub-consistent, with $\mu_{\tilde{P}} = M_{\tilde{D}}$, such that in $\forall \alpha \in [0,1]$ and in any $\theta \in (0,1)$, all (P^θ) of

$$(\tilde{P}^\theta) \qquad \begin{aligned} &\widetilde{\inf}\, \theta \tilde{g}_0(x) \\ &\text{s.t. } \theta \tilde{g}_i(x) \lesssim 1 \ (1 \leqslant i \leqslant p) \\ &\qquad x > 0 \end{aligned}$$

are super-consistent, so $(\tilde{P}^\theta)$ is fuzzy super-consistent.

Let $w > 0$ be a fuzzy feasible solution to $(\tilde{D})$. Since $(\tilde{D})$ has the same constraint condition as

$$(\tilde{D}^\theta) \qquad \begin{aligned} &\widetilde{\sup}\, \tilde{d}_0(w) = \tilde{d}(w)\theta^{\sum_{i=1}^{p} w_{i0}} \\ &\text{s.t. } w_{00} = 1 \\ &\qquad \Gamma^{\mathrm{T}} w = 0 \\ &\qquad w \geqslant 0 \end{aligned}$$

w is also a fuzzy feasible solution to $(\tilde{D}^\theta)$. From Theorem 2.4.2, we know that $(\tilde{P}^\theta)$ has a fuzzy optimal solution, i.e., there exists $x^\theta > 0$ satisfying

$$\theta\, \tilde{g}_i(x^\theta) \lesssim 1 \ (1 \leqslant i \leqslant p) \tag{3.43}$$

such that

$$\theta\, \tilde{g}_0(x^\theta) = M_{\tilde{P}^\theta} = M_{\tilde{D}^\theta}. \tag{3.44}$$

Because

$$(3.43) \Longleftrightarrow \theta g_i(x^\theta) \leqslant \tilde{B}_i^{-1}(\alpha),$$

$$\overset{g_i(x^\theta,\alpha)=g_i(x^\theta)/\tilde{B}_i^{-1}(\alpha)}{\Longleftrightarrow} \theta g_i(x^\theta, \alpha) \leqslant 1, \tag{3.45}$$

$$(3.44) \Longleftrightarrow \theta g_0(x^\theta, \alpha) = M^\theta_{P_\alpha} = M^\theta_{D_\alpha}, \tag{3.46}$$

there exists $x^\theta > 0$ satisfying (3.45), such that (3.46) holds. Again, θ has nothing to do with α. Similar to the proof in Theorem 2.5.4 in Reference [WY82] we can confirm the following:

When $\theta \to 1$, from (3.45) and (3.46), then

$$\theta g_i(x^*, \alpha^*) \leqslant 1 \ \text{ and } \ \theta g_0(x^*, \alpha^*) = M_{P_\alpha} = M_{D_\alpha}. \tag{3.47}$$

From the equivalence of (3.47) together with (3.43) and (3.44), we know

$$\tilde{g}_i(x^*) \lesssim 1 \text{ and } \tilde{g}_0(x^*) = M_{\tilde{P}} = M_{\tilde{D}}$$

holds. It naturally follows that x^* is a fuzzy feasible solution to $(\tilde{P})$ and a fuzzy optimal solution to $(\tilde{P})$.

By concluding 1^0, 2^0 and 3^0, it is easy to prove 4^0 and 5^0. □

This theorem hints that if $(\tilde{D})$ has a feasible solution $w > 0$, and an optimal solution w^*, then a primal problem $(\tilde{P})$ also contains a fuzzy optimal solution x^*, and x^* can be obtained from $\tilde{g}_0(x^*) = \tilde{d}(w^*)$, or from

$$\sum_{l=1}^{m} \tilde{\gamma}_{ikl} \log x_l = \begin{cases} \log(w_{0k}^* \tilde{d}(w^*)/\tilde{c}_{0k}, & (i = 0; 1 \leqslant k \leqslant J_0) \\ \log(w_{ik}^*/(w_{i0}^* \tilde{c}_{ik})), & (i \neq 0, w_{i0}^* \neq 0, 1 \leqslant k \leqslant J_i). \end{cases}$$

C) Degenerate Type of Nontotal degeneration.

Since $(\tilde{D}^*)$ is the dual fuzzy GP of $(\tilde{P}^*)$, then

Theorem 3.5.4 *Let $\tilde{A}$ be a monotonic and continuous function. If $(\tilde{P})$ and $(\tilde{D})$ are degenerate fuzzy PGP of nontotal degeneration, then*

1^0 *The fuzzy reduced programming of $(\tilde{P}^*)$ and $(\tilde{D}^*)$ are canonical.*

2^0 *$(\tilde{D})$ and $(\tilde{D}^*)$ are fuzzy consistent;*

3^0 *$(\tilde{P}^*)$ fuzzy consistent $\Longleftrightarrow$ the primal fuzzy PGP $(\tilde{P})$ is fuzzy sub-consistent, with $0 < \mu_{\tilde{P}} < \infty$, at this moment, $M_{\tilde{P}^*} = \mu_{\tilde{P}}$.*

4^0 *If $(\tilde{P})$ is fuzzy consistent, so is $(\tilde{P}^*)$, with $M_{\tilde{P}^*} = M_{\tilde{P}}$.*

5^0 *If $(\tilde{P})$ is fuzzy super-consistent, then there exists a fuzzy optimal solution $x^* \Longrightarrow x^*$ is also a fuzzy optimal solution to $(\tilde{P}^*)$.* ◊

Proof From Lemma 3.5.1, there exists a fuzzy feasible solution w satisfying (3.42) in $(\tilde{D})$. A feasible solution $w^* > 0$ to $(\tilde{D}^*)$ can be obtained after components in subscripts which do not belong to $\mathcal{I}$ in w are deleted, such that $(\tilde{P}^*)$ and $(\tilde{D}^*)$ are canonical with 1^0 and 2^0 holding.

3^0 Since $(\tilde{P}^*)$ fuzzy consistence $\Longleftrightarrow$

$$\begin{aligned} &\inf\ g_0^*(x, \alpha) \\ &\text{s.t. } \ g_i^*(x, \alpha) \leqslant 1 (1 \leqslant i \leqslant p) \\ &\qquad x > 0 \end{aligned} \tag{3.48}$$

consistence, (3.48) is consistent for an $\alpha \in [0, 1]$. And from (iii) in Theorem 2.5.5 in Reference [WY82], we know, (3.48) is equivalent to the primal parameter GP

$$\begin{aligned} &\inf\ g_0(x, \alpha) \\ &\text{s.t. } \ g_i(x, \alpha) \leqslant 1, (1 \leqslant i \leqslant p) \\ &\qquad x > 0 \end{aligned}$$

being sub-consistent, with $M_{P_\alpha^*} = \mu_{P_\alpha^*}$ for $0 < \mu_{P_\alpha^*} < \infty$. Therefore 3^0 holds.

4^0 Fuzzy consistence must be fuzzy sub-consistence, with $M_{\tilde{P}} = \mu_{\tilde{P}}$ (Theorem 3.2.2). From 3^0, we know only $0 < M_{\tilde{P}} < \infty$ needs proving. But since $(\tilde{D})$ is fuzzy consistent (from 2^0), the proved inequality holds from Theorem 3.1.2, therefore 4^0 holds.

5^0 The condition holds from the assumption that

$$\begin{aligned}
&\overset{\text{Th.2.4.1}}{\Longleftrightarrow} x^* \text{ is a fuzzy optimal solution of } (\tilde{P}) \\
&\Longrightarrow (\tilde{P}) \text{ fuzzy consistence} \\
&\overset{3^0}{\Longrightarrow} (\tilde{P}^*) \text{ fuzzy consistence} \\
&\overset{\text{Th.3.3.3;}2^0 3^0}{\Longrightarrow} (\tilde{P}^*) \text{ contains a fuzzy optimal solution } x^*.
\end{aligned}$$

□

Chapter 4

INITIAL STUDY OF FUZZY REVERSE PGP

In this Chapter, we extend a fuzzy PGP into a fuzzy reverse PGP, go initially into its determinacy method as well as its dual programming, study a Lagrange problem, and meanwhile advance a dual and a direct algorithm in solutions to a fuzzy reverse PGP and in the Lagrange problem, respectively.

4.1 Fuzzy Reverse PGP and Its Dual Form

Definition 4.1.1 We call a problem

$$\begin{aligned}&\widetilde{\min}\ \tilde{g}_0(x)\\ &\text{s.t. } \tilde{g}_i(x) \lesssim 1(1 \leqslant i \leqslant p)\\ &\qquad x > 0\end{aligned}$$

a primal-type fuzzy PGP, while we call

$$(\tilde{P})\qquad \begin{aligned}&\widetilde{\min}\ \tilde{g}_0(x)\\ &\text{s.t. } \tilde{g}_i(x) \lesssim 1(1 \leqslant i \leqslant p')\\ &\qquad \tilde{g}_i(x) \gtrsim 1(p'+1 \leqslant i \leqslant p)\\ &\qquad x > 0\end{aligned} \tag{4.1}$$

a fuzzy reverse PGP. Here, $x = (x_1, x_2, \cdots, x_m)^{\mathrm{T}}$ is an $m-$dimensional variable vector, "$\lesssim$" and "$\gtrsim$" correspond to " approximate $\leqslant$ " and " approximate $\geqslant$", respectively, and all

$$\tilde{g}_i(x) = \sum_{k=1}^{J_i} \tilde{v}_{ik}(x)(0 \leqslant i \leqslant p)$$

are fuzzy posynomials, whereas

$$\tilde{v}_{ik}(x) = \begin{cases} \tilde{c}_{ik} \prod\limits_{l=1}^{m} x_l^{\tilde{\gamma}_{ikl}}, & (1 \leqslant k \leqslant J_i; 0 \leqslant i \leqslant p') \\ \tilde{c}_{ik} \prod\limits_{l=1}^{m} x_l^{-\tilde{\gamma}_{ikl}}, & (1 \leqslant k \leqslant J_i; p'+1 \leqslant i \leqslant p) \end{cases}$$

where its coefficients $\tilde{c}_{ik} \geqslant 0$ and exponents $\tilde{\gamma}_{ikl}$ may be freely fixed in a closed value interval

$$[c_{ik}^-, c_{ik}^+], [\gamma_{ikl}^-, \gamma_{ikl}^+](0 \leq i \leqslant p, 1 \leqslant k \leqslant J_i, 1 \leq l \leq m),$$

here $c_{ik}^- < c_{ik}^+$ and $\gamma_{ikl}^- < \gamma_{ikl}^+$; c_{ik}^-, c_{ik}^+ and $\gamma_{ikl}^-, \gamma_{ikl}^+$ belonging to left and right endpoints in the intervals respectively are all real numbers. $-\tilde{\gamma}_{ikl}$ are exponents corresponding to each item x_l in the reversed direction inequality $\tilde{g}_i(x) \gtrsim 1$.

Theorem 4.1.1 *Suppose that fuzzy-valued functions $\tilde{A}_i = \tilde{B}_i \circ \tilde{\phi}_i$ are CSM functions, where $\tilde{\phi}_i = \tilde{c}_{ik} \circ \tilde{\gamma}_{ikl}(1 \leqslant k \leqslant J_i, 0 \leqslant i \leqslant p, 1 \leqslant l \leqslant m)$. A dual programming of $(\tilde{P})$ is*

$$(\tilde{D}) \ \widetilde{\max} \ \ \tilde{d}(w) = \left(\frac{\tilde{a}_{0k}}{w_{00}}\right)^{w_{00}} \prod_{i=0}^{p'} \prod_{k=1}^{J_i} \left(\frac{\tilde{c}_{ik}}{\tilde{a}_{ik} w_{ik}}\right)^{w_{ik}} \prod_{i=p'+1}^{p} \prod_{k=1}^{J_i} \left(\frac{\tilde{c}_{ik}\tilde{a}_{ik}}{w_{ik}}\right)^{-w_{ik}} \prod_{i=1}^{p'} w_{i0}^{w_{i0}} \prod_{i=p'+1}^{p} w_{i0}^{-w_{i0}}$$

$$\text{s.t.} \quad w_{00} = \sum_{k=1}^{J_0} w_{0k} = 1$$

$$\tilde{\Gamma}^{\mathrm{T}} w = 0$$

$$w \geqslant 0$$

where $\tilde{\Gamma}$ still represents a fuzzy exponent matrix, i.e.,

$$\tilde{\Gamma} = \begin{pmatrix} \tilde{\gamma}_{011} \cdots & \tilde{\gamma}_{01l} \cdots & \tilde{\gamma}_{01m} \\ \cdots & \cdots & \cdots \\ \tilde{\gamma}_{0J_0 1} \cdots & \tilde{\gamma}_{0J_0 l} \cdots & \tilde{\gamma}_{0J_0 n} \\ \cdots & \cdots & \cdots \\ \tilde{\gamma}_{p'J_{p'}1}^{-1} \cdots & \tilde{\gamma}_{p'J_{p'}l}^{-1} \cdots & \tilde{\gamma}_{p'J_{p'}m}^{-1} \\ -\tilde{\gamma}_{p'+1J_{p'+1}1}^{-1} \cdots & -\tilde{\gamma}_{p'+1J_{p'+1}l}^{-1} \cdots & -\tilde{\gamma}_{p'+1J_{p'+1}m}^{-1} \\ \cdots & \cdots & \cdots \\ -\tilde{\gamma}_{pJ_p 1} \cdots & -\tilde{\gamma}_{pJ_p l} \cdots & -\tilde{\gamma}_{pJ_p m} \end{pmatrix}$$

Here $w = (w_{00}, w_{01}, \ldots, w_{0J_0}, w_{p1}, \ldots, w_{pJ_p})^{\mathrm{T}}$ *is a J'−dimensional fuzzy variable vector* $(J' = 1 + J_0 + \cdots + J_p)$, *and* $w_{i0} = w_{i1} + w_{i2} + \cdots + w_{iJ_i}$;

$-w_{ik}$ and $-w_{i0}$ denote a reversed direction inequality $\tilde{g}_i(x)\gtrsim 1$ corresponding to factors $(\frac{\tilde{c}_{ik}}{w_{ik}})^{-w_{ik}}$ and $w_{i0}^{-w_{i0}}$ in the upper-right-corner exponent. ◊

Proof By (1.14),(1.17) and (1.18) in Section 1.6.3 of Chapter 1, $(\tilde{P})$ can be turned into

$$
\begin{aligned}
(\tilde{P})\quad & \max\alpha=\min(-\alpha)\\
& \text{s.t. } \exp\left\{-\frac{\sum_{k=1}^{J_0}\tilde{c}_{0k}^{-1}(\beta)\prod_{l=1}^{m}x_l^{\tilde{\gamma}_{0kl}^{-1}(\beta)}-\bar{g}_0}{d_0}\right\}\geqslant\alpha\\
& \exp\left\{-\frac{\sum_{k=1}^{J_i}\tilde{c}_{ik}^{-1}(\beta)\prod_{l=1}^{m}x_l^{\tilde{\gamma}_{ikl}^{-1}(\beta)}-1}{d_i}\right\}\geqslant\alpha\\
& (1\leqslant i\leqslant p')\\
& 1-\exp\left\{-\frac{\sum_{k=1}^{J_i}\tilde{c}_{ik}^{-1}(\beta)\prod_{l=1}^{m}x_l^{\tilde{\gamma}_{ikl}^{-1}(\beta)}-1}{d_i}\right\}\geqslant\alpha\\
& (p'+1\leqslant i\leqslant p),\alpha,\beta\in[0,1]\\
& x>0
\end{aligned}
\tag{4.2}
$$

where $\bar{g}_0=\sup_{x\in\Re^m}g_0(\bar{x}')$ is supremum of $\tilde{g}_0(x)$, such that

$$
\begin{aligned}
(4.2)\Leftrightarrow\quad & \min(-\alpha)\\
& \text{s.t. } g_0(\bar{x})=\frac{\sum_{k=1}^{J_0}\tilde{c}_{0k}^{-1}(\beta)\prod_{l=1}^{m}x_l^{\tilde{\gamma}_{0kl}^{-1}(\beta)}}{\tilde{B}_0^{-1}(\alpha)}\leqslant 1,\\
& g_i(\bar{x})=\frac{\sum_{k=1}^{J_i}\tilde{c}_{ik}^{-1}(\beta)\prod_{l=1}^{m}x_l^{\tilde{\gamma}_{ikl}^{-1}(\beta)}}{\tilde{B}_i^{-1}(\alpha)}\leqslant 1\\
& (1\leqslant i\leqslant p'),\\
& g_i(\bar{x})=\frac{\sum_{k=1}^{J_i}\tilde{c}_{ik}^{-1}(\beta)\prod_{l=1}^{m}x_l^{\tilde{\gamma}_{ikl}^{-1}(\beta)}}{\tilde{B}_i^{-1}(\alpha)}\geqslant 1\\
& (p'+1\leqslant i\leqslant p),\alpha,\beta\in[0,1],\\
& x>0,
\end{aligned}
\tag{4.3}
$$

where $\bar{x}=(x,\alpha,\beta)$, $\tilde{B}_i^{-1}(\alpha)(0\leqslant i\leqslant p)$ are defined as

$$
\begin{aligned}
&\tilde{B}_0^{-1}(\alpha)=\bar{g}_0-d_0\log\alpha\\
&\tilde{B}_i^{-1}(\alpha)=\begin{cases}1-d_i\log\alpha, & (1\leqslant i\leqslant p')\\ 1+d_i\log(1-\alpha), & (p'+1\leqslant i\leqslant p).\end{cases}
\end{aligned}
$$

It is obvious that they are greater than 0 for $\alpha \in [0,1]$.

Similar to the proof in Theorem 4.1 in Reference [Cao93a], we deduce the dual programming of (4.3) as follows:

$$\begin{aligned} \max & \left(\frac{-\alpha}{\bar{w}_{00}}\right)^{\bar{w}_{00}} \prod_{i=0}^{p'} \prod_{k=1}^{J_i} \left(\frac{\tilde{c}_{ik}^{-1}(\beta)}{\tilde{B}_i^{-1}(\alpha)\bar{w}_{ik}}\right)^{\bar{w}_{ik}} \prod_{i=p'+1}^{p} \prod_{k=1}^{J_i} \\ & \left(\frac{\tilde{c}_{ik}^{-1}(\beta)}{\tilde{B}_i^{-1}(\alpha)\bar{w}_{ik}}\right)^{-\bar{w}_{ik}} \prod_{i=1}^{p'} \bar{w}_{i0}^{\bar{w}_{i0}} \prod_{i=p'+1}^{p} \bar{w}_{i0}^{-\bar{w}_{i0}} \\ \text{s.t. } & \bar{w}_{00} = 1 \\ & \sum_{k=1}^{J_0} \bar{w}_{0k} = \bar{w}_{00} \\ & \bar{\Gamma}^{\mathrm{T}} \bar{w} = 0, \alpha, \beta \in [0,1] \\ & \bar{w} \geqslant 0 \end{aligned} \tag{4.4}$$

where $\bar{w} = (w, \alpha, \beta)$, and

$$\bar{\Gamma} = \begin{pmatrix} \tilde{\gamma}_{011}^{-1}(\beta)\cdots & \tilde{\gamma}_{01l}^{-1}(\beta)\cdots & \tilde{\gamma}_{01m}^{-1}(\beta) \\ \cdots & \cdots & \cdots \\ \tilde{\gamma}_{0J_01}^{-1}(\beta)\cdots & \tilde{\gamma}_{0J_0l}^{-1}(\beta)\cdots & \tilde{\gamma}_{0J_0m}^{-1}(\beta) \\ \cdots & \cdots & \cdots \\ \tilde{\gamma}_{p'J_{p'}1}^{-1}(\beta)\cdots & \tilde{\gamma}_{p'J_{p'}l}^{-1}(\beta)\cdots & \tilde{\gamma}_{p'J_{p'}m}^{-1}(\beta) \\ -\tilde{\gamma}_{p'+1J_{p'+1}1}^{-1}(\beta)\cdots & -\tilde{\gamma}_{p'+1J_{p'+1}l}^{-1}(\beta)\cdots & -\tilde{\gamma}_{p'+1J_{p'+1}m}^{-1}(\beta) \\ \cdots & \cdots & \cdots \\ -\tilde{\gamma}_{pJ_p1}^{-1}(\beta)\cdots & -\tilde{\gamma}_{pJ_pl}^{-1}(\beta)\cdots & -\gamma_{pJ_pm}^{-1}(\beta) \end{pmatrix}.$$

(4.4) is equivalent to $(\tilde{D})$. Therefore $(\tilde{D})$ denotes the fuzzy dual programming of $(\tilde{P})$. Overall the theorem holds. □

Theorem 4.1.2 *Suppose that fuzzy-valued functions $\tilde{A}_i$ $(0 \leqslant i \leqslant p)$ are CSM functions. Then $(\tilde{P})$ contains the same fuzzy optimal valued as its dual program $(\tilde{D})$.* ◊

Proof $(\tilde{P})$ is changed into (4.3), and formula (4.3)$\Longleftrightarrow$ formula (4.4). But from the proof of Theorem 4.1.1, we know $(\tilde{D})$ can be turned into (4.4), and from the equivalence of (4.4) and (4.3), we also know (4.4) has the same parameter optimal value as (4.3), such that the theorem holds. □

Obviously, the problems around nonfuzzification are of the same DDs.

In order to give a further discussion, we always write $(\tilde{P})$ and $(\tilde{D})$ in a more general form.

Definition 4.1.2 Let the general form of fuzzy reverse PGP $(\tilde{P})$ and its dual program $(\tilde{D})$ be [Cao98a,b]

$$(\tilde{P}_I) \quad \inf\{\tilde{g}_0(x)|\text{s.t. } \tilde{g}_i(x) \lesssim 1 \ (1 \leqslant i \leqslant p'); \\ \tilde{g}_i(x) \gtrsim 1 \ (p'+1 \leqslant i \leqslant p), x > 0\}$$

and

$$(\tilde{D}_S) \ \sup\{\tilde{d}(w)|\text{s.t. } w_{00} = \sum_{k=1}^{J_0} w_{0k} = 1, \tilde{\Gamma}^{\mathrm{T}} w = 0, w \geqslant 0\}.$$

When $(\tilde{P}_I)$ and $(\tilde{D}_S)$ are fuzzy consistent, $M_{\tilde{P}_I}$ and $M_{\tilde{D}_S}$ are the fuzzy constrained infimum of $(\tilde{P}_I)$ and the fuzzy constrained supremum of $(\tilde{D}_S)$, respectively.

Generally, a reversed PGP is nonconvex, so is the deformed substitution by $x_i = e^{y_i}$. Therefore, it does not keep nice properties [WY82] as a PGP. But we find that many results can be obtained in a fuzzy reverse PGP corresponding to a fuzzy PGP [Caoupd].

4.2 Duality Properties of Fuzzy Reverse PGP

Definition 4.2.1 Let $\tilde{A}_0^*(x)$ be an objective function defined on $X \subset \Re^m$, which is a fuzzy optimum valued set for $\{x|\tilde{g_0}(x) \lesssim g_0\}$ with respect to $\tilde{A}_0^*$. Again suppose that

$$\tilde{A}^*(x) = \bigwedge_{1\leqslant i\leqslant p'} \tilde{A}_i^{*1}(x) \bigwedge (\bigwedge_{p'+1\leqslant i\leqslant p} A_i^{*2}(x))$$

is a constrained function defined on $X \subset \Re^m$, which is a fuzzy feasible solution set for

$$\{x \in \Re^m|\tilde{g}_i(x) \lesssim 1 \ (1 \leqslant i \leqslant p'), x > 0\}$$

and

$$\{x \in \Re^m|\tilde{g}_i(x) \gtrsim 1 \ (p'+1 \leqslant i \leqslant p), x > 0\}$$

with respect to $\tilde{A}_i^{*1}$ and $\tilde{A}_i^{*2}$, respectively. Let $\tilde{Y}^* = \tilde{A}_0^* \bigcap \tilde{A}^*$, where $\tilde{A}^* = \tilde{A}_i^{*1} \cap \tilde{A}_i^{*2}$. Then $(\tilde{P})$ is written as

$$\max_{x\in\Re^+}\{\tilde{Y}^*(x) = \min_{0\leqslant i\leqslant p} \tilde{A}_i^{\ *}(x)\}. \tag{4.5}$$

Definition 4.2.2 The fuzzy reverse PGP $(\tilde{P})$ is fuzzy convex, i.e., (4.5) is defined as the fuzzy convex equivalent to $\forall\alpha \in (0,1)$,

$$\max_{x\in\Re^+}\{\min_{0\leqslant i\leqslant p} \tilde{A}_i^*(\bar{x})\}$$

is a convex programming.

Theorem 4.2.1 *If for all $i \in (0,p), \tilde{B_i}^*$ is a CSN function, $\tilde{c}_{ik}$, and $\tilde{\gamma}_{ikl}$ $(1 \leqslant k \leqslant J_i, 0 \leqslant i \leqslant p, 1 \leqslant l \leqslant m)$ are CSM, then the fuzzy reverse PGP $(\tilde{P})$ can be turned into a convex programming with respect to x.* ◊

Proof $(\tilde{P})$ can be changed into (4.2) by formula (1.14), (1.15) and (1.16) in Section 1.6.3 in Chapter 1, while (4.2) $\Leftrightarrow$ (4.3). In formula (4.3),

$$g_i(\bar{x}) = \sum_{k=1}^{J_i} \tilde{c}_{ik}^{-1}(\beta) \prod_{l=1}^{m} x_l^{\tilde{\gamma}_{ikl}^{-1}(\beta)} / \tilde{B}_i^{-1}(\alpha),$$

where we suppose $f_{ik}(\bar{x}) = \prod\limits_{l=1}^{m} x_l^{\tilde{\gamma}_{ikl}^{-1}(\beta)} / \tilde{B}_i(\alpha)$, so $g_i(\bar{x}) = \sum\limits_{k=1}^{J_i} \tilde{c}_{ik}^{-1}(\beta) f_{ik}(\bar{x})$. We might make $f_{ik}(\bar{x})$ nondecreasing and then it is easy to prove

$$f_{ik}[\lambda x_1 + (1-\lambda)x_2, \alpha, \beta] \leqslant \lambda f_{ik}(x_1, \alpha, \beta) + (1-\lambda) f_{ik}(x_2, \alpha, \beta)$$

for $\lambda \in [0,1]$, i.e., it is a convex function with respect to x for $\alpha, \beta \in [0,1]$. Again, from $\tilde{c}_{ik}^{-1}(\beta) \geqslant 0$, we know $g_i(\bar{x})$ is a convex function with respect to

x for α,β, formula (4.3), i.e., (4.2) is a convex GP with respect to x for α,β, such that the theorem holds. □

Theorem 4.2.2 *Let $\tilde{A}_i^*(0 \leqslant i \leqslant p)$ be a CSM function. Then the fuzzy local minimum solutions of the fuzzy reverse PGP $(\tilde{P})$ are also its fuzzy global minimum ones.* ◊

Proof From Theorem 4.2.1, we know $(\tilde{P})$ is fuzzy convex, then , $\forall\alpha,\beta\in[0,1]$,

$$\tilde{Y}^*(\bar{x}) = \tilde{A}_0^*(\bar{x})\bigwedge\Big(\bigwedge_{1\leqslant i\leqslant p'}\tilde{A}_i^{*1}(\bar{x})\Big)\bigwedge\Big(\bigwedge_{p'+1\leqslant i\leqslant p}\tilde{A}_i^{*2}(\bar{x})\Big)$$

is a convex fuzzy-valued function [Cao93a]. Therefore, the local minimum solutions of $\tilde{Y}^*(\bar{x})$ are its global minimum ones, and the theorem also holds. □

Corollary 4.2.1 *Let $\tilde{A}_i^*(0 \leqslant i \leqslant p)$ be a CSM function. Then the strictly fuzzy local minimum solutions to $(\tilde{P})$ are its strictly fuzzy global minimum ones.* ◊

Theorem 4.2.3 *Let $\tilde{A}_i^*(0 \leqslant i \leqslant p)$ be a CSM function. Then any fuzzy local minimum solutions to $(\tilde{P})$ are its fuzzy global minimum ones.* ◊

The theorems and the corollary above can be easily proved by means of corresponding Theorem 4.2.2 in Section 3 in Reference [Cao93a].

Definition 4.2.3 If $x > 0$, such that the conditions

$$\begin{cases} v(\tilde{g}_i(x)\tilde{<}1) = 1, & (1\leqslant i\leqslant p') \\ v(\tilde{g}_i(x)\tilde{>}1) = 1, & (p'+1\leqslant i\leqslant p) \end{cases}$$

or

$$\begin{cases} v(\tilde{g}_i(x)\lesssim 1) = 1, & (1\leqslant i\leqslant p') \\ v(\tilde{g}_i(x)\gtrsim 1) = 1, & (p'+1\leqslant i\leqslant p) \end{cases}$$

hold, respectively, then we call the fuzzy reverse PGP $(\tilde{P})$ a fuzzy super-consistence or fuzzy consistence, respectively, where $v(\cdot)$ denotes a degree of possibility, with

$$v(\tilde{g}_i(x)\tilde{\circledcirc}1) = \sup_{x,y|x\circledcirc y}\min\{\mu_{\tilde{g}_i}(x),\mu_1(y)\}(1\leqslant i\leqslant p),$$

here "$\tilde{\circledcirc}$" denotes "$<$" "$>$" "$\lesssim$" or "$\gtrsim$".

Lemma 4.2.1 *Let $\tilde{B}_i^*$ be the CSN function and $\tilde{\phi}_i = \tilde{c}_{ik}\circ\tilde{\gamma}_{ikl}(1\leqslant k\leqslant J_i, 0\leqslant i\leqslant p, 1\leqslant l\leqslant m)$ be a CSM function. Then to any fuzzy feasible solution x of the primal fuzzy reverse PGP $(\tilde{P})$ and any feasible solution w of dual program $(\tilde{D})$, we have*

$$\tilde{g}_0(x)\gtrsim\tilde{g}_0(x)\prod_{i=1}^{p'}(\tilde{g}_i(x))^{w_{i0}}\prod_{i=p'+1}^{p}(\tilde{g}_i(x))^{-w_{i0}}\gtrsim\tilde{d}(w) \tag{4.6}$$

and

$$\tilde{g}_0(x) = \tilde{d}(w) \tag{4.7}$$

iff

$$w_{ik} = \begin{cases} \tilde{v}_{0k}(x)/\tilde{g}_0(x), & (i = 0; 1 \leqslant k \leqslant J_0) \\ w_{i0}\tilde{v}_{ik}(x), & (i \neq 0; 1 \leqslant k \leqslant J_i) \end{cases} \tag{4.8}$$

holds. Now, x is a fuzzy optimum solution to $(\tilde{P})$ and w is an optimum one to $(\tilde{D})$. ◊

Proof Since x denotes a fuzzy feasible solution to $(\tilde{P})$, then $\bar{x}$ is a feasible one to (4.3). i.e., for $\tilde{B}_i^*, \tilde{c}_{ik}, \tilde{\gamma}_{ikl}(1 \leqslant k \leqslant J_i, 0 \leqslant i \leqslant p, 1 \leqslant l \leqslant m)$, we have $g_i(\bar{x}) = g_i(\bar{x}')/\tilde{B}_i^{-1}(\alpha) \leqslant 1 (0 \leqslant i \leqslant p')$ and $\bar{g}_i(\bar{x}) = \tilde{B}_i^{-1}(\alpha)/g_i(\bar{x}') \leqslant 1 (p'+1 \leqslant i \leqslant p)$. Again, $\bar{w}_{i0} \geqslant 0$, then

$$g_0(\bar{x}) \geqslant g_0(\bar{x}) \prod_{i=1}^{p'} (g_i(\bar{x}))^{\bar{w}_{i0}} \prod_{i=p'+1}^{p} ({g'}_i(\bar{x}))^{\bar{w}_{i0}}.$$

And because $g_i(\bar{x}) = \sum\limits_{k=1}^{J_i} v_{ik}(\bar{x})$, where

$${g_i}'(\bar{x}) = 1/g_i(\bar{x}), v_{ik}(\bar{x}) = v_{ik}(\bar{x}')/\tilde{B}_i^{-1}(\alpha).$$

Hence, from the ordinary geometric inequality, we have

$$(g_i(\bar{x}))^{\bar{w}_{i0}} \geqslant \prod_{k=1}^{J_i} \left(\frac{v_{ik}(\bar{x})}{\bar{w}_{ik}} \right)^{\bar{w}_{ik}} \bar{w}_{i0}^{\bar{w}_{i0}} (0 \leqslant i \leqslant p).$$

Whereas there must exist $\alpha, \beta \in [0, 1]$ for the feasible solutions $\bar{x}$ and $\bar{w}$ of (4.3) and (4.4), such that $\bar{w}_{00} = 1$ and $\Gamma^{\mathrm{T}}\bar{w} = 0$, we have

$$\begin{aligned} \alpha &\geqslant \alpha \prod_{i=0}^{p'} (g_i(\bar{x}))^{\bar{w}_{ik}} \prod_{i=p'+1}^{p} ({g'}_i(\bar{x}))^{\bar{w}_{ik}} \\ &\geqslant \alpha \prod_{i=0}^{p'} \prod_{k=1}^{J_i} \left(\frac{\tilde{c}_{ik}^{-1}(\beta)}{\tilde{B}_i^{-1}(\alpha)\bar{w}_{ik}} \right)^{\bar{w}_{ik}} \prod_{i=1}^{p'} \bar{w}_{i0}^{\bar{w}_{i0}} \prod_{i=p'+1}^{p} \prod_{k=1}^{J_i} \\ &\quad \left(\frac{\tilde{c}_{ik}^{-1}(\beta)}{\tilde{B}_i^{-1}(\alpha)\bar{w}_{ik}} \right)^{-\bar{w}_{ik}} \prod_{i=p'+1}^{p} \bar{w}_{i0}^{-\bar{w}_{i0}} \prod_{l=1}^{m} x_l^{\sum\limits_{i=0}^{p} \sum\limits_{k=1}^{J_i} \tilde{\gamma}_{ikl}^{-1}(\beta)\bar{w}_{ik}} \\ &= \alpha \prod_{i=0}^{p'} \prod_{k=1}^{J_i} \left(\frac{\tilde{c}_{ik}^{-1}(\beta)}{\tilde{B}_i^{-1}(\alpha)\bar{w}_{ik}} \right)^{\bar{w}_{ik}} \prod_{i=p'+1}^{p} \prod_{k=1}^{J_i} \\ &\quad \left(\frac{\tilde{c}_{ik}^{-1}(\beta)}{\tilde{B}_i^{-1}(\alpha)\bar{w}_{ik}} \right)^{-\bar{w}_{ik}} \prod_{i=1}^{p'} \bar{w}_{i0}^{\bar{w}_{i0}} \prod_{i=p'+1}^{p} \bar{w}_{i0}^{-\bar{w}_{i0}} \\ &= -d(\bar{w}). \end{aligned} \tag{4.9}$$

Therefore, $g_0(\bar{x}') \geqslant -d(w^*) = \tilde{B}_0^{-1}(\alpha)d(\bar{w})/(-\alpha)$, and

$$g_0(\bar{x}') = -d(w^*) \tag{4.10}$$

iff

$$\bar{w}_{ik} = \begin{cases} v_{00}(\bar{x}')/(-\alpha), & (i = 0; k = 0, \alpha \neq 0) \\ \bar{w}_{i0} v_{ik}(\bar{x}'), & (1 \leqslant k \leqslant J_i; 0 \leqslant i \leqslant p) \end{cases} \tag{4.11}$$

holds. Since (4.6) – (4.8) $\Leftrightarrow$ (4.9) – (4.11) holds in the light of arbitrariness of $\alpha, \beta \in [0, 1]$, the lemma holds. □

Theorem 4.2.4 (The first fuzzy dual theorem) *Let the fuzzy reverse PGP $(\tilde{P})$ be fuzzy super-consistent with fuzzy optimal solution x^* existing. Then there must exist a Lagrange multiplier $\mu^* = (\mu_1^*, \mu_2^*, \ldots, \mu_{p'}^*, \mu_{p'+1}^*, \ldots, \mu_p^*)^T \geqslant 0$ and CSM functions $\tilde{A}_i^*$ $(0 \leqslant i \leqslant p)$, such that*

$$\nabla \tilde{g}_0(x^*) + \sum_{i=1}^{p'} \mu_i^* \nabla \tilde{g}_i(x^*) + \sum_{i=p'+1}^{p} \mu_i^* \nabla(-\tilde{g}_i(x^*)) = 0, \tag{4.12}$$

$$\mu_i^*(\tilde{g}_i(x^*) - 1) = 0 \; (1 \leqslant i \leqslant p), \tag{4.13}$$

while w^* defined by

$$w_{ik}^* = \begin{cases} \tilde{v}_{0k}(x^*)/\tilde{g}_0(x^*), & (i = 0; 1 \leqslant k \leqslant J_0) \\ \mu_i^* \tilde{v}_{ik}(x^*)/\tilde{g}_0(x^*), & (i \neq 0; 1 \leqslant k \leqslant J_i) \end{cases} \tag{4.14}$$

is a fuzzy optimal solution of dual program $(\tilde{D})$, such that

$$\tilde{g}_0(x^*) = \tilde{d}(w^*). \tag{4.15}$$

◊

Proof As for all the CSM functions $\tilde{A}_i^*(0 \leqslant i \leqslant p)$, $(\tilde{P})$ fuzzy super-consistence having a fuzzy optimal solution x^* is equivalent to program (4.3) being super-consistent for all $\alpha, \beta \in [0, 1]$. Again, because (4.3) is convex with respect to x for α, β, satisfying the Slater condition, there exists an optimal parametric solution $\bar{x}^*$. Therefore, from Theorem 5.1 in Reference [Cao93a], we know there exists a Lagrange multiplier

$$\mu^* = (\mu_1^*, \mu_2^*, \ldots, \mu_{p'}^*, \mu_{p'+1}^*, \ldots, \mu_p^*)^{\mathrm{T}} \geqslant 0,$$

such that

$$\nabla g_0(\bar{x}^*) + \sum_{i=1}^{p'} \mu_i^* \nabla g_i(\bar{x}^*) + \sum_{i=p'+1}^{p} \mu_i^* \nabla(-g_i(\bar{x}^*)) = 0, \tag{4.16}$$

$$\mu_i^*(g_i(\bar{x}^*) - 1) = 0 \quad (1 \leqslant i \leqslant p), \tag{4.17}$$

while $\bar{w}^*_{ik}$ defined by

$$\bar{w}^*_{ik} = \begin{cases} v_{00}(x^*, \beta)/(-\alpha), & (i = 0; k = 0, \alpha \neq 0) \\ \mu^*_i v_{ik}(x^*, \beta)/g_0(\bar{x}^*), & (1 \leqslant k \leqslant J_i; 0 \leqslant i \leqslant p) \end{cases} \tag{4.18}$$

is an optimal parametric solution of dual program (4.4) with the determined $\alpha, \beta \in [0, 1]$, such that

$$g_0(\bar{x}^*) = d(\bar{w})\tilde{B}_i^{-1}(\alpha)/(-\alpha) = -d(\bar{w}^*).$$

Again, (4.16)–(4.18) hold by the arbitrariness of $\alpha, \beta \in [0, 1]$. It follows that the theorem holds. □

Corollary 4.2.2 *Let a general fuzzy reverse PGP and its dual programming be $(\tilde{P}_I)$ and $(\tilde{D}_S)$. If, to the CSM functions $\tilde{A}^*_i (0 \leqslant i \leqslant p)$, $(\tilde{P}_I)$ is fuzzy super-consistent, with $M_{\tilde{P}_I} > 0$, there must exist a Lagrange multiplier*

$$\mu^* = (\mu^*_1, \mu^*_2, \ldots, \mu^*_{p'}, \mu^*_{p'+1}, \ldots, \mu^*_p)^T \geqslant 0,$$

such that (4.12), (4.13) and (4.14) hold, and w^* is an optimal solution to a dual program $(\tilde{D}_S)$, with

$$M_{\tilde{D}_I} = \tilde{d}(w^*).$$

◊

Proof $(\tilde{P})$ has a minimum in Theorem 4.2.4 corresponding to $(\tilde{P}_I)$ being a constrained minimum which must be positive. The theorem holds in the same way as Theorem 4.2.4. □

All of the above conclusions assume that there exists a fuzzy optimal solution to fuzzy reverse PGP $(\tilde{P})$. When do we obtain a fuzzy optimal one in $(\tilde{P})$? Now we have the following:

Theorem 4.2.5 (the second fuzzy dual theorem) *Let $(\tilde{P})$ be fuzzy consistent and $(\tilde{D})$ have a fuzzy feasible solution with components being positive. Then to CSM functions $\tilde{B}_i$, $\tilde{c}_{ik}$ and $\tilde{\gamma}_{ikl}$ $(1 \leqslant k \leqslant J_i, 0 \leqslant i \leqslant p, 1 \leqslant l \leqslant m)$, there exists a fuzzy optimal solution to $(\tilde{P})$.* ◊

Proof For the CSM functions $\tilde{B}_i$, $\tilde{c}_{ik}$ and $\tilde{\gamma}_{ikl}$ $(1 \leqslant k \leqslant J_i, 0 \leqslant i \leqslant p, 1 \leqslant l \leqslant m)$, $(\tilde{P})$ can be changed into a convex GP (4.3), while the dual form of (4.3) is (4.4). Because $(\tilde{P})$ fuzzy consistence ⇔ (4.3) consistence. Again, the constraint condition of $(\tilde{D})$ is equivalent to that of (4.4), i.e., $(\tilde{D})$ contains a feasible solution with components being positive ⇔ (4.4) has a parameter feasible solution with components being positive.

Let $x_l = e^{z_l} (1 \leqslant l \leqslant m)$. We can get a deformed primal programming of (4.3), which is also consistent with points range $\{z^k\}$ existing, similar to the classical 2nd-dual theorem (Theorem 1.8.1 in Reference [WY82]), satisfying

$$G_i(\bar{z}^k) \leqslant 1 (1 \leqslant i \leqslant p'),$$
$$G_i^{-1}(\bar{z}^k) \leqslant 1 (p' + 1 \leqslant i \leqslant p)$$

such that

$$\lim_{k\to\infty} G_0(\bar{z}^k) = M_{\tilde{P}}$$

where $\bar{z}^k = (z^k, \alpha, \beta)$. Accordingly, there must exist α, $\beta \in [0,1]$ and then

$$\begin{aligned}
G_i(\bar{z}^*) &= \sum_{k=1}^{J_i} \tilde{c}_{ik}^{-1}(\beta)/(\tilde{B}_i^{-1}(\alpha)) \prod_{l=1}^{m} (e^{z_l^*})^{\sigma_i \tilde{\gamma}_{ikl}^{-1}(\beta)} \\
&= \sum_{k=1}^{J_i} \tilde{c}_{ik}^{-1}(\beta)/(\tilde{B}_i^{-1}(\alpha)) \prod_{l=1}^{m} x_l^{*\sigma_i \tilde{\gamma}_{ikl}^{-1}(\beta)} \\
&= \begin{cases} \sum_{k=1}^{J_i} \tilde{c}_{ik}^{-1}(\beta)/\tilde{B}_i^{-1}(\alpha) \prod_{l=1}^{m} x_l^{*\tilde{\gamma}_{ikl}^{-1}(\beta)} = \begin{cases} = M_{\tilde{P}}, & (i=0) \\ \leqslant 1, & (1 \leqslant i \leqslant p') \end{cases} \\ \sum_{k=1}^{J_i} \tilde{c}_{ik}^{-1}(\beta)/\tilde{B}_i^{-1}(\alpha) \prod_{l=1}^{m} x_l^{*-\tilde{\gamma}_{ikl}^{-1}(\beta)} \geqslant 1, \quad (p'+1 \leqslant i \leqslant p) \end{cases}
\end{aligned}$$

where $\bar{z}^* = (z^*, \alpha, \beta)$, $\bar{x}^* = (x^*, \alpha, \beta)$, $\bar{x}^k = (x^k, \alpha, \beta)$, $\bar{z}^*$ is a limit point of $\bar{z}^k$ and $\bar{x}^*$ a limit point of $\bar{x}^k$,

$$\sigma_i = \begin{cases} 1, & (0 \leqslant i \leqslant p') \\ -1, & (p'+1 \leqslant i \leqslant p) \end{cases}$$

i.e., $\bar{x}_l^*(1 \leqslant l \leqslant m)$ denotes an optimal solution to (4.3), therefore the theorem holds. □

4.3 Dual Algorithm for Fuzzy Reverse PGP

4.3.1 Algorithm Steps

It is simple and easy, in most cases, to find an analytic solution to a dual problem like fuzzy PGP. Therefore, similar to Section 2.6 in Chapter 2, we shall exhibit steps for fuzzy dual algorithms to the fuzzy reverse PGP $(\tilde{P})$ in the case of a DD $D > 0$:

Step 1 Change $(\tilde{P})$ into two determined reverse GP from formula (1.14) (1.15) and (1.16) (resp.(1.14) (1.17) and (1.18)) in Section 1.6.3, Chapter 1 as follows:

$$\begin{aligned} &\min\ g_0(\bar{x}') \\ &\text{s.t. } g_i(\bar{x}') \leqslant 1\,, (1 \leqslant i \leqslant p') \\ &\qquad g_i(\bar{x}') \geqslant 1\,, (p'+1 \leqslant i \leqslant p), \beta \in [0,1] \\ &\qquad x > 0 \end{aligned}$$

and

$$\begin{aligned} &\min\ g_0(\bar{x}') \\ &\text{s.t. } g_i(\bar{x}') \leqslant 1 + d_i,\ (1 \leqslant i \leqslant p') \\ &\qquad g_i(\bar{x}') \geqslant 1 + d_i, (p'+1 \leqslant i \leqslant p), \beta \in [0,1] \\ &\qquad x > 0 \end{aligned}$$

Step 2 Write out their dual form

$$\begin{aligned} \max\ d(\bar{w}) = &\left(\frac{-\alpha}{\bar{w}_{00}}\right)^{\bar{w}_{00}} \prod_{i=0}^{p'} \prod_{k=1}^{J_i} \left(\frac{\tilde{c}_{ik}^{-1}(\beta)/(1+d_i)}{\bar{w}_{ik}}\right)^{\bar{w}_{ik}} \\ &\prod_{i=p'+1}^{p} \prod_{k=1}^{J_i} \left(\frac{\tilde{c}_{ik}^{-1}(\beta)/(1+d_i)}{\bar{w}_{ik}}\right)^{-\bar{w}_{ik}} \prod_{i=1}^{p'} \bar{w}_{i0}^{\bar{w}_{i0}} \prod_{i=p'+1}^{p} \bar{w}_{i0}^{-\bar{w}_{i0}} \\ \text{s.t. } &\bar{w}_{00} = 1 \\ &\Gamma^{\mathrm{T}} \bar{w} = 0, \alpha, \beta \in [0,1] \\ &\bar{w} \geqslant 0 \end{aligned} \tag{4.19}$$

where d_i is selected as 0 or d_i.

Step 3 Find (4.19) (resp. (4.4)) for $d_i = 0$ or for $d_i = d_i$, and we can find two optimal values g_0^* and g_0^{*1}. As $D > 0$, we can find a feasible solution to (4.19) (resp.(4.4)) by means of optimizing methods such as a brief gradient, a gradient projection, feasible direction or variable elimination. Especially for $D = 1$, we apply search methods (or methods of 0.618, twice or three times in interpolation) into obtaining its optimal solution.

Step 4 Find an ordinary GP

$$\begin{aligned} &\max \alpha = \min(-\alpha) \\ &\text{s.t. } g_0(\bar{x}') \leqslant \tilde{B}_0^{-1}(\alpha) \\ &\quad g_i(\bar{x}') \leqslant \tilde{B}_i^{-1}(\alpha)(1 \leqslant i \leqslant p') \\ &\quad g_i(\bar{x}') \geqslant \tilde{B}_i^{-1}(\alpha)(p'+1 \leqslant i \leqslant p), \alpha, \beta \in [0,1] \\ &\quad x > 0 \end{aligned}$$

where $z_0 = g_0^{*1}$ (we had better suppose $g_0^{*1} < g_0^*$), $d_0 = g_0^{*1} - g_0^*$ and its dual programming denotes (4.19) (resp.(4.4)). And (4.19) optimal parameter solutions are obtained by using one method of Step 3 to (4.19) (resp.(4.4)); accordingly the value of α, β is determined by $d(\bar{w}) = -\alpha$.

Step 5 According to formula

$$\bar{w}_{ik} = \begin{cases} v_{00}(\bar{x})/(-\alpha), & (i = 0, k = 0, \alpha \neq 0) \\ \bar{w}_{i0} v_{ik}(\bar{x}), & (1 \leqslant k \leqslant J_i, 0 \leqslant i \leqslant p) \end{cases}$$

we find $\bar{x}$, i.e., it is the optimal solution to (4.19) (resp.(4.4)) with respect to parameters α, β, respectively. In application, we stop when we come to this step.

Step 6 $\forall \alpha, \beta \in [0,1]$, a fuzzy optimal solution x to $(\tilde{P})$ is compounded by $\bar{x}$.

4.3.2 Numerical Example

Example 4.1 Find a fuzzy reverse PGP

$$\begin{aligned} &\widetilde{\min}\ g_0(x) = 2x_1 + 3x_2 \\ &\text{s.t. } g_1(x) = x_1^2 + x_2^2 \gtrsim 1 \\ &\quad x_1, x_2 > 0 \end{aligned} \tag{4.20}$$

Solution By means of (4.3), we take $d_0 = 1, d_1 = 1/2$, whereas the classical type of (4.15) behaves as

$$\begin{aligned} &\min\ g_0(x) = 2x_1 + 3x_2 \\ &\text{s.t. } g_1(x) = x_1^2 + x_2^2 \geqslant 1 \\ &\quad x_1, x_2 > 0 \end{aligned}$$

which is nonconvex. Given $x_1 = e^u, x_2 = e^v$, its deformed programming is

$$\begin{aligned} &\min\ 2e^u + 3e^v \\ &\text{s.t. } e^{2u} + e^{2v} \geqslant 1 \end{aligned}$$

so, its deformed programming is nonconvex. For example, when $(0,-1)$ and $(-1,0)$ are taken for (u,v), the deformed programming satisfies the constraint condition, but their middle points $(-1/2, -1/2)$ fail to do that and their constraint infimum $M_P = 2$ is less than an objective function value $d(w) = 3$ of the dual programming [WY82].

Again, let $g_0(x) = d(w) = 3$ and we have $\alpha = 0.486$. This result is the same as the above one, then (4.20) can be turned into a determined type:

$$\begin{aligned}
\max\ & \alpha = \min(-\alpha) \\
\text{s.t.}\ & \frac{2x_1 + 3x_2}{3 - \log\alpha} \leqslant 1 \\
& \frac{x_1^2 + x_2^2}{1 + 2\log(1-\alpha)} \geqslant 1, \alpha \in (0,1) \\
& x_1, x_2 > 0
\end{aligned} \tag{4.21}$$

whose dual programming is

$$\begin{aligned}
\max\ & \left(\frac{-\alpha}{\bar{w}_{00}}\right)^{\bar{w}_{00}} \left(\frac{2}{(3-\log\alpha)\bar{w}_{01}}\right)^{\bar{w}_{01}} \left(\frac{3}{(3-\log\alpha)\bar{w}_{02}}\right)^{\bar{w}_{02}} \\
& \left(\frac{1}{(1+2\log(1-\alpha))\bar{w}_{11}}\right)^{-\bar{w}_{11}} \left(\frac{1}{(1+2\log(1-\alpha))\bar{w}_{12}}\right)^{-\bar{w}_{12}} \bar{w}_{10}^{-\bar{w}_{10}} \\
\text{s.t.}\ & \bar{w}_{00} = \bar{w}_{01} + \bar{w}_{02} = 1 \\
& \bar{w}_{01} - 2\bar{w}_{11} = 0 \\
& \bar{w}_{02} - 2\bar{w}_{12} = 0, \alpha \in (0,1) \\
& \bar{w} \geqslant 0
\end{aligned}$$

and its dual feasible solution behaves as several groups. Take $\bar{w}_{00} = 1, \bar{w}_{01} = 0, \bar{w}_{02} = 1, \bar{w}_{11} = 0, \bar{w}_{12} = 1/2$, substitute them for an objective function, then

$$d(\bar{w}) = (-\alpha)\frac{3}{3-\log\alpha}\left(\frac{2}{(1+2\log(1-\alpha))}\right)^{-1/2}\left(\frac{1}{2}\right)^{-1/2}.$$

From $d(w^*) = d(\bar{w})(3 - \log\alpha)/(-\alpha) = M_{\bar{P}}$, we get $\alpha = 0.243$. By means of (4.11), we have $x_1 = 0, x_2 = \frac{2}{3}$.

Therefore, $(x_1, x_2, \alpha) = (0, 2/3, 0.243)$ is an optimal solution to (4.21). Hence, the optimal value of the primal problem is equal to its dual optimal value as follows:

$$M_{\bar{P}} = -d(w^*) = \left.\frac{d(\bar{w})(3-\log\alpha)}{(-\alpha)}\right|_{\substack{\bar{w} = \bar{w}^0 \\ \alpha = 0.243}} = 2.$$

Definition 4.3.1 Let the general form of the deformed primal programming in the fuzzy reverse PGP (4.1) be

$$(\tilde{G})\ \{\min \tilde{G}_0(z),\ \text{s.t.}\ \tilde{G}_i(z) \lesssim 1\ (1 \leqslant i \leqslant p');\ \tilde{G}_i(z) \gtrsim 1\ (p'+1 \leqslant i \leqslant p)\}.$$

When it is fuzzy consistent, $M_{\tilde{G}}$ denotes the fuzzy constrained infimum of $(\tilde{G})$. And $(\tilde{G})$ does not mean a fuzzy convex programming of z, hence many good properties in the fuzzy PGP fail to remain in the fuzzy reverse PGP [Cao98b][Caoupd], which is more difficult for us to study. Here, we exhibit an example with a dent made in it as follows.

Let us consider the deformed primal parameter programming of (4.21) as

$$\begin{aligned} &\min(-\alpha) \\ &\text{s.t.}\ \frac{2e^u + 3e^v}{3 - \log\alpha} \leqslant 1 \\ &\quad\ \ \frac{e^{2u} + e^{2v}}{1 + 2\log(1-\alpha)} \geqslant 1, \alpha \in (0,1) \end{aligned} \tag{4.22}$$

$(0, -1, 0.243)$,$(-1, 0, 0.243)$ and $(-1/2, -1/2, 0.243)$ are selected for points (u, v, α), respectively, which all satisfy its constraint condition. Therefore, the deformed program (4.22) can be changed into a convex programming with respect to u, v, and α.

Example 4.2 In Example 4.1, let a membership function be as (1.17) and (1.18) in Section 1.6.3. Then the inequalities $2x_1 + 3x_2 \gtrsim 3$ and $x_1^2 + x_2^2 \gtrsim 1$ can be turned into

$$\begin{cases} \dfrac{1}{1 + \exp\{-(x_1^2 + x_2^2 - 1) - \frac{1}{2}\}} \geqslant \alpha \\ \dfrac{1}{1 + \exp\{2x_1 + 3x_2 - 3\}} \geqslant \alpha \end{cases}$$

we find (4.20) before $\alpha = 0.486$ is gained by steps 1-6 above, whose results are the same as that of Example 4.1.

Attention: Here $\alpha \neq 0$ or 1, otherwise, the primal fuzzy reverse PGP is degenerated into a classical reverse PGP or a nonconsistent reverse PGP. Besides, when exponents of the problem are also fuzzy numbers, an optimal solution can be obtained by the method above.

4.3.3 Conclusion

This section changes the fuzzy reverse PGP into a parameter convex GP with respect to α,β, so that the result of the former differs from that of the reverse PGP. Meanwhile, policymakers in applications decide an optimal distribution scheme with respect to α,β in terms of an objective value, which leaves flexible room for their decision.

4.4 Lagrange Problem of Fuzzy Reverse PGP

We consider a Lagrange problem of a fuzzy reverse PGP (4.1) in this section.

4.4.1 Lagrange Problem in Fuzzy Reverse PGP

Definition 4.4.1 Suppose $\tilde{A}_i(0 \leqslant i \leqslant p)$ to be a CSM function; we call $\tilde{g}_i(x^*) = 1$ the fuzzy equality, if its membership is $\tilde{A}_i(\tilde{g}_i(x) - 1) = 1$. We call the problem a fuzzy Lagrange problem for (4.1), i.e., finding a fuzzy feasible solution x^* to (4.1) and $\mu^* = (\mu_1^*, \mu_2^*, \ldots, \mu_p^*)^{\mathrm{T}} \geqslant 0$ satisfying $\mu_i^*(\tilde{g}_i(x^*) - 1) = 0 (1 \leqslant i \leqslant p)$, such that a fuzzy Lagrange function

$$\tilde{L}(x, \mu) = \tilde{g}_0(x) + \sum_{i=1}^{p'} \mu_i(\tilde{g}_i(x) - 1) + \sum_{i=p'+1}^{p} \mu_i(1 - \tilde{g}_i(x))$$

satisfies $\nabla_x \tilde{L}(x^*, \mu^*) = 0$.

Theorem 4.4.1 *Let x^* be a fuzzy feasible solution to (4.1). If $\mathcal{I} = \{i | \tilde{g}_i(x^*) = 1, 0 \leqslant i \leqslant p\}$ denotes a subscript set of fuzzy effective constraints and if $\tilde{A}_i(0 \leqslant i \leqslant p)$ is a CSM function, then there exists μ^* such that (x^*, μ^*) becomes a fuzzy solution to the Lagrange problem $\Longleftrightarrow$ to all vectors there exists $x > 0$ satisfying*

$$\sum_{l=1}^{m} \tilde{\Gamma}_{il}(\log x_l - \log x_l^*) \lesssim 0 \ (i \in \mathcal{I}); \tag{4.23}$$

then

$$\tilde{g}_0(x^*) \lesssim \tilde{g}_0(x). \tag{4.24}$$

Here

$$\tilde{\Gamma}_{il} = \sum_{k=1}^{J_i} \tilde{\gamma}_{ikl} \tilde{v}_{ik}(x^*) \ (i \in \mathcal{I}, 1 \leqslant l \leqslant m). \tag{4.25}$$

◊

Proof Let $\tilde{A}_i(0 \leqslant i \leqslant p)$ [Cao93a] be a CSM function. Then (4.1) $\Longleftrightarrow$

$$\begin{aligned} &\min \ \tilde{A}_0(\tilde{g}_0(x)) \\ &\text{s.t. } \tilde{A}_i(\tilde{g}_i(x) - 1) \geqslant \alpha \ (1 \leqslant i \leqslant p') \\ &\quad \tilde{A}_i(1 - \tilde{g}_i(x)) \geqslant \alpha \ (p' + 1 \leqslant i \leqslant p) \\ &\quad x > 0, \alpha \in [0, 1] \end{aligned} \tag{4.26}$$

while

$$\mathcal{I} \Longleftrightarrow \mathcal{I}' = \{i | \tilde{A}_i(\tilde{g}_i(x^*) - 1) = 0 \ (1 \leqslant i \leqslant p)\}, \tag{4.27}$$

$$(4.23) \Longleftrightarrow \tilde{A}_i[\Gamma_{il}(\log x_l - \log x_l^*)] \geqslant \alpha(i \in \mathcal{I}'), \tag{4.28}$$

$$(4.24) \Longleftrightarrow \tilde{A}_0[\tilde{g}_0(x^*) - \tilde{g}_0(x)] \geqslant \alpha. \tag{4.29}$$

□

From the condition of the theorem, we know that x^* is a fuzzy feasible solution to (4.1), which is equivalent to $\bar{x}^*$ being a parameter feasible solution to (4.26) [Cao93a]. Therefore, for subscript set $\mathcal{I}'$ and any $\alpha, \beta \in [0,1]$ there exists μ^* enabling $(\bar{x}^*, \mu^*)$ to be a Lagrange problem solution with parameters $\alpha, \beta \Longleftrightarrow$ all vectors $x > 0$ satisfying (4.28) can satisfy (4.29) from Theorem 1.4.5.

The theorem holds due to the arbitrariness of α, β on $[0,1]$. □

Proposition 4.4.1 *Let $\tilde{A}_i(0 \leqslant i \leqslant p)$ be the CSM function. On the assumption of a constraint complete lattice, a local fuzzy optimum solution to fuzzy reverse PGP (4.1) must be its part of a fuzzy solution to the Lagrange problem.* ◊

The reverse proposition is not necessarily true.

Proposition 4.4.2 *Let $\tilde{A}_i(0 \leqslant i \leqslant p)$ be the CSM function. If x^* is part of a fuzzy solution to the Lagrange problem, x^* denotes a global fuzzy optimum solution to (4.1) when (4.1) is fuzzy convex or $p' = p$. But x^* is not necessarily a global fuzzy optimum solution to (4.1), not even a local fuzzy optimum one for $p' \neq p$.* ◊

It is easy to prove the propositions in a similar way to Theorem 4.4.1.

In spite of that, x^* is still a fuzzy optimum solution to some extent (More details can be found in Example 4.3 below).

From Theorem 4.4.1, we know that x^* is still a fuzzy optimum for all x satisfying (4.23), which is called the tangentially optimal nature of fuzziness, though x^* is not a local fuzzy optimum one.

4.4.2 Direct Algorithm

There exist three algorithms in the fuzzy reverse PGP: 1) a direct algorithm, 2) a dual algorithm, 3) a balanced-solution algorithm. For algorithm 3, we use a fuzzy harmonic mean inequality and the properties of fuzzy balanced solution, change (4.1) into a fuzzy GP $(\tilde{P}(\alpha))$ with weight vector $\alpha(>0)$, and then find it, a step which remains to be investigated in the future. Here we only study algorithm 1 and 2.

Lemma 4.4.1 [TOA73][Cao87a,b] *If constraint $\tilde{A}$ is a strictly convex fuzzy set on $\Re^m$, i.e.,*

$$\forall \lambda \in (0,1), x^1, x^2 \in X, x^1 \neq x^2,$$
$$\tilde{A}[\lambda x^1 + (1-\lambda)x^2] > \tilde{A}(x^1) \wedge \tilde{A}(x^2),$$

then

$$\tilde{M} \triangleq \sup_{x \in A_\alpha} |\tilde{A}(\tilde{g}_0(x))|$$

is said to be continuous, where $\alpha \in [0,1]$. ◊

Lemma 4.4.2 *If $\tilde{M}$ is differentiable for arbitrary $\alpha \in [0,1]$ as for $x \in X \subset \Re^m$, then $\tilde{M}$ is continuous on X.* ◊

It can be proved that there exists a unique fixed point

$$\bar{\alpha} = \tilde{M}(\overline{\alpha}) = \sup_{x \in A_{\bar{\alpha}}} |\tilde{A}_0(\tilde{g}_0(x))|$$

if $\tilde{M}$ is a continuous function on [0,1].

Since $\tilde{g}_i(x)(0 \leqslant i \leqslant p)$ is a differentiable function, we can get a direct algorithm step for fuzzy reverse PGP (4.1)

1^0 Let $k = 1$, and determine α_1 and h by $1 - hd = \alpha_1$.

2^0 Calculate

$$\tilde{M}^{(k)} = \sup_{x \in A_{\alpha_k}} |\tilde{A}_0(\tilde{g}_0(x))|.$$

3^0 Calculate $\varepsilon_k = \alpha_k - \tilde{M}^{(k)}$ for $\forall \varepsilon > 0$. If $|\varepsilon_k| > \varepsilon$, we go to step 2^0, otherwise to step 5^0.

4^0 Select $r_k \in [0,1]$ properly. Suppose $\alpha_{k+1} = \alpha_k - r_k \varepsilon_k$, let $k : k+1$ go to 2^0.

5^0 For $\overline{\alpha} = \alpha_k$, find x such that

$$\sup_{x \in A_{\alpha_k}} |\tilde{A}_0(\tilde{g}_0(x))| = \tilde{M}^{(k)}(x).$$

According to experience, it is proper to take $\alpha_1 \in [0.9, 1]$ when $\tilde{g}_0(x)$ strictly increases monotonicity, otherwise to select $\alpha_1 \in [0.75, 0.9]$. If $b(> 0)$ is very large, large, smaller, very small, it is proper to take 0.02, 0.2, 2 and 20. To r_k selection, for $\varepsilon_1 \gg \varepsilon_2$, we may choose $r_k = 0.5$. If $\varepsilon_1 \searrow \varepsilon_2$ changes a little, we can take $r_k \in [0.618, 1]$. If $\varepsilon_1 \ll \varepsilon_2$, take $r_k \in [0.382, 0.4]$ properly, otherwise we might obtain an irrational result.

We introduce this method with a numerical example which is used to confirm Theorem 4.4.1.

4.4.3 Numerical Example

Example 4.3 Find Example 4.1 by a direct method

$$\begin{aligned} &\widetilde{\min}\ 2x_1 + 3x_2 \\ &\text{s.t. } x_1^2 + x_2^2 \gtrsim 1 \\ &\quad x_1, x_2 > 0 \end{aligned} \tag{4.30}$$

Solution. Consider

$$\tilde{M}(x) = \sup_{x \in \tilde{A}_\alpha} |\tilde{A}_0(\tilde{g}_0(x))|,$$

$$\tilde{A}_1(x) = \{x_1^2 + x_2^2 \gtrsim 1, x_1, x_2 > 0\},$$

where $\tilde{M} : M \to [0,1], \tilde{M}(x) = \dfrac{1}{\eta} g_0(x)$. The Lagrange solution of (4.30) is $x_1^* = \dfrac{2}{\sqrt{13}}, x_2^* = \dfrac{3}{\sqrt{13}}$, such that

$$\eta = \max_{x \in X} g_0(x) = 2x_1^* + 3x_2^* = \sqrt{13}.$$

Therefore, $\tilde{M}(x) = \dfrac{1}{\sqrt{13}}\{2x_1 + 3x_2\}$.

1^0 Let $d_1 = 0.25$. Turn (4.30) into

$$\begin{aligned} &\min\{\frac{2}{\sqrt{13}}x_1 + \frac{3}{\sqrt{13}}x_2\} \\ &\text{s.t. } x_1^2 + x_2^2 \geq 1 + 0.25 \\ &\quad x_1, x_2 > 0 \end{aligned}$$

2^0 Let $\alpha_1 = 0.9$, $\varepsilon = 0.001$. By taking $x_1 = 1, x_2 = 1/2$, we have

$$\begin{aligned} &\tilde{M} = \frac{3.5}{\sqrt{13}} \approx 0.9707, \\ &|\varepsilon_1| = |0.9 - 0.9707| = 0.0707 > \varepsilon = 0.001. \end{aligned}$$

Because $1 - 0.2d_1' = 0.9707 \Rightarrow d_1' = 0.1465$, then $\alpha_2 = 0.9 - 0.5(-0.0707) = 0.93535$.

3^0 Find

$$\begin{aligned} &\min\ \{\frac{2}{\sqrt{13}}x_1 + \frac{3}{\sqrt{13}}x_2\} \\ &\text{s.t. } \ x_1^2 + x_2^2 \geqslant 1.1465 \\ &\quad x_1, x_2 > 0 \end{aligned}$$

for $\alpha_2 = 0.93535$ and, by repeating Step 2^0, we get $x_1 = 0.915683, x_2 = 0.555, \tilde{M} = 0.969717, |\varepsilon_2| = 0.00098 < \varepsilon = 0.001$ and an objective function $S \approx 3.496$.

Obviously, $x^* = (x_1^*, x_2^*) = (1.07075, 0)$ is a fuzzy minimum solution to (4.29); the fuzzy constraint infimum is $M_{\tilde{P}} = g_0(x^*) = g_0(1.07075, 0) = 2.1415$, whereas $\hat{x} = (\hat{x}_1, \hat{x}_2) = (0.915683, 0.555)$ is not a global fuzzy optimum point of (4.29), not even a local fuzzy optimal point. Therefore, Proposition 4.4.2 is confirmed.

In spite of that, $\hat{x}$ is a fuzzy optimal point to (4.29) within a certain range.

For example, in Example 4.2, the formula corresponding to (4.23) behaves as

$$\begin{aligned} \Gamma_{11} &= -2(\hat{x}_1)^2 \\ &= -2(0.915683)^2 \approx -1.67695, \\ \Gamma_{12} &= -2(\hat{x}_2)^2 \\ &= -2(0.555)^2 \approx -0.61605, \end{aligned}$$

hence, for all of x_1 and x_2 satisfying

$$\begin{array}{ll} & -1.67695(\log x_1 - \log \hat{x}_1) - 0.61605(\log x_2 - \log \hat{x}_2) \leqslant 0 \\ \Longrightarrow & -1.67695(\log x_1 + 0.088) - 0.61605(\log x_2 + 0.484) \leqslant 0 \\ & x_1^{-1.677} x_2^{-0.616} \leqslant 0.44572 \end{array}$$

i.e.,

$$x_1^{1.677} x_2^{0.616} \geqslant 2.2436,$$

and so

$$2x_1 + 3x_2 \geqslant 3.496366.$$

such that

$$2\tilde{x}_1 + 3\tilde{x}_2 \gtrsim 3.496366.$$

Theorem 4.4.1 is confirmed by the numerical example.

It can be proved that, if $\tilde{A}_i (0 \leqslant i \leqslant p)$ is a fuzzy convex function, then all of the results in References [Cao93a][Cao96b] can be transplanted.

Chapter 5

GEOMETRIC PROGRAMMING WITH FUZZY COEFFICIENTS

In this chapter we recount the building of four GP models, encompassing models with an interval-valued and a fuzzy-valued coefficient [Cao93c], a type $(\cdot, c)$ one [Cao95a], and an L-R [Cao94a] and a flat fuzzy one [Cao92a][Cao00] by extending a classical GP. Such models play an important part in application of engineering design and budget, mechanical analysis and economical management, etc.

5.1 Properties of Five Types of Numbers

In the section, we shall discuss properties of five types of numbers including interval, fuzzy, type $(\cdot, c)$, L-R and flat fuzzy numbers.

5.1.1 Definition and Properties of Interval and Fuzzy Numbers

Definition 5.1.1 Let $\Re$ denote a real number set. We call c, d interval numbers, written as $c, d \in \mathcal{I}_\Re$, where $\mathcal{I}_\Re = \{[c_i, d_i] | c_i < d_i, c_i, d_i \in \Re, (i = 1, 2)\}$ is a set consisting of all interval numbers.

If $c = [c_1, c_2], d = [d_1, d_2]$. The operation of defined interval numbers is as follows:

$$\begin{aligned}
&c + d = [c_1 + d_1, c_2 + d_2] \\
&c - d = [(c_1 - d_1) \wedge (c_2 - d_2), (c_1 - d_1) \vee (c_2 - d_2)] \\
&c \cdot d = [c_1 d_1 \wedge c_1 d_2 \wedge c_2 d_1 \wedge c_2 d_2, c_1 d_1 \vee c_1 d_2 \vee c_2 d_1 \vee c_2 d_2] \\
&c \div d = [c_1/d_1 \wedge c_1/d_2 \wedge c_2/d_1 \wedge c_2/d_2, c_1/d_1 \vee c_1/d_2 \vee c_2/d_1 \vee c_2/d_2] \\
&c \vee d = [c_1 \vee d_1, c_2 \vee d_2] \\
&c \wedge d = [c_1 \wedge d_1, c_2 \wedge d_2]
\end{aligned}$$

Theorem 5.1.1 *Given $c, d \in \mathcal{I}_{\Re}$, then $c * d \in \mathcal{I}_{\Re}$, where " $*$ " denotes algebra operations $\{+, -, \cdot, \div, \vee, \wedge\}$ on $\Re$.* $\Diamond$

Fuzzy numbers are obtained by applying the extension principle. From now on, $\mathcal{F}(\Re)$ represents the set of real fuzzy numbers.

Definition 5.1.2 Given $\tilde{c}, \tilde{d}$ denote fuzzy numbers, written as $\tilde{c}, \tilde{d} \in \mathcal{F}(\Re)$, with α cuts of $\tilde{c}$ and $\tilde{d}$ being

$$c_\alpha = [c_1(\alpha), c_2(\alpha)],\ d_\alpha = [d_1(\alpha), d_1(\alpha)], \alpha \in [0,1]$$

respectively, so that the operations of a fuzzy number are defined as follows:

$$\tilde{c} + \tilde{d} = \bigcup_{\alpha\in[0,1]} \alpha(c_\alpha + d_\alpha) = \bigcup_{\alpha\in[0,1]} \alpha[c_1(\alpha) + d_1(\alpha), c_2(\alpha) + d_2(\alpha)]$$

$$\begin{aligned}\tilde{c} - \tilde{d} &= \bigcup_{\alpha\in[0,1]} \alpha(c_\alpha - d_\alpha)\\ &= \bigcup_{\alpha\in[0,1]} \alpha[(c_1(\alpha) - d_1(\alpha)) \wedge (c_2(\alpha) - d_2(\alpha)),\\ &\qquad (c_1(\alpha) - d_1(\alpha)) \vee (c_2(\alpha) - d_2(\alpha))]\end{aligned}$$

$$\begin{aligned}\tilde{c} \cdot \tilde{d} &= \bigcup_{\alpha\in[0,1]} \alpha(c_\alpha \cdot d_\alpha)\\ &= \bigcup_{\alpha\in[0,1]} \alpha[c_1(\alpha)d_1(\alpha) \wedge c_1(\alpha)d_2(\alpha) \wedge c_2(\alpha)d_1(\alpha) \wedge c_2(\alpha)d_2(\alpha),\\ &\qquad c_1(\alpha)d_1(\alpha) \vee c_1(\alpha)d_2(\alpha) \vee c_2(\alpha)d_1(\alpha) \vee c_2(\alpha)d_2(\alpha)]\end{aligned}$$

$$\begin{aligned}\tilde{c} \div \tilde{d} &= \bigcup_{\alpha\in[0,1]} \alpha(c_\alpha \div d_\alpha)\\ &= \bigcup_{\alpha\in[0,1]} \alpha[c_1(\alpha)/d_1(\alpha) \wedge c_1(\alpha)/d_2(\alpha) \wedge c_2(\alpha)/d_1(\alpha) \wedge c_2(\alpha)/d_2(\alpha),\\ &\qquad c_1(\alpha)/d_1(\alpha) \vee c_1(\alpha)/d_2(\alpha) \vee c_2(\alpha)/d_1(\alpha) \vee c_2(\alpha)/d_2(\alpha)]\end{aligned}$$

$$\tilde{c} \vee \tilde{d} = \bigcup_{\alpha\in[0,1]} \alpha[c_1(\alpha) \vee d_1(\alpha), c_2(\alpha) \vee d_2(\alpha)]$$

$$\tilde{c} \wedge \tilde{d} = \bigcup_{\alpha\in[0,1]} \alpha[c_1(\alpha) \wedge d_1(\alpha), c_2(\alpha) \wedge d_2(\alpha)]$$

Theorem 5.1.2 *Let $\tilde{c}, \tilde{d} \in \mathcal{F}(\Re)$. Then $\tilde{c} * \tilde{d} \in \mathcal{F}(\Re)$.* $\Diamond$

It is easy to prove the two above theorems in a manner similar to corresponding theorems in Reference [Lu92][Luo84a,b].

Definition 5.1.3 Suppose that $\tilde{c} \in \mathcal{F}(\Re)$ is called a fuzzy number, where $\Re$ represents the set of whole real numbers. If

i) $\tilde{c}$ is normal, i.e., $x_0 \in \Re$ exists, such that $\tilde{c}(x_0) = 1$.

ii) $\forall\alpha \in (0,1]$, c_α is a closed interval.

Theorem 5.1.3 *Let $\tilde{c} \in \mathcal{F}(\Re)$ be a fuzzy number. Then*

i) $\tilde{c}$ is fuzzy convex.

ii) If $\tilde{c}(x_0) = 1$, then $\tilde{c}(x)$ is nondecreasing for $x \leqslant x_0$ and $\tilde{c}(x)$ nonincreasing for $x \geqslant x_0$. ◊

Proof Because $c_\alpha(\alpha \in (0,1])$ is the closed interval $c_0 = \Re$, i.e., $\forall \alpha \in [0,1], c_\alpha$ is a convex set. $\tilde{c}$ can be proved to be fuzzy convex according to Theorem 1.2.3.

Now, take $x_1 < x_2 \leqslant x_0$ and let $\alpha = \tilde{c}(x_1)$. Since $\tilde{c}(x_0) = 1$, then

$$[x_1, x_0] \subset c_\alpha,$$

hence $x_2 \in c_\alpha$, such that $\tilde{c}(x) \geqslant \alpha$, i.e., $\tilde{c}(x_1) \leqslant \tilde{c}(x_2)$.

Similarly, $\tilde{c}(x_2) \leqslant \tilde{c}(x_1)$ can be proved if $x_0 \leqslant x_1 < x_2$.

Overall the theorem holds. □

Theorem 5.1.4 *Let $\tilde{c} \in \mathcal{F}(\Re)$, $\sup \tilde{c}$ be bounded. Then $\tilde{c}$ is a fuzzy number $\Leftrightarrow \exists$: interval $[c_1, c_2]$, such that*

$$\tilde{c}(x) = \begin{cases} 1, & x \in [c_1, c_2] \neq \phi \\ L(x), & x < c_1 \\ R(x), & x > c_2 \end{cases}$$

where $L(x)$ represents a continuously increasing function of the right $(0 \leqslant L(x) < 1)$; $R(x)$ represents a continuously decreasing one of the left $(0 \leqslant R(x) < 1)$. ◊

Proof *Necessity*. Let $\tilde{c} \in \mathcal{F}(\Re)$.

(1) Because $\tilde{c}_1^*$ is a closed convex set, $\tilde{c}_1^* = [c_1, c_2]$ and $\tilde{c}(x) = 1$ on $[c_1, c_2]$. It is obvious that $\tilde{c}(x) < 1$ for $x \notin [c_1, c_2]$.

(2) Because $\tilde{c} \in \mathcal{F}(\Re), \forall \alpha \in [0,1], c_\alpha$ is a closed interval, we assume $c_\alpha = [c_{1\alpha}, c_{2\alpha}] \subset [0,1]$, then

$$\tilde{c} = \bigcup_{\alpha \in (0,1]} \alpha c_\alpha = \bigcup_{\alpha \in (0,1]} \alpha [c_{1\alpha}, c_{2\alpha}].$$

As for $x < c_1$,

$$\begin{aligned} L(x) = \tilde{c}(x) &= \bigvee_{\alpha \in (0,1]} \alpha \wedge \chi_{[c_{1\alpha}, c_{2\alpha}]}(x) \\ &= \bigvee_{\alpha \in (0,1]} \{\alpha | x \in [c_{1\alpha}, c_{2\alpha})\} \\ &= \bigvee_{\alpha \in (0,1]} \{\alpha | c_{1\alpha} \leqslant x < c_{2\alpha}\} \end{aligned}$$

where χ represents a characteristic function. Therefore, $0 \leqslant L(x) < 1$.

If $x_1 < x_2 \leqslant c_1$, then $\tilde{c}(x_1) \leqslant \tilde{c}(x_2)$, otherwise, $\tilde{c}(x_1) > \tilde{c}(x_2)$. Again, $x_1 < x_2 \leqslant c_1 \Rightarrow x_2 \in (x_1, c_1 + \varepsilon) \Rightarrow \exists : \alpha \in [0,1]$, such that

$$x_2 = \alpha x_1 + (1 - \alpha)(c_1 + \varepsilon),\ c_1 + \varepsilon \in (c_1, c_2).$$

Since $\tilde{c}$ represents a convex fuzzy set on $[c_1, c_2]$,

$$\begin{aligned}\tilde{c}(x_2) &= \tilde{c}(\alpha x_1 + (1-\alpha)(c_1+\varepsilon)) \geqslant \tilde{c}(x_1) \wedge \tilde{c}(c_1+\varepsilon),\\ &\geqslant \tilde{c}(x_1) > \tilde{c}(x_2)\end{aligned}$$

which is a contradiction. Therefore, $L(x)$ is an increasing function.

(3) $L(x)$ continues on the right, otherwise there exists $x < c_1, x_n \hookrightarrow x$, then

$$\lim_{x_n \to x^*} L(x_n) = \alpha > L(x).$$

Since $x_n \in c_\alpha$ and c_α is closed, then $x \in c_\alpha$, such that $\tilde{c}(x) = L(x) \geqslant \alpha$, a contradiction.

For the same reason, $\tilde{c}(x) = R(x)$ is a continuously decreasing function on the left for $x > c_2$, with $0 \leqslant R(x) < 1$.

Sufficiency. Let $\tilde{c}$ satisfy the condition in the theorem.

(1) $\tilde{c}$ is obviously normal.

(2) Prove $c_\alpha = [c_{1\alpha}, c_{2\alpha}], \forall \alpha \in (0, 1]$.

$\tilde{c}(x) = L(x)$ for $x < c_1$, so we select $c_{1\alpha} = \min\{x | L(x) \geqslant \alpha\}$ and $\tilde{c}(x) = R(x)$ for $x > c_2$, such that we select $c_{2\alpha} = \max\{x | R(x) \geqslant \alpha\}$.

Obviously, $c_\alpha \subset [c_{1\alpha}, c_{2\alpha}]$. Now, prove $[c_{1\alpha}, c_{2\alpha}] \subset c_\alpha$, and we prove only $[c_{1\alpha}, c_1) \subset c_\alpha$(because we can prove $(c_2, c_{2\alpha}] \subset c_\alpha$ for the same reason). Again, we prove only $c_{1\alpha} \in c_\alpha$ due to the monotonicity of $L(x)$.

Select $x_n \hookrightarrow c_{1\alpha}$, then

$$L(c_{1\alpha}) = \lim_{x_n \to c_{1\alpha}} (L(x_n)) \geqslant \alpha,$$

such that $c_{1\alpha} \in c_\alpha$. □

5.1.2 Definition and Properties of Type $(\cdot, c), L - R$ and Flat Fuzzy Numbers

Definition 5.1.4 $\tilde{c} = (\alpha, c)$ is defined as a type $(\cdot, c)$ fuzzy number on a product space $\alpha_1 \times \alpha_2 \times \cdots \times \alpha_J$, where $\alpha = (a_1, a_2, \ldots, a_J)^{\mathrm{T}}, c = (c_1, c_2, \ldots, c_J)^{\mathrm{T}}$. Its membership function is

$$\tilde{c}(a) = \min_j [\tilde{c}_j(a_j)]$$

$$\tilde{c}(a_j) = \begin{cases} 1 - \dfrac{|\alpha_j - a_j|}{c_j}, & \alpha_j - c_j \leqslant a_j \leqslant \alpha_j + c_j \\ 0, & \text{otherwise,} \end{cases}$$

where α denotes the center of $\tilde{c}$, c the extension of $\tilde{c}$, with $c_j > 0$.

Special cases:

Definition 5.1.5 L is called a reference function of fuzzy numbers, if L satisfies

(i) $L(x) = L(-x)$; (ii) $L(0) = 1$;

(iii) $L(x)$ is a nonincreasing and piecewise continuous function at $[0, +\infty)$.

Definition 5.1.6 Let L, R be reference functions of fuzzy numbers, and call fuzzy number $\tilde{c}$ a type L-R fuzzy number. If

$$\tilde{c}(x) = \begin{cases} L\left(\dfrac{c-x}{\alpha}\right), & x \leqslant c, \alpha > 0 \\ R\left(\dfrac{x-c}{\beta}\right), & x \geqslant c, \beta > 0, \end{cases}$$

we write $\tilde{c} = (c, \alpha, \beta)_{LR}$, where c is a mean value and α and β are called the left and the right spreads of $\tilde{c}$, respectively. L is called a left reference and R a right reference.

Definition 5.1.7 If L and R are functions satisfying

$$T(x) = \begin{cases} 1 - |x|, & \text{if } -1 \leqslant x \leqslant 1 \\ 0, & \text{otherwise,} \end{cases}$$

then we call $\tilde{c} = (x, \underline{\xi}, \overline{\xi})_T$ T-fuzzy numbers, $T(\Re)$ representing T-fuzzy number sets.

Definition 5.1.8 Given that L, R are reference functions, the quadruple $\tilde{c} = (c^-, c^+, \alpha, \beta)_{LR}$ is called a type L-R flat fuzzy number. Then we have

$$\tilde{c}(x) = \begin{cases} L\left(\dfrac{c^- - x}{\alpha}\right), & x \leqslant c^-, \alpha > 0 \\ R\left(\dfrac{x - c^+}{\beta}\right), & x \geqslant c^+, \beta > 0 \\ 1, & \text{otherwise} \end{cases}$$

satisfying $\exists : (c^-, c^+) \in \Re, c^- < c^+, \tilde{c}(x) = 1, \forall x \in [c^-, c^+]$.

Especially, $\tilde{c} = (c^-, c^+, \sigma_c^-, \sigma_c^+)$ is said to be a flat fuzzy number, where

$$\tilde{c}(x) = \begin{cases} 1 - \dfrac{c^- - x}{\sigma_c^-}, & \text{if } c^- - \sigma_c^- \leqslant x \leqslant c^- \\ 1, & \text{if } c^- < x < c^+ \\ 1 - \dfrac{x - c^+}{\sigma_c^+}, & \text{if } c^+ \leqslant x \leqslant c^+ + \sigma_c^+ \\ 0, & \text{otherwise} \end{cases}$$

Definition 5.1.9 Suppose " $*$ " represents an arbitrary ordinary binary operation in $\Re$, such that $\forall \tilde{c}, \tilde{d} \in \mathcal{F}(\Re)$ and we define

$$\tilde{c} * \tilde{d} = \int_{x,y\in\Re} \tilde{c}(x) \wedge \tilde{d}(y)/x * y.$$

That is to say, $\forall z \in \Re$,

$$\tilde{c} * \tilde{d}(z) = \bigvee_{x*y=z} (\tilde{c}(x) \wedge \tilde{d}(y)),$$

here " $*$ " represents the arithmetic operations $+, -, \cdot, \div$.

Accordingly, we can define the operations of type $(\cdot, c)$, L-R and flat fuzzy numbers.

Fuzzy number operations in type L-R:

Let $\tilde{c} = (c, \underline{c}, \overline{c})_{LR}, \tilde{d} = (d, \underline{d}, \overline{d})_{LR}, \tilde{p} = (p, \underline{p}, \overline{p})_{RL}$ be an L-R fuzzy number. Then

1) $\tilde{c} + \tilde{d} = (c + d, \underline{c} + \underline{d}, \overline{c} + \overline{d})_{LR}$.

2) $k \cdot \tilde{c} = \begin{cases} (kc, k\underline{c}, k\overline{c})_{LR}, & \text{when } k \geqslant 0, \\ (kc, -k\overline{c}, -k\underline{c})_{LR}, & \text{when } k < 0. \end{cases} \quad (k \in \Re)$

Let $(-1)\tilde{c} = -\tilde{c}$ for $k = -1$. Then $-\tilde{c} = (-c, \overline{c}, \underline{c})_{RL}$.

3) $\tilde{c} - \tilde{p} = (c - p, \underline{c} + \overline{p}, \overline{c} + \underline{p})_{LR}$ for $L = R$.

4) $\tilde{c} \cdot \tilde{d} \approx (cd, c\underline{d} + d\underline{c}, c\overline{d} + d\overline{c})_{LR}$.

5) $\tilde{c} \div \tilde{p} \approx (\frac{c}{p}, \frac{\overline{p}c + \underline{c}p}{p^2}, \frac{\underline{p}c + \overline{c}p}{p^2})_{LR}$,

$p \neq 0$, $\tilde{c}$ and $\tilde{p}$ can not be divided for $L \neq R$.

6) $\widetilde{\max}(\tilde{c}, \tilde{d}) \approx (c \vee d, \underline{c} \wedge \underline{d}, \overline{c} \vee \overline{d})_{LR}$, $\widetilde{\min}(\tilde{c}, \tilde{d}) \approx (c \wedge d, \underline{c} \vee \underline{d}, \overline{c} \wedge \overline{d})_{LR}$.

7) $\tilde{c} \leqslant \tilde{d} \Longleftrightarrow c \leqslant d, \underline{c} \geqslant \underline{d}, \overline{c} \leqslant \overline{d}$; $\tilde{c} \subseteq \tilde{d} \Longleftrightarrow c + \overline{c} \leqslant d - \overline{d}$, or $\tilde{c} = \tilde{d}$.

The operation of type flat fuzzy numbers:

Let $\tilde{c} = (c^-, c^+, \sigma_c^-, \sigma_c^+)$ and $\tilde{d} = (d^-, d^+, \sigma_d^-, \sigma_d^+)$ be flat fuzzy numbers. Then

1) $\tilde{c} + \tilde{d} = (c^- + d^-, c^+ + d^+, \sigma_c^- + \sigma_d^-, \sigma_c^+ + \sigma_d^+)$.

2) $k \cdot \tilde{c} = \begin{cases} (kc^-, kc^+, k\sigma_c^-, k\sigma_c^+), & \text{for } k > 0 \\ (kc^+, kc^-, -k\sigma_c^-, -k\sigma_c^+), & \text{for } k \leqslant 0. \end{cases}$

It is easy to deduce operation properties of type $(\cdot, c)$ fuzzy numbers since they are extended over flat fuzzy ones.

The proof can be found in References [Dia87][Cao90a][Cao95a,b][DP80] [LW92][Zim91] about the operation properties of an interval number and a fuzzy one in this section.

5.2 GP with Interval and Fuzzy Valued Coefficients

5.2.1 GP with Interval or Fuzzy Coefficients in Constraint Conditions

Consider

$$\begin{aligned} \min & \sum_{k=1}^{J_0} c_{0k} \prod_{l=1}^{m} x_l^{\gamma_{0kl}} \\ \text{s.t.} & \sum_{k=1}^{J_i} c_{ik} \prod_{l=1}^{m} x_l^{\gamma_{ikl}} \leqslant 1 \ (1 \leqslant i \leqslant p) \\ & x > 0 \end{aligned} \tag{5.1}$$

I. Interval Coefficients

Suppose c_{ik} and 1 to be interval numbers, writing

$$\bar{c}_{ik} = [c_{ik}^-, c_{ik}^+](1 \leqslant k \leqslant J_i, 0 \leqslant i \leqslant p), \ \bar{1} = [1^-, 1^+],$$

and we can write (5.1) as

$$\begin{aligned} \min & \sum_{k=1}^{J_0} c_{0k} \prod_{l=1}^{m} x_l^{\gamma_{0kl}} \\ \text{s.t.} & \left[\sum_{k=1}^{J_i} c_{ik}^- \prod_{l=1}^{m} x_l^{\gamma_{ikl}}, \sum_{k=1}^{J_i} c_{ik}^+ \prod_{l=1}^{m} x_l^{\gamma_{ikl}}\right] \subseteq [1^-, 1^+] \\ & x > 0 (1 \leqslant i \leqslant p) \end{aligned} \tag{5.2}$$

Theorem 5.2.1 *The programming problem (5.2) is equivalent to*

$$\begin{aligned} \min & \sum_{k=1}^{J_0} c_{0k} \prod_{l=1}^{m} x_l^{\gamma_{0kl}} \\ s.t. & \sum_{k=1}^{J_i} c_{ik}^- \prod_{l=1}^{m} x_l^{\gamma_{ikl}} \leqslant 1^- \qquad (1 \leqslant i \leqslant p) \\ & \sum_{k=1}^{J_i} c_{ik}^+ \prod_{l=1}^{m} x_l^{\gamma_{ikl}} \leqslant 1^+ \\ & x > 0 \end{aligned} \tag{5.3}$$

and the optimal solution to (5.3) is that to (5.2). ◊

Proof The theorem is easy to prove by means of properties of an interval number. □

II. Fuzzy Coefficients

Suppose c_{ik} and 1 to be fuzzy numbers, writing

$$\tilde{c}_{ik} = \bigcup_{\alpha \in [0,1]} \alpha[c_{ik}^-(\alpha), c_{ik}^+(\alpha)](1 \leqslant k \leqslant J_i, 0 \leqslant i \leqslant p),$$

$$\tilde{1} = \bigcup_{\alpha\in[0,1]} \alpha[1^-(\alpha), 1^+(\alpha)],$$

and we can write (5.1) as

$$\begin{aligned}
&\min \sum_{k=1}^{J_0} c_{0k} \prod_{l=1}^{m} x_l^{\gamma_{0kl}} \\
&\text{s.t. } \bigcup_{\alpha\in[0,1]} \alpha\Big[\sum_{k=1}^{J_i} c_{ik}^-(\alpha) \prod_{l=1}^{m} x_l^{\gamma_{ikl}}, \sum_{k=1}^{J_i} c_{ik}^+(\alpha) \prod_{l=1}^{m} x_l^{\gamma_{ikl}}\Big] \\
&\quad \subseteq \bigcup_{\alpha\in[0,1]} \alpha[1^-(\alpha), 1^+(\alpha)]\ (1 \leqslant i \leqslant p) \\
&\quad x > 0
\end{aligned} \tag{5.4}$$

Theorem 5.2.2 $\forall\alpha \in [0,1]$, *problem (5.4) is equivalent to*

$$\begin{aligned}
&\min \sum_{k=1}^{J_0} c_{0k} \prod_{l=1}^{m} x_l^{\gamma_{0kl}} \\
&s.t. \sum_{k=1}^{J_i} c_{ik}^-(\alpha) \prod_{l=1}^{m} x_l^{\gamma_{ikl}} \leqslant 1^-(\alpha) \\
&\quad \sum_{k=1}^{J_i} c_{ik}^+(\alpha) \prod_{l=1}^{m} x_l^{\gamma_{ikl}} \leqslant 1^+(\alpha)\ (1 \leqslant i \leqslant p) \\
&\quad x > 0
\end{aligned} \tag{5.5}$$

If $x(\alpha)$ is an optimal solution to (5.5), then $x = \bigcup_{\alpha\in[0,1]} \alpha x(\alpha)$ is a fuzzy optimal solution to (5.4). ◊

Proof For $\alpha \in [0,1]$, it is not difficult to prove the equivalence of (5.4) and (5.5) by the cut-set properties of a fuzzy number operation and Theorem 5.2.1, such that $x = \bigcup_{\alpha\in[0,1]} \alpha x(\alpha)$ means a fuzzy optimal solution to (5.4). □

5.2.2 GP with Interval or Fuzzy Coefficients in Objective Function

I. Interval Coefficients

Let coefficients of an objective function in programming problem (5.1) be an interval number $\bar{c}_{0k} = [c_{0k}^-, c_{0k}^+](1 \leqslant k \leqslant J_0)$. Then (5.1) can be written as

$$\begin{aligned}
&\min\{\bar{g}_0(x) = \sum_{k=1}^{J_0} \bar{c}_{0k} \prod_{l=1}^{m} x_l^{\gamma_{0kl}}\} \\
&\text{s.t. } \sum_{k=1}^{J_i} c_{ik} \prod_{l=1}^{m} x_l^{\gamma_{ikl}} \leqslant 1\ (1 \leqslant i \leqslant p) \\
&\quad x > 0
\end{aligned} \tag{5.6}$$

Theorem 5.2.3 *Problem (5.6) is equivalent to*

$$\begin{aligned}&\min \sum_{k=1}^{J_0} c_{0k}^- \prod_{l=1}^{m} x_l^{\gamma_{0kl}}\\ &\text{s.t.}\ \sum_{k=1}^{J_i} c_{ik} \prod_{l=1}^{m} x_l^{\gamma_{ikl}} \leqslant 1\ (1 \leqslant i \leqslant p)\\ &x > 0\end{aligned} \tag{5.7}$$

and

$$\begin{aligned}&\min \sum_{l=1}^{J_0} c_{0k}^+ \prod_{l=1}^{m} x_l^{\gamma_{0kl}}\\ &\text{s.t.}\ \sum_{k=1}^{J_i} c_{ik} \prod_{l=1}^{m} x_l^{\gamma_{ikl}} \leqslant 1\ (1 \leqslant i \leqslant p)\\ &x > 0\end{aligned} \tag{5.8}$$

If $x^- = (x_1^-, x_2^-, \ldots, x_m^-)$ *and* $x^+ = (x_1^+, x_2^+, \ldots, x_m^+)$ *represent optimal solutions to (5.7) and to (5.8), respectively, then* $\bar{g}_0 = [g_0^-(x), g_0^+(x)]$ *represents an interval optimal solution to (5.6), where*

$$\begin{aligned}g_0^-(x) &= \min\{g_0^-(x^-) \wedge g_0^-(x^+), g_0^+(x^-) \wedge g_0^+(x^+)\},\\ g_0^+(x) &= \max\{g_0^-(x^-) \wedge g_0^-(x^+), g_0^+(x^-) \wedge g_0^+(x^+)\}.\end{aligned} \tag{5.9}$$

◊

Proof By using properties of interval numbers operations, it is easy to confirm the equivalence of (5.6) and (5.7) with (5.8).

Again, let (5.6) and (5.7) with (5.8) contain the same feasible solutions. If they represent optimal solutions to (5.7) and to (5.8), x^- and x^+ must be feasible solutions to (5.6). Now, we substitute them simultaneously for objective functions in (5.6), such that $\bar{g}_0 = \min[g_0^-(x), g_0^+(x)]$, where $g_0^-(x), g_0^+(x)$ are shown as in formula (5.9). Hence $\bar{g}_0(x)$ is an interval valued optimal solution to (5.6). □

II. Fuzzy Coefficients

Let an objective coefficient $\tilde{c}_{0k}$ of program (5.1) be a fuzzy number, i.e.,

$$\tilde{c}_{0k} = \bigcup_{\alpha\in[0,1]} \alpha[c_{ik}^-(\alpha), c_{ik}^+(\alpha)](1 \leqslant k \leqslant J_0).$$

Then (5.1) can be denoted by:

$$\begin{aligned}&\min\{\tilde{g}_0(x) = \sum_{k=1}^{J_0} \tilde{c}_{0k} \prod_{l=1}^{m} x_l^{\gamma_{0kl}}\}\\ &\text{s.t.}\ \sum_{k=1}^{J_i} c_{ik} \prod_{l=1}^{m} x_l^{\gamma_{ikl}} \leqslant 1\ (1 \leqslant i \leqslant p)\\ &x > 0\end{aligned} \tag{5.10}$$

Theorem 5.2.4 *Program (5.10) is equivalent to* $\forall\alpha\in[0,1]$, *find*

$$\begin{aligned} &\min \sum_{k=1}^{J_0} c_{0k}^-(\alpha)\prod_{l=1}^{m} x_l^{\gamma_{0kl}}\\ &\text{s.t. } \sum_{k=1}^{J_i} c_{ik}\prod_{l=1}^{m} x_l^{\gamma_{ikl}}\leqslant 1\ (1\leqslant i\leqslant p)\\ &\quad x>0 \end{aligned} \tag{5.11}$$

and

$$\begin{aligned} &\min \sum_{k=1}^{J_0} c_{0k}^+(\alpha)\prod_{l=1}^{m} x_l^{\gamma_{0kl}}\\ &\text{s.t. } \sum_{k=1}^{J_i} c_{ik}\prod_{l=1}^{m} x_l^{\gamma_{ikl}}\leqslant 1\ (1\leqslant i\leqslant p)\\ &\quad x>0 \end{aligned} \tag{5.12}$$

If

$$x^-(\alpha)=(x_1^-(\alpha),x_2^-(\alpha),\ldots,x_m^-(\alpha)),x^+(\alpha)=(x_1^+(\alpha),x_2^+(\alpha),\ldots,x_m^+(\alpha))$$

represent optimal solutions to (5.11) and to (5.12), respectively, then

$$\tilde{g}_0(x)=\bigcup_{\alpha\in[0,1]}[g_0^-(x(\alpha)),g_0^+(x(\alpha))]$$

is a fuzzy optimal value of (5.10), where

$$\begin{aligned} &g_0^-(x(\alpha))\\ &=\min\{g_0^-(x^-(\alpha))\wedge g_0^-(x^+(\alpha)),g_0^+(x^-(\alpha))\wedge g_0^+(x^+(\alpha))\},\\ &g_0^+(x(\alpha))\\ &=\max\{g_0^-(x^-(\alpha))\wedge g_0^-(x^+(\alpha)),g_0^+(x^-(\alpha))\wedge g_0^+(x^+(\alpha))\}. \end{aligned} \tag{5.13}$$

◊

Proof By means of cut set operation properties of fuzzy numbers,

$$\begin{aligned} &(5.10)\ \Leftrightarrow\\ &\min \bigcup_{\alpha\in[0,1]}\alpha\Big[\sum_{k=1}^{J_0} c_{0k}^-(\alpha)\prod_{l=1}^{m} x_l^{\gamma_{0kl}},\sum_{k=1}^{J_0} c_{0k}^+(\alpha)\prod_{l=1}^{m} x_l^{\gamma_{0kl}}\Big]\\ &\text{s.t. } \sum_{k=1}^{J_i} c_{ik}\prod_{l=1}^{m} x_l^{\gamma_{ikl}}\leqslant 1\ (1\leqslant i\leqslant p)\\ &\quad x>0 \end{aligned} \tag{5.14}$$

According to Theorem 5.2.3, for a certain $\alpha\in[0,1]$, finding the results of program (5.11) and (5.12) means getting an optimal value of (5.14). Again $\overline{g}_{0\alpha}(x)=\min[g_0^-(x(\alpha)),\ g_0^+(x(\alpha))]$, that is to say (5.14) corresponds to an optimal value of a certain α, i.e., for a certain α, if x^- and x^+ are optimal solutions to (5.11) and to (5.12), respectively, then $\overline{g}_{0\alpha}(x)$ is an optimal value of (5.14).

Now the theorem holds by the arbitrariness of $\alpha \in [0,1]$. □

5.2.3 GP Mixed with Interval or Fuzzy Coefficients in Objective and Constraints

I. Mixed with Interval Coefficients

Let $\overline{c}_{ik}(1 \leqslant k \leqslant J_i, 0 \leqslant i \leqslant p)$ and $\overline{1}$ be all interval numbers. Then (5.1) is written as

$$\begin{aligned} &\min \ \{\bar{g}_0(x) = \sum_{k=1}^{J_0} \overline{c}_{0k} \prod_{l=1}^{m} x_l^{\gamma_{0kl}}\} \\ &\text{s.t.} \ \sum_{k=1}^{J_i} \overline{c}_{ik} \prod_{l=1}^{m} x_l^{\gamma_{ikl}} \leqslant \overline{1} \ (1 \leqslant i \leqslant p \\ &\quad x > 0 \end{aligned} \tag{5.15}$$

Theorem 5.2.5 *Program (5.15) is equivalent to finding*

$$(P_1) \qquad \begin{aligned} &\min \sum_{k=1}^{J_0} c_{0k}^- \prod_{l=1}^{m} x_l^{\gamma_{0kl}} \\ &\text{s.t.} \sum_{k=1}^{J_i} c_{ik}^- \prod_{l=1}^{m} x_l^{\gamma_{ikl}} \leqslant 1^- \\ &\quad \sum_{k=1}^{J_i} c_{ik}^+ \prod_{l=1}^{m} x_l^{\gamma_{ikl}} \leqslant 1^+ \ (1 \leqslant i \leqslant p) \\ &\quad x > 0 \end{aligned}$$

and

$$(P_2) \qquad \begin{aligned} &\min \sum_{k=1}^{J_0} c_{0k}^+ \prod_{l=1}^{m} x_l^{\gamma_{0kl}} \\ &\text{s.t.} \sum_{k=1}^{J_i} c_{ik}^- \prod_{l=1}^{m} x_l^{\gamma_{ikl}} \leqslant 1^- \\ &\quad \sum_{k=1}^{J_i} c_{ik}^+ \prod_{l=1}^{m} x_l^{\gamma_{ikl}} \leqslant 1^+ \ (1 \leqslant i \leqslant p) \\ &\quad x > 0 \end{aligned}$$

If x_{P_1} and x_{P_2} stand for optimal solutions to (P_1) and to (P_2), respectively, then $\bar{g}_0(x) = [g_0^-(x), g_0^+(x)]$ denotes an optimal value of (5.15), where

$$g_0^-(x) = \min\{g_0^-(x_{P_1}) \wedge g_0^-(x_{P_2}), g_0^+(x_{P_1}) \wedge g_0^+(x_{P_2})\},$$
$$g_0^+(x) = \max\{g_0^-(x_{P_1}) \wedge g_0^-(x_{P_2}), g_0^+(x_{P_1}) \wedge g_0^+(x_{P_2})\}.$$

◊

Proof By means of Theorem 5.2.1 and Theorem 5.2.3, (5.15) is equivalent to (P_1) and (P_2). Again, x_{P_1} and x_{P_2} represent feasible solutions to (5.15), which are substituted for an objective function in (5.15), and the minimum objective value $\bar{g}_0(x)$ is gained in the problem, such that the theorem is true. □

II. Mixed with Fuzzy Coefficients

When coefficients $\tilde{c}_{ik}(1 \leqslant k \leqslant J_i, 0 \leqslant i \leqslant p)$ in objective and constraints are all fuzzy numbers, (5.1) is written as

$$\begin{aligned} &\min\{\tilde{g}_0(x) = \sum_{k=1}^{J_0} \tilde{c}_{0k} \prod_{l=1}^{m} x_l^{\gamma_{0kl}}\} \\ &\text{s.t. } \sum_{k=1}^{J_i} \tilde{c}_{ik} \prod_{l=1}^{m} x_l^{\gamma_{ikl}} \subseteq \tilde{1} \\ &\quad x > 0 \, (1 \leqslant i \leqslant p) \end{aligned} \tag{5.16}$$

Theorem 5.2.6 *Program (5.16)$\Longleftrightarrow \forall\alpha \in [0,1]$, then*

$$(\overline{P}_1) \qquad \begin{aligned} &\min \sum_{k=1}^{J_0} c_{0k}^-(\alpha) \prod_{l=1}^{m} x_l^{\gamma_{0kl}} \\ &\text{s.t. } \sum_{k=1}^{J_i} c_{ik}^-(\alpha) \prod_{l=1}^{m} x_l^{\gamma_{ikl}} \leqslant 1^-(\alpha) \\ &\quad \sum_{k=1}^{J_i} c_{ik}^+(\alpha) \prod_{l=1}^{m} x_l^{\gamma_{ikl}} \leqslant 1^+(\alpha) \, (1 \leqslant i \leqslant p) \\ &\quad x > 0 \end{aligned}$$

and

$$(\overline{P}_2) \qquad \begin{aligned} &\min \sum_{k=1}^{J_0} c_{0k}^+(\alpha) \prod_{l=1}^{m} x_l^{\gamma_{0kl}} \\ &\text{s.t. } \sum_{k=1}^{J_i} c_{ik}^-(\alpha) \prod_{l=1}^{m} x_l^{\gamma_{ikl}} \leqslant 1^-(\alpha) \\ &\quad \sum_{k=1}^{J_i} c_{ik}^+(\alpha) \prod_{l=1}^{m} x_l^{\gamma_{ikl}} \leqslant 1^+(\alpha) \, (1 \leqslant i \leqslant p) \\ &\quad x > 0 \end{aligned}$$

If $x_{\overline{P}_1}$ and $x_{\overline{P}_2}$ represent optimal solutions to $(\overline{P}_1)$ and to $(\overline{P}_2)$, respectively, then

$$\tilde{g}_0(x) = \bigcup_{\alpha\in[0,1]} \alpha\bar{g}_0(x(\alpha)) = \bigcup_{\alpha\in[0,1]} \alpha[g_0^-(x(\alpha)), g_0^+(x(\alpha))]$$

represent an optimal value of (5.16), where

$$\begin{aligned} &g_0^-(x(\alpha)) \\ &\quad = \min\{g_0^-(x_{\overline{P}_1}(\alpha)) \wedge g_0^-(x_{\overline{P}_2}(\alpha)), g_0^+(x_{\overline{P}_1}(\alpha)) \wedge g_0^+(x_{\overline{P}_2}(\alpha))\}, \\ &g_0^+(x(\alpha)) \\ &\quad = \max\{g_0^-(x_{\overline{P}_1}(\alpha)) \wedge g_0^-(x_{\overline{P}_2}(\alpha)), g_0^+(x_{\overline{P}_1}(\alpha)) \wedge g_0^+(x_{\overline{P}_2}(\alpha))\}. \end{aligned}$$

◇

Proof By means of operation properties for a fuzzy number cut set, we know

$$(5.16) \iff$$

$$\begin{aligned} \min \quad & \bigcup_{\alpha\in[0,1]} \alpha\Big[\sum_{k=1}^{J_0} c_{0k}^-(\alpha) \prod_{l=1}^{m} x_l^{\gamma_{0kl}}, \sum_{k=1}^{J_0} c_{0k}^+(\alpha) \prod_{l=1}^{m} x_l^{\gamma_{0kl}}\Big] \\ \text{s.t.} \quad & \bigcup_{\alpha\in[0,1]} \alpha\Big[\sum_{k=1}^{J_i} c_{ik}^-(\alpha) \prod_{l=1}^{m} x_l^{\gamma_{ikl}}, \sum_{k=1}^{J_i} c_{ik}^+(\alpha) \prod_{l=1}^{m} x_l^{\gamma_{ikl}}\Big] \\ & \subseteq \bigcup_{\alpha\in[0,1]} \alpha\big[1^-(\alpha), 1^+(\alpha)\big] \ (1 \leqslant i \leqslant p) \\ & x > 0 \end{aligned} \tag{5.17}$$

Again, according to Theorem 5.2.2 and Theorem 5.2.4, finding (5.17) is equivalent to finding $(\overline{P}_1)$ and $(\overline{P}_2)$ $\forall\alpha \in [0, 1]$. Therefore, for a certain $\alpha \in [0, 1]$, Theorem 5.2.5 holds and from the arbitrariness of $\alpha \in [0, 1]$ the theorem holds, too. □

5.3 GP with Type $(\cdot, c)$ Fuzzy Coefficients

5.3.1 Properties of Type $(\cdot, c)$ Fuzzy Functions

Consider a general fuzzy GP[Cao95a]

$$\begin{aligned} &\widetilde{\min}\ \tilde{g}_0(x) \\ &\text{s.t. } \tilde{g}_i(x) \otimes \tilde{1}\ (1 \leqslant i \leqslant p) \\ &\quad\ x > 0 \end{aligned} \tag{5.18}$$

where $\tilde{g}_i(x) = \sum\limits_{k=1}^{J_i} \tilde{c}_{ik} \prod\limits_{l=1}^{m} x_l^{\gamma_{ikl}} (1 \leqslant i \leqslant p)$ is type $(\cdot, c)$ fuzzy function, here $\tilde{c}_{ik} > 0$ and $\tilde{1}$ are type $(\cdot, c)$ fuzzy numbers, sign "$\otimes$" the aggregation of "$\lesssim$" or "$\gtrsim$", and "$\otimes$" is taken to be "$\lesssim$" for $1 \leqslant i \leqslant p'$ and to be "$\gtrsim$" for $p'+1 \leqslant i \leqslant p$ respectively. We call (5.18) a fuzzy GP with type $(\cdot, c)$ fuzzy coefficients.

In order to find (5.18), we would better first introduce the definitions and concepts as follows:

Definition 5.3.1 Fuzzy function $\tilde{g}_i : x \to \mathcal{F}(g_i), \tilde{g}_i = \tilde{g}_i(x, \tilde{c})$, where $\mathcal{F}(g_i)$ is a fuzzy subset on g_i, and the membership function of the fuzzy set $\tilde{g}_i$ is defined as follows

$$\mu_{\tilde{g}_i}(g_i) = \begin{cases} \max\limits_{\{a | g_i = g_i(x,a)\}} \mu_{\tilde{c}}(a), & \{a | g_i = g_i(x, a)\} \neq \phi \\ 0, & \text{otherwise,} \end{cases}$$

where $\tilde{c}$ is a fuzzy set on product space consisting of $[c_{i1} \times c_{i2} \times \cdots \times c_{iJ_i}]$, the membership function of $\tilde{c}$ is denoted by $\mu_{\tilde{c}}(a)$.

It is assumed that fuzzy parameters are defined by fuzzy sets in Definition 5.1.4.

Definition 5.3.2 $\tilde{g}_i^* \gtrsim 0$ represents 'almost positive', which can be defined by applying equivalent form $\mu_{\tilde{g}_i^*}(0) \leqslant 1 - h, \alpha_i^{*\mathrm{T}} G_i \geqslant 0$, where

$$\tilde{g}_i^* = \begin{cases} \tilde{g}_i, & 1 \leqslant i \leqslant p' \\ -\tilde{g}_i, & p'+1 \leqslant i \leqslant p; \end{cases}$$

h stands for the degree of $\tilde{g}_i^* \gtrsim 0$. The larger h is, the stronger the meaning of 'almost positive' is.

5.3.2 Nonfuzzification

If an objective expectation value $\tilde{b}_0$ is presented by decision makers, then (5.18) can be turned into the following form:

$$
\begin{array}{lll}
\text{objection} & \tilde{g}_0 = \tilde{b}_0 G_{00} - \sum\limits_{k=1}^{J_0} \tilde{c}_{0k} G_{0k}(x) \gtrsim 0 & \\
\text{constraint} & \begin{cases} \tilde{g}_i = G_{00} - \sum\limits_{k=1}^{J_i} \tilde{c}_{ik} G_{ik}(x) \gtrsim 0 & (1 \leqslant i \leqslant p') \\ -\tilde{g}_i = -G_{00} + \sum\limits_{k=1}^{J_i} \tilde{c}_{ik} G_{ik}(x) \gtrsim 0 & (p'+1 \leqslant i \leqslant p) \\ x > 0 & \end{cases} & \\
\text{where} & G_{00} = 1, G_{ik}(x) = \prod_{l=1}^{m} x_l^{\gamma_{ikl}} & (0 \leqslant i \leqslant p)
\end{array}
$$

Theorem 5.3.1 *Given that fuzzy coefficients are denoted by $\tilde{c} = (\alpha, c)$, where $\alpha = (a_{i1}, a_{i2}, \ldots, a_{iJ_i})^T$, $c = (c_{i1}, c_{i2}, \ldots, c_{iJ_i})^T (0 \leqslant i \leqslant p)$, and fuzzy functions are denoted by*

$$
\begin{aligned}
\tilde{g}_i(x) &= \tilde{c}_{i1} \prod_{l=1}^{m} x_l^{\gamma_{i1l}} + \tilde{c}_{i2} \prod_{l=1}^{m} x_l^{\gamma_{i2l}} + \ldots + \tilde{c}_{iJ_i} \prod_{l=1}^{m} x_l^{\gamma_{ikl}} \\
&= \tilde{c}G = (\alpha^T G_i, c^T G_i);
\end{aligned}
$$

here $G_i = (\prod_{l=1}^{m} x_l^{\gamma_{i1l}}, \prod_{l=1}^{m} x_l^{\gamma_{i2l}}, \ldots, \prod_{l=1}^{m} x_l^{\gamma_{iJ_il}})^T$ and its membership function is

$$
\mu_{\tilde{g}_i}(g_i) = \begin{cases} 1 - \frac{|g_i - \alpha^{\mathrm{T}} G_i|}{c^{\mathrm{T}}|G_i|}, & G_i \neq 0 \\ 1, & G_i = 0, g_i = 0 \\ 0, & G_i = 0, g_i \neq 0 \end{cases}
$$

where $|G_i| = (|G_{i1}|, \ldots, |G_{iJ_i}|)^T$ and $\mu_{\tilde{g}_i}(g_i) = 0$ for $c^T|G_i| \leqslant |g_i - \alpha^T G_i|$. ◊

Proof We prove only the case to be true for $G_i \neq 0$, but the other cases are self-evident.

Because $\triangle ABD \sim \triangle AEF$ (shown in Figure 5.1), then $\frac{V}{c_{ik_i}^{\mathrm{T}}} = \frac{1-h}{1} \Rightarrow$ $V = c_{ik_i}^{\mathrm{T}}(1-h)$, while

$$
\begin{aligned}
K &= |\frac{g_i - \Sigma\alpha_{i(k_i-1)}^{\mathrm{T}} G_{i(k_i-1)}}{G_{ik_i}} - \alpha_{ik_i}^{\mathrm{T}}| - V \\
&= \frac{|g_i - \Sigma\alpha_{i(k_i-1)}^{\mathrm{T}} G_{i(k_i-1)} - \alpha_{ik_i}^{\mathrm{T}} G_{ik_i}|}{|G_{ik_i}|} - c_{ik_i}^{\mathrm{T}}(1-h) \\
&= \frac{|g_i - \Sigma\alpha_{ik_i}^{\mathrm{T}} G_{ik_i}| - c_{ik_i}^{\mathrm{T}}|G_i|(1-h)}{|G_{ik_i}|}.
\end{aligned}
$$

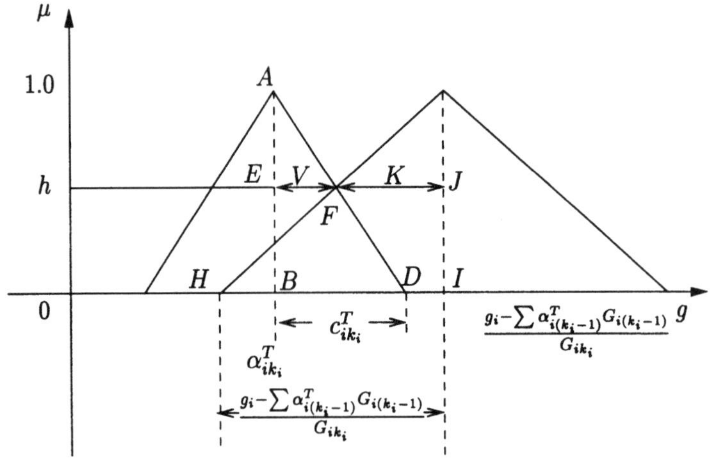

Figure 5.1. Illustration of expectation value and fuzzy constraint function.

Applying similarity of right triangles, we have

$$
\begin{aligned}
1-h &= \frac{K}{\Sigma c^{\mathrm{T}}_{i(k_i-1)}|G_{i(k_i-1)}|/|G_{ik_i}|}\\
\Rightarrow 1-h &= \frac{|g_i-\Sigma\alpha^{\mathrm{T}}_{ik_i}G_{ik_i}|-c^{\mathrm{T}}_{ik_i}|G_{ik_i}|(1-h)}{\Sigma c^{\mathrm{T}}_{i(k_i-1)}|G_{i(k_i-1)}|}\\
&\Rightarrow (1-h)\Sigma c^{\mathrm{T}}_{i(k_i-1)}|G_{i(k_i-1)}|+(1-h)c^{\mathrm{T}}_{ik_i}|G_{ik_i}|\\
&= |g_i-\Sigma\alpha^{\mathrm{T}}_{ik_i}G_{ik_i}|\\
\Rightarrow 1-h &= \frac{|g_i-\alpha^{\mathrm{T}}G_i|}{\Sigma c^{\mathrm{T}}_{ik_i}|G_{ik_i}|}\\
\Rightarrow h &= 1-\frac{|g_i-\alpha^{\mathrm{T}}G_i|}{c^{\mathrm{T}}_{ik}|G_i|}\\
&(c^{\mathrm{T}}_{ik}>0)
\end{aligned}
$$

□

Theorem 5.3.2 *If a fuzzy coefficient is known to be* $\tilde{c}=(\alpha,c)$, *where* $\alpha=(\alpha_{i1},\ldots,\alpha_{iJ_i})^T$, $c=(c_{i1},\ldots,c_{iJ_i})^T$, *then*

$$\tilde{g}^*_i \gtrsim 0 \Longleftrightarrow$$

$$(\alpha^*_{ik}-hc^*_{ik})^{\mathrm{T}}G_i \geqslant 0. \tag{5.19}$$

◊

Proof From Definition 5.3.1, we know

$$\begin{array}{lll}\tilde{g}_0^* \gtrsim 0 & \Longleftrightarrow \mu_{\tilde{g}_0^*}(0) & = 1 - \frac{\alpha_{0k}^{*\mathrm{T}} G_i}{c_{0k}^{*\mathrm{T}} G_i} \leqslant 1 - h, \quad \alpha_{0k}^{*\mathrm{T}} G_i \geqslant 0 \\ \vdots & \vdots & \vdots \\ \tilde{g}_i^* \gtrsim 0 & \Longleftrightarrow \mu_{\tilde{g}_i^*}(0) & = 1 - \frac{\alpha_{ik}^{*\mathrm{T}} G_i}{c_{ik}^{*\mathrm{T}} G_i} \leqslant 1 - h, \quad \alpha_{ik}^{*\mathrm{T}} G_i \geqslant 0\end{array}$$

where $G_i > 0$; then

$$\begin{array}{lll}c_{0k}^{*\mathrm{T}} G_i - \alpha_{0k}^{*\mathrm{T}} G_i & \leqslant (1-h) c_{0k}^{*\mathrm{T}} G_i \Longrightarrow & (\alpha_{0k}^{*\mathrm{T}} - h c_{0k}^{*\mathrm{T}}) G_i \geqslant 0 \\ \vdots & \vdots & \vdots \\ c_{ik}^{*\mathrm{T}} G_i - \alpha_{ik}^{*\mathrm{T}} G_i & \leqslant (1-h) c_{ik}^{*\mathrm{T}} G_i \Longrightarrow & (\alpha_{ik}^{*\mathrm{T}} - h c_{ik}^{*\mathrm{T}}) G_i \geqslant 0.\end{array}$$

Hence the theorem holds. □

Theorem 5.3.3 (5.18) $\Longleftrightarrow$

$$\begin{array}{l}\max h \\ \text{s.t. } (\alpha_{ik}^* - h c_{ik}^*)^{\mathrm{T}} G_i(x) \geqslant 0 \; h \in [0,1] \\ \qquad x > 0 \; (1 \leqslant k \leqslant J_i, 0 \leqslant i \leqslant p)\end{array} \tag{5.20}$$

where $(\alpha_{ik}^* - h c_{ik}^*)^T G_i(x) = g_i^*(x,h)$. ◊

Proof According to References [Cao87a,b], [Lu92], what we want is to let

$$\max_x \{\mu_{\tilde{D}}(x) = \min_x \{\mu_{\tilde{g}_0^*}(x), \bigwedge_{i=1}^{p} \mu_{\tilde{g}_i^*}(x)\}\}.$$

Here, $\mu_{\tilde{g}_0^*}(x)$ and $\mu_{\tilde{g}_i^*}(x)$ represent fuzzy objective and fuzzy constraint functions of (5.18), respectively, which is equivalent to causing the height of membership intersection to be highest between objective and constraints. The theorem is proved from Theorem 5.3.2. □

Theorem 5.3.4 *Let $\mathcal{X}$ be a feasible solution set of (5.19) with h being $\mathcal{X}_h$. Then*

$$h_1 < h_2 \Longrightarrow \mathcal{X}_{h_1}^* \supset \mathcal{X}_{h_2}^*.$$

◊

The theorem is proved by means of (5.19) without difficulty.

According to this theorem, we can choose a better constraint under the level of $h + r\triangle$ ($\triangle > 0$ is a small increment) by means of (5.19) and we might as well suppose it to be the i-th constraint, where the left inequality is regarded as a new objective function of the problem, such that problem (5.20) can be changed into finding

$$\begin{array}{l}\max\{\alpha_{i0}^* G_0(x) + \cdots + \alpha_{ip}^* G_p(x)\} \\ \text{s.t. } (\alpha_{ik}^* - (h + r\triangle) c_{ik}^*)^{\mathrm{T}} G_i(x) \geqslant 0 \; (0 < h < 1) \\ \qquad x > 0 \; (1 \leqslant k \leqslant J_i, 0 \leqslant i \leqslant p)\end{array}$$

whose solution x^* denotes an approximate solution to (5.20).

5.3.3 Algorithm and Numerical Example

Based on the theory mentioned above, we build the algorithms to (5.18). Because of (5.18) $\Longleftrightarrow$ (5.20), we have

Algorithm I. Choose the i-th constraint inequality in (5.19), solve h and substitute it for the objective function and for the remaining constraints in (5.20) and obtain a determined GP. Again, find its optimal solution by a direct algorithm in References [Cao87a,b], [Cao93a].

Algorithm II. Turn problem (5.19) into (5.20) before writing the dual form of (5.20); solve the optimal solution to its dual problem by a dual algorithm in References [Cao89a], [Cao93a] and we get an optimal solution in (5.19).

Algorithm III. [Cao92a][RT91] In Algorithm III, we advance steps for solution below:

1^0 Define the lower and the upper bounds for h and we suppose $h_0^- = 0, h_0^+ = 1$ for $\theta = 0$ in (5.20).

2^0 Fix $h_{\theta+1}$ and let $h_{\theta+1}$ = small end+(big end − small end) × 0.618, where small end and big end mean left and right endpoint value in the interval we refer to. If $|h_\theta^+ - h_\theta^-| < \varepsilon$ (ε is a sufficiently small positive number), then we take $h^* = h_{\theta+1}$. It ends, otherwise, we go on to 3^0.

3^0 If there exists a feasible solution set $\mathcal{X}$ for $h = h_{\theta+1}$, then go ahead 4^0. Otherwise, back to 2^0, and let $h_{\theta+1}^+ = h_\theta, h_{\theta+1}^- = h_\theta^-$.

4^0 Let $x^* \in \mathcal{X}$. We define

$$\bar{h}^n = \min\{g_0(x,h), \max_p g_i^*(x,h)\}.$$

We take $h_{\theta+1}^- = \bar{h}^n, h_{\theta+1}^+ = h_\theta^+$ and turn back to 2^0.

Continuing like this, we can find an approximate optimal solution to (5.20).

It is easy to compose an approximate fuzzy optimal value for (5.18) after the optimal solution converges to (5.20) by the three algorithms mentioned above.

Example 5.1 Find a fuzzy PGP

$$\begin{aligned} \min\ & \widetilde{37}x_1^{-1}x_2^{-1}x_3^{-1} + \widetilde{38.5}x_2x_3 \\ \text{s.t. } & \widetilde{1.5}x_1x_3 + \widetilde{0.9}x_1x_2 \lesssim \widetilde{4.5} \\ & x_1, x_2, x_3 > 0 \end{aligned} \tag{5.21}$$

where $\widetilde{37} = (37,6), \widetilde{38.5} = (38.5,3), \widetilde{1.5} = (1.5,1), \widetilde{0.9} = (0.9,0.2), \widetilde{4.5} = (4.5,1)$, and suppose the expected objective value is $\widetilde{64} = (64,8)$.

Solution Turn the object and constraint of (5.21) into

$$\begin{aligned} & \widetilde{64} - \widetilde{37}x_1^{-1}x_2^{-1}x_3^{-1} - \widetilde{38.5}x_2x_3 \gtrsim 0 \\ & \widetilde{4.5} - \widetilde{1.5}x_1x_3 - \widetilde{0.9}x_1x_2 \gtrsim 0 \\ & x_1, x_2, x_3 > 0; \end{aligned} \tag{5.22}$$

here, we suppose fuzzy sets $\tilde{c}_i (i = 0, 1)$ to be

$$\tilde{c}_0 = \{\alpha_0^* = (64, -37, -38.5)^{\mathrm{T}},\ c_0^* = (8, 6, 4)^{\mathrm{T}}\},$$
$$\tilde{c}_1 = \{\alpha_1^* = (4.5, -1.5, -0.9)^{\mathrm{T}},\ c_1^* = (1, 1, 0.2)^{\mathrm{T}}\}.$$

According to formula (5.20) in Theorem 5.3.3, (5.22) can be changed into

$$\begin{aligned}
&\max\ h \\
&\text{s.t. } 64 - 8h + (-37 - 6h)x_1^{-1}x_2^{-1}x_3^{-1} + (-38.5 - 3h)x_2x_3 \geqslant 0 \\
&\qquad 4.5 - h + (-1.5 - h)x_1x_3 + (-0.9 - 0.2h)x_1x_2 \geqslant 0, h \in [0, 1] \\
&\qquad x_1, x_2, x_3 > 0
\end{aligned}$$

Different optimal solutions can be obtained for different levels h. A decision maker may select k levels and compare their obtained k-group optimal solutions, among which the best is the most satisfactory solution.

In the example, if we choose $h = 0.5$, we can obtain the unique feasible solution by applying a dual algorithm, which means a dual optimal solution

$$w^* = (w_{01}, w_{02}, w_{11}, w_{12}; h)^{\mathrm{T}} = (\frac{2}{3}, \frac{1}{3}, \frac{1}{3}, \frac{1}{3}; 0.5)^{\mathrm{T}}$$

such that a feasible solution can be obtained corresponding to the primal problem, i.e., an optimal solution:

$$x^* = (x_1, x_2, x_3; h)^{\mathrm{T}} = (2, 1, \frac{1}{2}; \frac{1}{2})^{\mathrm{T}}$$

and the optimal value is 60.

5.3.4 Concluding Remarks

A series of results can be concluded from the discussion above.

i) Any fuzzy GP (5.18) with $(\cdot, c)$-coefficients can be completely turned into an ordinary GP (5.20) with a parameter h.

ii) Programming problem (5.18) has the same DD as (5.20).

iii) When exponent in (5.18) stands for a fuzzy number as in formula (1.14) in Section 1.6.3 in Chapter 1, the GP with fuzzy parametric $(\cdot, c)$ is

$$\begin{aligned}
\min\ & \sum_{k=1}^{J_0} \tilde{c}_{0k} \prod_{l=1}^{m} x_l^{\tilde{\gamma}_{0kl}} \\
\text{s.t. } & \sum_{k=1}^{J_i} \tilde{c}_{ik} \prod_{l=1}^{m} x_l^{\tilde{\gamma}_{ikl}} \otimes \tilde{1}\ (1 \leqslant i \leqslant p) \\
& x > 0
\end{aligned}$$

And then, it can be changed into a GP with h and β being parameters as follows:

If "$\otimes$" is taken to be "$\lesssim$", we have

$$\begin{aligned} &\min\ g_0(x,h,\beta) \\ &\text{s.t.}\ \ g_i(x,h,\beta) \leqslant 1 (1 \leqslant i \leqslant p) \\ &\qquad x > 0, h, \beta \in [0,1] \end{aligned}$$

If "$\otimes$" is taken to be "$\lesssim$" for $1 \leqslant i \leqslant p'$ and to be "$\gtrsim$" for $p'+1 \leqslant i \leqslant p$, we have

$$\begin{aligned} &\min\ g_0(x,h,\beta) \\ &\text{s.t.}\ \ g_i(x,h,\beta) \leqslant 1\ (1 \leqslant i \leqslant p') \\ &\qquad g_i(x,h,\beta) \geqslant 1\ (p'+1 \leqslant i \leqslant p) \\ &\qquad x > 0, h, \beta \in [0,1] \end{aligned}$$

where $g_i(x,h,\beta) = \sum\limits_{k=1}^{J_i} \tilde{c}_{ik}^{-1}(h) \prod\limits_{l=1}^{m} x_l^{\tilde{\gamma}_{ikl}^{-1}(\beta)} (0 \leqslant i \leqslant p)$.

Again, the most satisfactory solution can be found by methods mentioned above.

5.4 GP with L-R Fuzzy Coefficients

In this section, we will introduction GP with L-R fuzzy coefficients.[cao94a]

5.4.1 GP Model with L-R Coefficients

From Theorem 5.1.3 and Theorem 5.1.4, we call $\tilde{c} = (c, \alpha, \beta)_{LR}$ an L-R fuzzy number. Especially, $\tilde{c} = (c, 0, 0)_{LR}$ is an ordinary real number.

Definition 5.4.1 A matrix with element $\tilde{c}_{ij}$ being an L-R fuzzy number is said to be an L-R matrix, written $\tilde{c} = (\tilde{c}_{ij})$.

Definition 5.4.2 Define

$$\widetilde{\log c} = (\log c, \log \underline{c}, \log \overline{c})_{LR};\ \widetilde{e^c} = (e^c, e^{\underline{c}}, e^{\overline{c}})_{LR}$$

as an L-R logarithm and as an L-R exponent, and define

$$\begin{aligned} \min \tilde{f} &\Longleftrightarrow (\min f, \max \underline{f}, \min \overline{f}) \\ &(\text{resp.} \max \tilde{f} \Longleftrightarrow (\max f, \min \underline{f}, \max \overline{f})) \end{aligned}$$

and

$$\tilde{f} \leqslant \tilde{b} \Longleftrightarrow f \leqslant b, \underline{f} \geqslant \underline{b}, \overline{f} \leqslant \overline{b}.$$

Consider a PGP with L-R fuzzy coefficients as follows [Cao94a]:

$$\begin{aligned} &\widetilde{\min}\ \tilde{g}_0(x) \\ &\text{s.t. } \tilde{g}_i(x) \leqslant \tilde{1}\ (1 \leqslant i \leqslant p) \\ &\qquad x > 0 \end{aligned} \tag{5.23}$$

where $x = (x_1, x_2, \ldots, x_m)^{\mathrm{T}}$ is a variable vector,

$$\begin{aligned} \tilde{g}_i(x) &= \sum_{k=1}^{J_i} \tilde{c}_{ik} \prod_{l=1}^{m} x_l^{\gamma_{ikl}} \\ &= (\sum_{k=1}^{J_i} c_{ik} \prod_{l=1}^{m} x_l^{\gamma_{ikl}}, \sum_{k=1}^{J_i} \underline{c}_{ik} \prod_{l=1}^{m} x_l^{\gamma_{ikl}}, \sum_{k=1}^{J_i} \overline{c}_{ik} \prod_{l=1}^{m} x_l^{\gamma_{ikl}})_{LR} \\ &(0 \leqslant i \leqslant p) \end{aligned}$$

is a monomial; $\tilde{c}_{ik} = (c_{ik}, \underline{c}_{ik}, \overline{c}_{ik})_{LR}$ and $\tilde{1} = (1, \underline{1}, \overline{1})_{LR}$ are all L-R fuzzy numbers.

Because $\widetilde{\min}\ \tilde{g}_0(x)$ is equivalent to $\min g_0(x)$, $\max \underline{g}_0(x)$ and $\min \overline{g}_0(x)$, while $\tilde{g}_i(x) \leqslant \tilde{1}$ is equivalent to $g_i(x) \leqslant 1, \underline{g}_i(x) \geqslant \underline{1}$ and $\overline{g}_i(x) \leqslant \overline{1}$, and note that for $\min f = \max(-f)$, the nonfuzzification form of (5.23) means

$$\begin{aligned}
\min\ & g_0(x) = \sum_{k=1}^{J_0} c_{0k} \prod_{l=1}^{m} x_l^{\gamma_{0kl}} \\
\max\ & \underline{g}_0(x) = \sum_{k=1}^{J_0} \underline{c}_{0k} \prod_{l=1}^{m} x_l^{\gamma_{0kl}} \\
\min\ & \bar{g}_0(x) = \sum_{k=1}^{J_0} \bar{c}_{0k} \prod_{l=1}^{m} x_l^{\gamma_{0kl}} \\
\text{s.t.}\ & g_i(x) \leqslant 1 \\
& \underline{g}_i(x) \geqslant \underline{1} \\
& \overline{g}_i(x) \leqslant \overline{1}\ (1 \leqslant i \leqslant p) \\
& x > 0
\end{aligned} \tag{5.24}$$

Let $x_l = e^{z_l}(1 \leqslant l \leqslant m)$. Then $\tilde{g}_i(x)$ is deformed into

$$\tilde{G}_i(z) = \sum_{k=1}^{J_i} \tilde{c}_{ik}\exp(\sum_{l=1}^{m} \gamma_{ikl} z_l)(0 \leqslant i \leqslant p),$$

such that (5.23) can be changed into

$$\begin{aligned}
& \min\ \tilde{G}_0(z) \\
& \text{s.t. } \tilde{G}_i(z) \leqslant \tilde{1}\ (1 \leqslant i \leqslant p)
\end{aligned} \tag{5.25}$$

It is easy to prove the following theorems and corollaries by combining Theorems 2.2.2, 2.2.3 and 2.2.6 in Section 2.2 in Chapter 2 from the definition of L-R fuzzy numbers.

Theorem 5.4.1 *$\tilde{G}_i(z)$ serves as a fuzzy convex function for all $i(0 \leqslant i \leqslant p)$, so the deformed PGP (5.25) with L-R fuzzy coefficients is a fuzzy convex programming with its fuzzy local minimum solution being its fuzzy global minimum one.* ◇

Corollary 5.4.1 *Any strict fuzzy local minimum solution in (5.25) is its fuzzy global minimum one.* ◇

Theorem 5.4.2 *Let (5.25) be a strongly fuzzy convex programming problem. Then its fuzzy local minimum solution is its unique fuzzy global minimum one.* ◇

Theorem 5.4.3 *Given $\tilde{c}_k > \tilde{0}$ $(1 \leqslant k \leqslant J)$, then*

$$\widetilde{\log G}(z) = \log \sum_{k=1}^{J} \tilde{c}_k \exp\{\sum_{l=1}^{m} \gamma_{kl} z_l\}$$

means a fuzzy convex function of z. ◊

Proof From Definition 5.4.3 and Theorem 5.4.1, we can prove the result as we did in Theorem 2.3 in Reference [Cao87a,b]. □

Besides, (5.24) is a multi-objective GP problem. Its calculation is to be exhibited in Section 7.3.1 in Chapter 7.

5.4.2 PGP of Constraint with L-R Fuzzy Coefficients

Suppose (5.23) only with constraint coefficients being L-R fuzzy numbers. Obviously, (5.23) is equivalent to

$$\begin{aligned} &\min\ x_0 \\ &\text{s.t. } x_0^{-1} g_0(x) \leqslant 1 \\ &\qquad \tilde{g}_i(x) \leqslant \tilde{1}\ (1 \leqslant i \leqslant p) \\ &\qquad x_0 > 0, x > 0 \end{aligned}$$

which still represents a PGP having L-R fuzzy coefficients, with objective functions being very simple. Now, we might as well suppose $g_0(x) = x_1$. And practically, we can estimate the range in fuzzy optimal solutions, such that we can consider the fuzzy PGP with variables subject to lower and upper bounds

$$\begin{aligned} &\min\ x_1 \\ &\text{s.t. } \tilde{g}_i(x) \leqslant \tilde{1}\ (1 \leqslant i \leqslant p) \\ &\qquad 0 < x_L \leqslant x \leqslant x_U \end{aligned} \tag{5.26}$$

For all $x^0 > 0$, let

$$\varepsilon_{ik} = c^*_{ik} \prod_{l=1}^{m} x_l^{0\gamma_{ikl}} / g^*_i(x^0)\ (1 \leqslant k \leqslant J_i; 1 \leqslant i \leqslant p)$$

where $c^*_{ik} = c_{ik} + (\underline{c}_{ik} + \overline{c}_{ik})/2, g^*_i(x^0) = g_i(x^0) + (\underline{g}_i(x^0) + \overline{g}_i(x^0))/2$. Then $\sum_{k=1}^{J_i} \varepsilon_{ik} = 1$, such that we have the following lemma.

Lemma 5.4.1 *Let* $\tilde{c}^*_i > \tilde{0}, x_l > 0, \gamma^*_{il} > 0$. *Then*

$$\tilde{g}^*_i(x, x^0) = \tilde{c}^*_i \prod_{l=1}^{m} x_l^{\gamma^*_{il}} \leqslant \tilde{c}^*_i \sum_{l=1}^{m} \gamma^*_{il} x_l = \tilde{g}^*_i(x) \tag{5.27}$$

where

$$\tilde{c}^*_i = \prod_{k=1}^{J_i} (\tilde{c}^*_{ik}/\varepsilon_{ik})^{\varepsilon_{ik}},\ \gamma^*_{il} = \sum_{k=1}^{J_i} \gamma_{ikl}\varepsilon_{ik},\ \sum_{k=1}^{J_i} \varepsilon_{ik} = 1\ (0 \leqslant i \leqslant p, 1 \leqslant l \leqslant m).$$

◊

Proof Since

$$\tilde{c}_i^* \prod_{l=1}^{m} x_l^{\gamma_{il}^*} - \tilde{c}_i^* \sum_{l=1}^{m} \gamma_{il}^* x_l = \tilde{c}_i^* (\prod_{l=1}^{m} x_l^{\gamma_{il}^*} - \sum_{l=1}^{m} \gamma_{il}^* x_l),$$

and when $\gamma_{il}^* > 0, x_l > 0$, due to ordinary geometric inequality,

$$\prod_{l=1}^{m} x_l^{\gamma_{il}^*} - \sum_{l=1}^{m} \gamma_{il}^* x_l \leqslant 0$$

holds.

But $\tilde{c}_i^* > \tilde{0}$, and from the definition of L-R fuzzy numbers and their operation properties 8 in Section 5.1 in the chapter, we have

$$\tilde{c}_i^* (\prod_{l=1}^{m} x_l^{\gamma_{il}^*} - \sum_{l=1}^{m} \gamma_{il}^* x_l) \leqslant \tilde{0}.$$

Therefore, (5.27) holds. □

Since (5.27) holds, then (5.26) is equivalent to a monomial PGP with L-R fuzzy coefficients where variables are limited by lower and upper bounds

$$\begin{aligned} &\min\ x_1 \\ &\text{s.t. } x \in \tilde{F}^0 \\ &\quad \tilde{F}^0 = \{x | \tilde{g}_i^*(x, x^0) \leqslant \tilde{1}\ (1 \leqslant i \leqslant p), x_L \leqslant x \leqslant x_U\} \end{aligned} \tag{5.28}$$

Theorem 5.4.4 *If there exists a minimum solution x^k in (5.28) for a determinate k, then x^k must denote a fuzzy optimal solution to (5.26) after finite steps, otherwise any limit point of $\{x^k\}$ is a fuzzy optimal solution to (5.26).* ◊

Proof The fuzzy PGP (5.26) is equivalent to

$$\begin{aligned} &\min\ x_1 \\ &\text{s.t. } x \in F_j\ (j = 1, 2, 3) \end{aligned} \tag{5.29}$$

where

$$\begin{aligned} F_1 &= \{x | g_i(x) \leqslant 1\ (1 \leqslant i \leqslant p),\ 0 < x_L \leqslant x \leqslant x_U\}, \\ F_2 &= \{x | \underline{g}_i(x) \geqslant \underline{1}\ (1 \leqslant i \leqslant p),\ 0 < x_L \leqslant x \leqslant x_U\}, \\ F_3 &= \{x | \overline{g}_i(x) \leqslant \overline{1}\ (1 \leqslant i \leqslant p),\ 0 < x_L \leqslant x \leqslant x_U\}. \end{aligned}$$

But (5.29) is equivalent to a monomial fuzzy PGP by means of (5.27) [WY82] as follows:

$$\begin{aligned} &\min\ x_1 \\ &\text{s.t. } x \in F_j^0\ (j = 1, 2, 3) \\ &\quad F_1^0 = \{x | g_i^*(x, x^0) \leqslant 1\ (1 \leqslant i \leqslant p),\ x_L \leqslant x \leqslant x_U\}, \\ &\quad F_2^0 = \{x | \underline{g}_i^*(x, x^0) \geqslant \underline{1}\ (1 \leqslant i \leqslant p),\ x_L \leqslant x \leqslant x_U\}, \\ &\quad F_3^0 = \{x | \overline{g}_i^*(x, x^0) \leqslant \overline{1}\ (1 \leqslant i \leqslant p),\ x_L \leqslant x \leqslant x_U\}, \end{aligned}$$

and its optimal solution x^k must be that to (5.29), otherwise any limit point in a point range $\{x^k\}$ must be an optimal solution to (5.29)[Shi81] and the theorem is true. □

This indicates that any multinomial PGP with L-R fuzzy coefficients can be turned into a monomial one. Now, we consider a monomial PGP with only L-R fuzzy coefficients

$$\begin{aligned}
&\min\ g_0(x) = c_0 \prod_{l=1}^{m} x_l^{\gamma_{0l}} \\
&\text{s.t.}\ \tilde{g}_i(x) = \tilde{c}_i \prod_{l=1}^{m} x_l^{\gamma_{il}} \leqslant \tilde{1}\ (1 \leqslant i \leqslant p) \\
&\quad\ x > 0
\end{aligned} \tag{5.30}$$

We substitute $z_l = \log x_l$, such that

$$\begin{aligned}
&\min\ \sum_{l=1}^{m} \gamma_{0l} z_l \\
&\text{s.t.}\ \sum_{l=1}^{m} \gamma_{il} z_l + \widetilde{\log c_i} \leqslant \tilde{0}\ (1 \leqslant i \leqslant p)
\end{aligned} \tag{5.31}$$

Theorem 5.4.5 *The fuzzy optimal solution of program (5.30) is also that of (5.31).* ◊

Proof Because

$$\begin{aligned}
&\tilde{c}_i \prod_{l=1}^{m} x_l^{\gamma_{il}} \leqslant \tilde{1} \\
\Longleftrightarrow\ & c_i \prod_{l=1}^{m} x_l^{\gamma_{il}} \leqslant 1,\ \underline{c}_i \prod_{l=1}^{m} x_l^{\gamma_{il}} \geqslant \underline{1},\ \overline{c}_i \prod_{l=1}^{m} x_l^{\gamma_{il}} \leqslant \overline{1}.
\end{aligned}$$

Take logarithms from formulas on both sides on the right end above, respectively, and we have

$$\sum_{l=1}^{m} \gamma_{il} z_l + \log c_i \leqslant 0,\ \sum_{l=1}^{m} \gamma_{il} z_l + \log \underline{c}_i \geqslant \underline{0},\ \sum_{l=1}^{m} \gamma_{il} z_l + \log \overline{c}_i \leqslant \overline{0}$$

which is equivalent to $\sum_{l=1}^{m} \gamma_{il} z_l + \widetilde{\log c_i} \leqslant \tilde{0}$.

Then the theorem holds in a similar way as the proof of Theorem 5.4.4. □

5.4.3 Numerical Example

Example 5.2 Suppose we make about a $400m^3$ box having a bottom without a cover which is used to transport chemical raw materials. The bottom and both sides of the box are made of $4m^2$ special materials with negligible cost.

The materials of both ends cost ¥20/m^2, and the transporting expense of each box is a little more than ¥0.1. How much does it cost to transport about 400m^3 of chemical raw materials?

Solution Let x_1, x_2, x_3 represent the length, width and height of a box, respectively. Then its cost is equal to the total cost of transportation and material for the sides of the box, i.e.,

$$\tilde{g}_0(x) = \tilde{40}x_1^{-1}x_2^{-1}x_3^{-1} + 40x_2x_3.$$

Again, because the area sum of its bottom and both sides is less than $4m$, we have

$$g_1(x) = 2x_1x_3 + x_1x_2 \leqslant 4.$$

This problem can be concluded as a problem to find an answer to the fuzzy PGP

$$\begin{aligned} &\widetilde{\min}\ (\tilde{40}x_1^{-1}x_2^{-1}x_3^{-1} + 40x_2x_3) \\ &\text{s.t. } 2x_1x_3 + x_1x_2 \leqslant 4 \\ &\quad\ x_1, x_2, x_3 > 0 \end{aligned} \tag{5.32}$$

where $\tilde{40} = (40, 1, 9)_{LR} (= \widetilde{400} \cdot \widetilde{0.1} = (400, 10, 10)_{LR} \cdot (0.1, 0, 0.02)_{LR})$.

Let $\varepsilon_{01} = \frac{2}{3}, \varepsilon_{02} = \frac{1}{3}, \varepsilon_{11} = \varepsilon_{12} = \frac{1}{2}$. Then (5.32) can be turned into the following form by means of (5.27):

$$\begin{aligned} &\widetilde{\min}\ \left(\frac{\tilde{40}x_1^{-1}x_2^{-1}x_3^{-1}}{2/3}\right)^{2/3}\left(\frac{40x_2x_3}{1/3}\right)^{1/3} \\ &\text{s.t. } \left(\frac{2x_1x_3}{1/2}\right)^{1/2}\left(\frac{x_1x_2}{1/2}\right)^{1/2} \leqslant 4 \\ &\quad\ x_1, x_2, x_3 > 0 \end{aligned}$$

i.e.

$$\begin{aligned} &\widetilde{\min}(\tilde{40})^{2/3}(270)^{1/3}x_1^{-\frac{2}{3}}x_2^{-\frac{1}{3}}x_3^{-\frac{1}{3}} \\ &\text{s.t. } x_1x_2^{1/2}x_3^{1/3} \leqslant \sqrt{2} \\ &\quad\ x_1, x_2, x_3 > 0 \end{aligned}$$

$$\overset{z_i = \log x_i}{\Longleftrightarrow}$$

$$\begin{aligned} &\widetilde{\min}\{\frac{2}{3}\log\tilde{40} + \frac{1}{3}\log 270 - \frac{2}{3}z_1 - \frac{1}{3}z_2 - \frac{1}{3}z_3\} \\ &\text{s.t. } z_1 + \frac{1}{2}z_2 + \frac{1}{2}z_3 \leqslant \log\sqrt{2} \end{aligned}$$

We obtain

$$z_1 = \log\sqrt{2}, z_2 = z_3 = 0,$$

or

$$z_1 = \log 2, z_2 = 0, z_3 = \log\frac{1}{2}$$

and then

$$x_1 = 2, x_2 = x_3 = 1,$$

or

$$x_1 = 2, x_2 = 1, x_3 = \frac{1}{2}$$

are optimal solutions and a fuzzy optimal value is

$$\tilde{g}_0^* = (\widetilde{40})^{2/3} \cdot (270)^{1/3} \cdot 0.7937.$$

If we change 39 to $\widetilde{40}$, then $g_0^* = 58.996$ and if we change 49 to $\widetilde{40}$, then $g_0^* = 68.692$, so it will cost between ¥58.996 and ¥68.692 for transporting materials something like $400m^3$. If the weight ε is properly chosen, the budget expense is superiorly obtained.

5.5 GP with Flat Fuzzy Coefficients

We consider the fuzzy PGP as

$$\begin{aligned} \min\ \tilde{g}_0(x) &= \sum_{k=1}^{J_0} \tilde{c}_{0k} \prod_{l=1}^{m} x_l^{\gamma_{0kl}} \\ \text{s.t.}\ \ \tilde{g}_i(x) &= \sum_{k=1}^{J_i} \tilde{c}_{ik} \prod_{l=1}^{m} x_l^{\gamma_{ikl}} \circledast \tilde{1} \\ x &> 0 \end{aligned} \tag{5.33}$$

where $x = (x_1, x_2, \ldots, x_m)^{\mathrm{T}}$, "T" is a transpose sign, $\tilde{c}_{ik}(\gtrsim 0)$ and $\tilde{1}$ all flat fuzzy numbers, with sign "$\circledast$" denoting the pool of "$\subseteq$", "$\leqslant_\theta$", "$\geqslant_\theta$" and "$\approx_\theta$". We call (5.33) a PGP with flat fuzzy coefficients.

5.5.1 Change of Fuzzy Objective Function

In (5.33), an objective function is denoted by

$$\begin{aligned} \tilde{g}_0(x) &= \sum_{k=1}^{J_0} \tilde{c}_{0k} \prod_{l=1}^{m} x_l^{\gamma_{0kl}} = (g_0^-(x), g_0^+(x), \sigma^-_{g_0(x)}, \sigma^+_{g_0(x)}) \\ &= (\sum_{k=1}^{J_0} c_{0k}^- \prod_{l=1}^{m} x_l^{\gamma_{0kl}}, \sum_{k=1}^{J_0} c_{0k}^+ \prod_{l=1}^{m} x_l^{\gamma_{0kl}}, \\ &\quad \sum_{k=1}^{J_0} \sigma^-_{c_{0k}} \prod_{l=1}^{m} x_l^{\gamma_{0kl}}, \sum_{k=1}^{J_0} \sigma^+_{0k} \prod_{l=1}^{m} x_l^{\gamma_{0kl}}). \end{aligned} \tag{5.34}$$

An expected object is written as $\tilde{F} = (\overline{F}, \overline{F}, 0, \sigma^+_{\overline{F}})$; then let $\tilde{g}_0(x)$ intersect with an expected objective function $\tilde{F}$ pictured in the following figure:

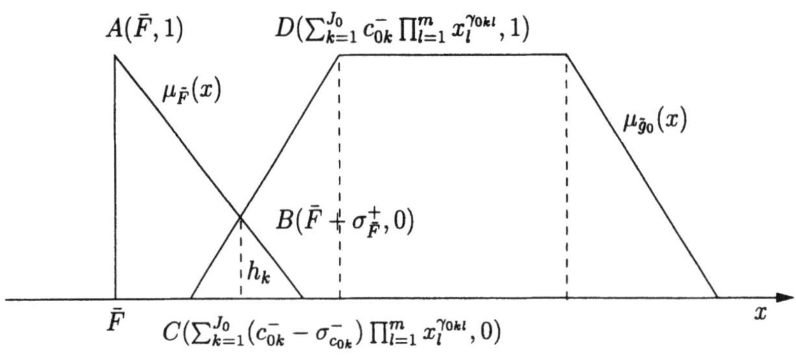

Figure 5.2. Relationship between $\tilde{g}_0(x)$ and $\tilde{F}$.

Theorem 5.5.1 *Given that $\tilde{g}_0(x)$ is like (5.34), it intersects an expected object $\tilde{F} = (\overline{F}, \overline{F}, 0, \sigma^{+}_{\overline{F}})$, then*

$$\min \tilde{g}_0(x)$$
$$\Longleftrightarrow \min \frac{\sum_{k=1}^{J_0} c^{-}_{0k} \prod_{l=1}^{m} x_l^{\gamma_{0kl}} - \overline{F}}{\sum_{k=1}^{J_0} \sigma_{c^{-}_{0k}} \prod_{l=1}^{m} x_l^{\gamma_{0kl}} + \sigma^{+}_{\overline{F}}}. \tag{5.35}$$

◊

Proof Let the equation of AB, CD denote (shown in Figure 5.2)

$$h_k - 1 = -\frac{1}{\sigma^{+}_{F}}(x - \overline{F}), \tag{5.36}$$

and

$$h_k - 1 = \frac{x - \sum_{k=1}^{J_0} c^{-}_{0k} \prod_{l=1}^{m} x_l^{\gamma_{0kl}}}{\sum_{k=1}^{J_0} \sigma^{-}_{0k} \prod_{l=1}^{m} x_l^{\gamma_{0kl}}}, \tag{5.37}$$

respectively.

Find (5.36) and (5.37) and then

$$h_k = 1 - \frac{\sum_{k=1}^{J_0} c^{-}_{0k} \prod_{l=1}^{m} x_l^{\gamma_{0kl}} - \overline{F}}{\sum_{k=1}^{J_0} \sigma_{c^{-}_{0k}} \prod_{l=1}^{m} x_l^{\gamma_{0kl}} + \sigma^{+}_{\overline{F}}}.$$

Since

$$\begin{aligned} \mathrm{PD}(\tilde{F}, \tilde{g}) &\triangleq \max\nolimits_{x,y|x>y} \min\{\mu_{\overline{F}}(x), \mu_{\tilde{g}}(y)\} \\ &= \min\{1, \mathrm{hgt}(\inf \tilde{g} \cap \sup \tilde{F})\} \end{aligned}$$

where, $\mathrm{hgt}(\inf \tilde{g} \cap \sup \tilde{F})$ stands for the nonnegative height in the intersection of a right decreased end side for $\mu_{\tilde{F}}(x)$ and a left increased one for $\mu_{\tilde{g}}(y)$, we have $\mathrm{PD}(\tilde{F}, \tilde{g}) = h_k$.

According to the judgment criterion, $\min \tilde{g}_0(x)$ means making h_k as high as possible, i.e., $\max h_k$, which is equivalent to the truth of (5.35). □

5.5.2 Determination of Fuzzy Constraints

Given

$$\tilde{g}_i(x) = (g_i^-(x), g_i^+(x), \sigma^-_{g_i(x)}, \sigma^+_{g_i(x)}), \tilde{1} = (1^-, 1^+, \sigma_{\tilde{1}}^-, \sigma_{\tilde{1}}^+),$$

according to the method in References [Dia87][Cao89b][Cao00][RT91], we may prove:

$$\tilde{g}_i(x) \subseteq \tilde{1} \Longleftrightarrow \begin{cases} g_i^+(x) \leqslant 1^+ \\ g_i^+(x) + \sigma^+_{g_i(x)} \leqslant 1^+ + \sigma_{\tilde{1}}^+ \\ g_i^-(x) - \sigma^-_{g_i(x)} \geqslant 1^- - \sigma_{\tilde{1}}^- \\ g_i^-(x) \geqslant 1^-. \end{cases} \tag{5.38}$$

Definition 5.5.1 $I(\tilde{g}_i(x), \alpha) = \{x | \mu_{\tilde{g}_i}(x) \geqslant \alpha, \alpha \in [0,1]\}$ is called an $\alpha-$ level set for $\tilde{g}_i(x)$.

The level set in Definition 5.5.1 denotes an open interval where on a real axis are embodied an initial end $\inf\limits_x I(\tilde{g}_i(x), \alpha)$ and a final end $\sup\limits_x I(\tilde{g}_i(x), \alpha)$. By means of monotonicity of $\mu_{\tilde{g}_i(x)}$ and $\mu_{\tilde{1}}$, (5.38) is equivalent to

$$\tilde{g}_i(x) \subseteq \tilde{1} \Longleftrightarrow \begin{cases} \sup\limits_x I(\tilde{g}_i(x), \alpha) \leqslant \sup\limits_x I(\tilde{1}, \alpha), \ \forall \alpha \in [0,1] \\ \inf\limits_x I(\tilde{g}_i(x), \alpha) \leqslant \inf\limits_x I(\tilde{1}, \alpha), \ \forall \alpha \in [0,1]. \end{cases}$$

If height $h_k = \text{hgt}(\inf\limits_x \tilde{1} \cap \sup\limits_x \tilde{g}_i(x)) > 0$, where the left end side for $\mu_{\tilde{1}}(x)$ increases, whereas the right end side for $\mu_{\tilde{g}_i}(x)$ decreases, then

$$\begin{aligned} \text{hgt}(\inf_x \tilde{1} \cap \sup_x \tilde{g}_i(x)) &= \max\{\frac{g_i^+ - 1^-}{\sigma^+_{g_i(x)}(x) + \sigma_{\tilde{1}}^-} + 1, 0\} \\ &= \begin{cases} \geqslant 1, & \text{if } g_i^+(x) \geqslant 1^- \\ < 1, & \text{if } g_i^+(x) < 1^- \end{cases} \end{aligned} \tag{5.39}$$

such that the degree of possibility of $\tilde{g}_i(x)$ superior to $\tilde{1}$ is denoted by

$$\begin{aligned} \text{PD}(\tilde{g}_i(x), \tilde{1}) &= \max\nolimits_{x,y|x>y} \min\{\mu_{\tilde{g}_i(x)}(x), \mu_{\tilde{1}}(y)\} \\ &= \min\{1, \text{hgt}(\inf \tilde{1} \cap \sup \tilde{g}_i(x))\}. \end{aligned}$$

Definition 5.5.2 Let $\theta \in [0,1]$ be an expected level. Then

$$\begin{aligned} &\tilde{g}_i(x) \geqslant_\theta \tilde{1} \Longleftrightarrow \text{no } \tilde{g}_i(x) <_\theta \tilde{1}, \\ &\tilde{g}_i(x) \approx_\theta \tilde{1} \Longleftrightarrow \text{no } \tilde{g}_i(x) \leqslant_\theta \tilde{1} \text{ and } \tilde{g}_i(x) \geqslant_\theta \tilde{1}, \end{aligned}$$

by which

$$\begin{cases} \tilde{g}_i(x) <_\theta \tilde{1} \Longleftrightarrow \mathrm{PD}(\tilde{g}_i(x), \tilde{1}) < \theta \\ \tilde{g}_i(x) \geqslant_\theta \tilde{1} \Longleftrightarrow \mathrm{PD}(\tilde{g}_i(x), \tilde{1}) \geqslant \theta \\ \tilde{g}_i(x) \approx_\theta \tilde{1} \Longleftrightarrow \mathrm{PD}(\tilde{g}_i(x), \tilde{1}) \geqslant \theta \\ \qquad\qquad\qquad \text{and } \mathrm{PD}(\tilde{1}, \tilde{g}_i(x)) \geqslant \theta \end{cases} \tag{5.40}$$

can be deduced.

Theorem 5.5.2

$$\tilde{g}_i(x) \leqslant_\theta \tilde{1} \Longleftrightarrow g_i^-(x) - (1-\theta)\sigma^-_{g_i(x)} \leqslant 1^+ + (1-\theta)\sigma^+_{\tilde{1}},$$
$$\tilde{g}_i(x) >_\theta \tilde{1} \Longleftrightarrow g_i^+(x) + (1-\theta)\sigma^+_{g_i(x)} > 1^- - (1-\theta)\sigma^-_{\tilde{1}},$$

$$\tilde{g}_i(x) \approx_\theta \tilde{1} \Longleftrightarrow \begin{cases} g_i^-(x) - (1-\theta)\sigma^-_{g_i(x)} \leqslant 1^+ + (1-\theta)\sigma^+_{\tilde{1}} \\ g_i^+(x) + (1-\theta)\sigma^+_{g_i(x)} \geqslant 1^- - (1-\theta)\sigma^-_{\tilde{1}}. \end{cases}$$

◊

Proof From (5.40), (5.39) and Dubois' proof [DP80], we have

$$\begin{aligned} \tilde{g}_i(x) \leqslant_\theta \tilde{1} &\Longleftrightarrow \frac{1^+ - g_i^-(x)}{\sigma^+_{\tilde{1}} + \sigma^-_{g_i(x)}} + 1 \geqslant \theta \\ &\Longleftrightarrow g_i^-(x) - (1-\theta)\sigma^-_{g_i(x)} \leqslant 1^+ + (1-\theta)\sigma^+_{\tilde{1}}, \end{aligned} \tag{5.41}$$

$$\begin{aligned} \tilde{g}_i(x) >_\theta \tilde{1} &\Longleftrightarrow \frac{g_i^+(x) - 1^-}{\sigma^+_{g_i(x)} + \sigma^+_{\tilde{1}}} + 1 > \theta \\ &\Longleftrightarrow 1^- - (1-\theta)\sigma^-_{\tilde{1}} < g_i^+(x) + (1-\theta)\sigma^+_{g_i(x)}. \end{aligned} \tag{5.42}$$

Combining the two cases mentioned above, $\tilde{g}_i(x) \approx_\theta \tilde{1}$ proves to be true. □

5.5.3 Equivalent Form

I. When "⊛" in (5.33) represents sign "⊆", by means of (5.35) and (5.38), then

$$(5.33) \Longleftrightarrow \min \frac{\sum\limits_{k=1}^{J_0} c^-_{0k} \prod\limits_{l=1}^{m} x_l^{\gamma_{0kl}} - \overline{F}}{\sum\limits_{k=1}^{J_0} \sigma_{c^-_{0k}} \prod\limits_{l=1}^{m} x_l^{\gamma_{0kl}} + \sigma^+_{\overline{F}}} \tag{5.43}$$

$$\left.\begin{array}{l} \text{s.t. } \sum_{k=1}^{J_i} c_{ik}^{+} \prod_{l=1}^{m} x_l^{\gamma_{ikl}} \leqslant 1^{+} \\ \sum_{k=1}^{J_i} (c_{ik}^{+} + \sigma_{c_{ik}}^{+}) \prod_{l=1}^{m} x_l^{\gamma_{ikl}} \leqslant 1^{+} + \sigma_1^{+} \\ \sum_{k=1}^{J_i} (c_{ik}^{-} - \sigma_{c_{ik}}^{-}) \prod_{l=1}^{m} x_l^{\gamma_{ikl}} \geqslant 1^{-} - \sigma_1^{-} \\ \sum_{k=1}^{J_i} c_{ik}^{-} \prod_{l=1}^{m} x_l^{\gamma_{ikl}} \geqslant 1^{-} \ (1 \leqslant i \leqslant p) \\ x_1, x_2, \ldots, x_m > 0 \end{array}\right\} \tag{5.44}$$

but

$$(5.43) \quad \Longleftrightarrow \quad \max\left\{1 - \frac{\sum_{k=1}^{J_0} c_{0k}^{-} \prod_{l=1}^{m} x_l^{\gamma_{0kl}} - \overline{F}}{\sum_{k=1}^{J_0} \sigma_{c_{0k}}^{-} \prod_{l=1}^{m} x_l^{\gamma_{0kl}} + \sigma_{\overline{F}}^{+}}\right\}$$

Programming (5.43),(5.44) above is equivalent to

$$\begin{array}{l} \max \ \theta \\ \text{s.t. } \sum_{k=1}^{J_0} [c_{0k}^{-} - (1-\theta)\sigma_{c_{0k}}^{-}] \prod_{l=1}^{m} x_l^{\gamma_{ikl}} \leqslant \overline{F} + (1-\theta)\sigma_{\overline{F}}^{+} \\ \quad (5.44), \theta \in [0,1] \end{array} \tag{5.45}$$

II. When "⊛" in (5.33) is selected as $\leqslant$ for $1 \leqslant i \leqslant p'$ and $\geqslant$ for $p'+1 \leqslant i \leqslant p$, then from (5.39) and (5.40), we know

$$\begin{array}{rl} (5.33) \Longleftrightarrow & \max \ \mathrm{PD}(\tilde{F}, g_0(x)) \\ & \text{s.t. } \mathrm{PD}(\tilde{1}, \tilde{g}_i(x)) \geqslant \theta \ (1 \leqslant i \leqslant p') \\ & \quad \mathrm{PD}(\tilde{g}_i(x), \tilde{1}) \geqslant \theta \ (p'+1 \leqslant i \leqslant p) \\ & \quad \theta \in [0,1], x > 0 \end{array}$$

$$\Longleftrightarrow$$

min (5.43)

$$\left.\begin{array}{l} \text{s.t. } \sum_{k=1}^{J_i} [c_{ik}^{-} - (1-\theta)\sigma_{c_{ik}}^{-}] \prod_{l=1}^{m} x_i^{\gamma_{ikl}} \leqslant 1^{+} + (1-\theta)\sigma_1^{+} \\ \qquad (1 \leqslant i \leqslant p') \\ \sum_{k=1}^{J_i} [c_{ik}^{+} + (1-\theta)\sigma_{c_{ik}}^{+}] \prod_{l=1}^{m} x_i^{\gamma_{ikl}} \geqslant 1^{-} - (1-\theta)\sigma_1^{-} \\ \qquad (p'+1 \leqslant i \leqslant p) \\ \theta \in [0,1] \\ x_1, x_2, \ldots, x_m > 0 \end{array}\right\} \tag{5.46}$$

$$\Longleftrightarrow \max\ \theta$$

$$\text{s.t.}\ \sum_{k=1}^{J_i}[c_{0k}^{-}-(1-\theta)\sigma_{c_{0k}}^{-}]\prod_{l=1}^{m}x_l^{\gamma_{0kl}} \leqslant \overline{F}+(1-\theta)\sigma_{\overline{F}}^{+} \tag{5.47}$$

(5.46)

$\theta \in [0,1]$

Comparing I to II, (5.45) contains only $3p$ constraints more than (5.47). Accordingly, we take only (5.47) into consideration. In order to handle a negative term in constraints, we introduce a sign

$$\delta_{ik}=\begin{cases}\delta_i, & \text{for } 1\leqslant k\leqslant S_i\\ -\delta_i, & \text{for } S_{i+1}\leqslant k\leqslant J_i\end{cases} \qquad (1\leqslant k\leqslant J_i, 0\leqslant i\leqslant p)$$

where S_i denotes all items with the same sign as a constraint function sign δ_i. If the constraint with negative items denotes the $i-$th one, the constraint can be uniquely written as

$$\begin{aligned}\overline{g}_i(x) &= \delta_i[1^{+}+(1-\theta)\sigma_1^{+}-\delta_i\sum_{k=1}^{J_i}\delta_{ik}\left(c_{ik}^{-}(1-\theta)\sigma_{c_{ik}}^{-}\right)\prod_{l=1}^{m}x_i^{\gamma_{ikl}}]\\ &= \delta_i[1^{+}+(1-\theta)\sigma_1^{+}-P_i+N_i]\geqslant 0\\ &\Longleftrightarrow\begin{cases}\delta_i[1^{+}+(1-\theta)\sigma_1^{+}+x_{i0}^{-1}P_i]\geqslant 0\\ \delta_i[1^{+}+(1-\theta)\sigma_1^{+}-x_{i0}^{-1}-x_{i0}^{-1}N_i]\geqslant 0\end{cases}\end{aligned}$$

where

$$P_i=\sum_{k=1}^{S_i}[c_{ik}^{-}-(1-\theta)\sigma_{c_{ik}}^{-}]\prod_{l=1}^{m}x_l^{\gamma_{0kl}}>0,$$
$$N_i=\sum_{k=S_i+1}^{J_i}[c_{ik}^{-}-(1-\theta)\sigma_{c_{ik}}^{-}]\prod_{l=1}^{m}x_l^{\gamma_{ikl}}>0,$$

x_{i0} is a new non-tentative-value variable, such that an inequality constraint polynomial with an arbitrary sign coefficient can be turned into a monomial. In (5.47), let

$$b_0=\overline{F}+(1-\theta)\sigma_{\overline{F}}^{+}>0,$$
$$b_1=1^{+}+(1-\theta)\sigma_1^{+}>0,$$
$$b_2=1^{-}-(1-\theta)\sigma_1^{-}>0.$$

Then (5.47) can be turned into an ordinary reverse PGP

$$\begin{aligned} &\max\theta=\min(-\theta)\\ &\text{s.t. } \frac{1}{b_0}\sum_{k=1}^{J_0}[c_{0k}^- -(1-\theta)\sigma_{c_{0k}}^-]\prod_{l=1}^{m}x_l^{\gamma_{0kl}}\leqslant 1\\ &\frac{1}{b_1}\sum_{k=1}^{J_i}[c_{ik}^- -(1-\theta)\sigma_{c_{ik}}^-]\prod_{l=1}^{m}x_l^{\gamma_{ikl}}\leqslant 1\ (1\leqslant i\leqslant p')\\ &\frac{1}{b_2}\sum_{k=1}^{J_i}[c_{ik}^+ +(1-\theta)\sigma_{c_{ik}}^+]\prod_{l=1}^{m}x_l^{\gamma_{ikl}}\geqslant 1\ (p'+1\leqslant i\leqslant p)\\ &\theta\in[0,1]\\ &x_1,x_2,\ldots,x_m>0 \end{aligned} \tag{5.48}$$

such that we can obtain the following.

Theorem 5.5.3 *There exists a fuzzy optimal solution to the fuzzy PGP (5.33) which is equivalent to existence of a parameter optimal solution to a reverse PGP (5.48) with parameter θ.* ◊

5.5.4 Algorithm

We have several solution methods to (5.48), for example, we can turn (5.33) into (5.48) to solve. But here, we only introduce a new algorithm, with steps as follows:

1^0 Define the lower and the upper bounds for θ and we suppose $\theta_0^- = 0, \theta_0^+ = 1$ for $l = 0$.

2^0 Fix θ_{l+1} and let θ_{l+1} = small end+(big end − small end) × 0.618, where the small one and the big one mean left and right endpoint values in the interval we refer to. If $|\theta_l^+ - \theta_l^-| < \varepsilon$ (ε means a sufficiently small positive number), then we take $\theta^* = \theta_{l+1}$. It ends, otherwise, go on to 3^0.

3^0 If there exists a feasible solution set $\mathcal{X}$ for $\theta = \theta_{l+1}$, then move on to 4^0, otherwise, back to 2^0, and let $\theta_{l+1}^+ = \theta_l, \theta_{l+1}^- = \theta_l^-$.

4^0 Let $x^* \in \mathcal{X}$ and we define

$$\theta^0=\min\{\mathrm{PD}(\tilde{F},\tilde{g}_0(x)),\min_{1\leqslant i\leqslant p}\mathrm{PD}(\tilde{g}_i(x),\tilde{1}),\min_{p'+1\leqslant i\leqslant p}\mathrm{PD}(\tilde{1},\tilde{g}_i(x))\}.$$

If we fix $\theta_{l+1}^- = \theta^0, \theta_{l+1}^+ = \theta_l^+$, turn to 2^0.

Continue like this and we can find an approximate optimal solution to (5.47) before we obtain an approximate fuzzy optimal value for (5.33).

Finally, we point out that after (5.33) is turned into (5.43), (5.44) or (5.48), we can solve it by the primal or the dual algorithm in Chapter 2. Therefore, (5.33) is solved.

Chapter 6

PROGRAMMING WITH FUZZY VARIABLES

This chapter is intended to deal with a special fuzzy variable (FV) including T–FV and trapezoidal FV applied in a programming, build-up LP [Cao96a] and GP [Cao97a] models with T–FV, and GP with trapezoidal FV, respectively. Because this is a new research direction, we propose one method of variable determination in anticipation of many new research achievements constantly being produced. Some ideas introduced in the chapter will give rise to many more valuable ones.

It falls into six parts. Section 1 embodies necessary knowledge for T–FV and trapezoidal FV and for determination of LP, GP with T–FV and GP with trapezoidal FV. Sections 2,3 and 4 involve their solution methods, respectively. Section 5 and 6 are devoted to development of a dual method for programming problems with FV [MTM00] and a fuzzy genetic algorithm [LZ98], respectively.

6.1 Basic Knowledge

Definition 6.1.1 [Dia87][Cao89b][Cao93a] Let $\tilde{x}_l = (\tilde{x}_{l1}, \tilde{x}_{l2}, \cdots, \tilde{x}_{ln})^{\mathrm{T}}$ $(1 \leqslant l \leqslant m)$ denote a trapezoidal FV. For each partition, we divide the set of natural numbers $\{1, 2, \cdots, n\}$ into two exhaustive, mutually exclusive subsets $\mathcal{T}(+)$ and $\mathcal{T}(-)$, one of which may be an empty set ϕ. Each partition associates a binary multi-index $\mathcal{T} = (\mathcal{T}_1, \mathcal{T}_2, \cdots, \mathcal{T}_n)$ defined by $\mathcal{T}_l = \{0, \quad \text{if } l \in \mathcal{T}(+); \quad 1, \quad \text{if } l \in \mathcal{T}(-)\}$.

Especially, $\mathcal{T}_0 = (0, 0, \cdots, 0), \quad \mathcal{T}_1 = (1, 1, \cdots, 1)$.

Definition 6.1.2 Let $\mathcal{F}(\Re)$ denote the set of normal fuzzy numbers and a quadruple parameter denote $\tilde{x} = (x^-, x^+, \underline{\xi}, \overline{\xi})$. Then we call $\tilde{x}$ a trapezoidal

FV, where interval (x^-, x^+) is the mean value of $\tilde{x}$, and $\underline{\xi}$, $\overline{\xi}$ stand for left and right spreads of $\tilde{x}$, respectively, with membership function of $\tilde{x}$,

$$\mu_{\tilde{x}}(x) = \begin{cases} 1 - \dfrac{x^- - x}{\underline{\xi}}, & \text{for } x^- - \underline{\xi} \leqslant x \leqslant x^- \\ 1, & \text{for } x^- < x < x^+ \\ 1 - \dfrac{x - x^+}{\overline{\xi}}, & \text{for } x^+ \leqslant x \leqslant x^+ + \overline{\xi} \\ 0, & \text{otherwise.} \end{cases} \tag{6.1}$$

Especially, suppose $\tilde{x} = (x, \alpha, \beta)_{LR}$ to be an L-R FV, where x represents the mean value of $\tilde{x}$, and α,β denote the left and the right spreads of $\tilde{x}$, respectively. If reference functions L and R are functions

$$T(u) = \begin{cases} 1 - |u|, & \text{if } \quad -1 \leqslant u \leqslant 1 \\ 0, & \text{if } \quad \text{otherwise,} \end{cases} \tag{6.2}$$

then $\tilde{x} = (x, \underline{\xi}, \overline{\xi})_T$ is called $T-$FV with the whole represented by $\mathcal{T}(\Re)$.

Note 6.1 Now, we can define $T-$FV as in Definition 6.1.1 [Cao89b,c] [Cao90a][Dia87].

Definition 6.1.3 $\forall x, y, z \in \Re$ with $x \leqslant y \leqslant z$ satisfying

$$\begin{aligned} &i) \quad \mu_{\tilde{x}}(y) \geqslant \mu_{\tilde{x}}(x) \wedge \mu_{\tilde{x}}(z), \\ &ii) \quad \max_{x \in \Re} \mu_{\tilde{x}}(x) = 1, \end{aligned}$$

we call $\tilde{x}$ a convex normalized fuzzy variable.

We also call $x_\alpha = \{x | \mu_{\tilde{x}} \geqslant \alpha, 0 < \alpha \leqslant 1\}$ an α-cut set of a trapezoidal FV (resp. $T-$FV). Especially, $A_0 = \{x | \mu_{\tilde{x}}(x) > 0\}$ is called a platform of trapezoidal FV $\tilde{x}$ (resp. a cone when $\tilde{x}$ is $T-$FV).

Let $P(\Re)$ denote the sets of all trapezoidal FV. If $\tilde{x} = (x^-, x^+, \underline{\xi}, \overline{\xi})$, $\tilde{x}_1 = (x_1^-, x_1^+, \underline{\xi}_1, \overline{\xi}_1) \in P(\Re)$ represent trapezoidal FV, then, by an extension principle, we have

$$\begin{aligned} &i) \quad \tilde{x} + \tilde{x}_1 = (x^- + x_1^-, x^+ + x_1^+, \underline{\xi} + \underline{\xi}_1, \overline{\xi} + \overline{\xi}_1), \\ &ii) \quad k\tilde{x} = \begin{cases} (kx^-, kx^+, k\underline{\xi}, k\overline{\xi}), & \text{for } k \geqslant 0 \\ (kx^+, kx^-, -k\overline{\xi}, -k\underline{\xi}), & \text{for } k < 0. \end{cases} \end{aligned}$$

Similarly, if $\tilde{x}$, $\tilde{x}_1$ behaves as $T-$FV $\tilde{x} = (x, \underline{\xi}, \overline{\xi})_T$, $\tilde{x}_1 = (x_1, \underline{\xi}_1, \overline{\xi}_1)_T$; then

$$\begin{aligned} &i) \quad \tilde{x} + \tilde{x}_1 = (x + x_1, \underline{\xi} + \underline{\xi}_1, \overline{\xi} + \overline{\xi}_1)_T, \\ &ii) \quad k\tilde{x} = \begin{cases} (kx, k\underline{\xi}, k\overline{\xi})_T, & \text{for } k \geqslant 0 \\ (kx, -k\overline{\xi}, -k\underline{\xi})_T, & \text{for } k < 0. \end{cases} \end{aligned}$$

Proposition 6.1.1 *The trapezoidal FV (resp. $T-$FV) $\tilde{x}$ is convex iff all of $x_\alpha (0 < \alpha \leqslant 1)$ are intervals.* $\Diamond$

Proof "$\Longrightarrow$" If $\tilde{x}$ is convex, from $i)$ in Definition 6.1.3, we know $y \in x_\alpha$. Again from the arbitrary x, y and z, we know x_α must be interval.

"$\Longleftarrow$" If $\alpha \in (0, 1)$, x_α is interval and we consider x, $z \in \Re$, with $\alpha_0 = \mu_{\tilde{x}}(x) \bigwedge \mu_{\tilde{x}}(z)$, then α_0 must be interval. It is easy to know $y \in x_{\alpha_0}$ from $x \leqslant y \leqslant z$, such that $\mu_{\tilde{x}}(y) \geqslant \alpha_0$, i.e., $\tilde{x}$ is convex. $\Box$

From Definition 6.1.3, we know the trapezoidal FV (resp. $T-$FV) must satisfy

$$\max_{x \in \Re} \mu_{\tilde{x}}(x) = 1,$$

therefore it is a fuzzy subset of a normal convex set.

Definition 6.1.4 Suppose that the space of nonempty compact convex subsets in $\Re^n$ can be embedded in a Banach space by identification with support functions and an L_2-metric defined. If $a = [\underline{a}, \overline{a}]$ and $b = [\underline{b}, \overline{b}]$ are two compact intervals, then this metric is reduced to

$$D_2(a, b)^2 = (\underline{a} - \underline{b})^2 + (\overline{a} - \overline{b})^2.$$

Define a metric d on $\mathcal{T}(\Re)$ by

$$d(\tilde{x}, \tilde{y})^2 = D_2(\text{supp}\tilde{x}, \text{supp}\tilde{y})^2 + (m(\tilde{x}) - m(\tilde{y}))^2,$$

where supp$(\cdot)$ denotes the interval support of $(\cdot)$ and $m(\cdot)$ is its modal value.

Proposition 6.1.2 *$(P(\Re), d)$ is complete.* $\Diamond$

Proof Because $\Re$ is complete, for an arbitrary Cauchy sequence $\{\tilde{x}_l\}$, where $\tilde{x}_l = (x_l^-, x_l^+, \underline{\xi}_l, \overline{\xi}_l)$, $\exists : \tilde{y} = (y^-, y^+, \underline{\eta}, \overline{\eta}) \in (P(\Re), d)$, we have $d(\tilde{x}_l, \tilde{y})^2 < \epsilon$, such that the proposition holds. $\Box$

Similarly, we can prove the case when x, y behave as $T-$FV.

If $\tilde{x} = (x^-, x^+, \underline{\xi}, \overline{\xi}), \tilde{y} = (y^-, y^+, \underline{\eta}, \overline{\eta}) \in P(\Re)$ are trapezoidal FV, then we define a metric d between $\tilde{x}, \tilde{y} \in P(\Re)$ as

$$\begin{aligned} d(\tilde{x}, \tilde{y})^2 =& [x^- - y^- - (\underline{\xi} - \underline{\eta})]^2 + [x^+ - y^+ + (\overline{\xi} - \overline{\eta})]^2 \\ &+ (x^- - y^-)^2 + (x^+ - y^+)^2, \ \tilde{x}, \tilde{y} \in P(\Re). \end{aligned} \tag{6.3}$$

If $\tilde{x} = (x, \underline{\xi}, \overline{\xi})_T, \tilde{y} = (y, \underline{\eta}, \overline{\eta})_T \in P(\Re)$ are $T-$FV, then we define a metric d between $\tilde{x}, \tilde{y} \in P(\Re)$ as

$$\begin{aligned} d(\tilde{x}, \tilde{y})^2 = [x - y - (\underline{\xi} - \underline{\eta})]^2 + [x - y + (\overline{\xi} - \overline{\eta})]^2 + (x - y)^2, \\ \tilde{x}, \tilde{y} \in P(\Re). \end{aligned}$$

Definition 6.1.5 Let $P(\Re)$ be that subspace of $\mathcal{T}(\Re)$, consisting of all those elements with nonnegative support: for each trapezoidal FV $(x^-, x^+, \underline{\xi}, \overline{\xi}) \in P(\Re)$, $x^- - \underline{\xi} \geqslant 0, x^+ - \overline{\xi} \geqslant 0$. Then $P(\Re)$ is a platform in $\mathcal{T}(\Re)$ and a closed convex subset of $\mathcal{T}(\Re)$ with respect to the topology induced by d.

Especially, for each $T-$FV $(x, \underline{\xi}, \overline{\xi}) \in P(\Re)$, $x - \underline{\xi} \geqslant 0$. Then $P(\Re)$ is a cone in $\mathcal{T}(\Re)$ and a closed convex subset of $\mathcal{T}(\Re)$ with respect to the topology induced by d.

Definition 6.1.6 If $\tilde{x} > 0$, such that

$$g_i(\tilde{x}) \leqslant \tilde{1}(< \tilde{1})(1 \leqslant i \leqslant p),$$

then primal PGP $(\tilde{P})$ with the trapezoidal FV (resp. $T-$FV) is called a fuzzy consistence (or fuzzy super-consistence). If $z(\mathcal{T}) > 0$ (resp. $z(\mathcal{J}) > 0$), such that

$$g_i(z(\mathcal{T})) \leqslant 1(< 1) \text{ (resp. } g_i(z(\mathcal{J})) \leqslant 1(< 1)) \ (1 \leqslant i \leqslant p),$$

then primal PGP depending on platform index $\mathcal{T}$ (resp. cone index $\mathcal{J}$) is called a consistence (or super-consistence).

We always see Model $(\tilde{P})$ in cybernetics and management decision, but there appear difficulties in its solution as follows:

1^0. Obtainment of the trapezoidal FV (resp. $T-$FV). 2^0. Determination method.

There are many methods to obtain the trapezoidal FV (resp. $T-$FV):

1) Direct obtainment. Write down the data from tests or measures characteristically as the trapezoidal FV (resp. $T-$FV).

2) Fitting. Fit the chosen data with spread functions of existing trapezoidal fuzzy variate (resp. $T-$fuzzy variate), and we find an answer to its extension value by chosen approximate functions and determine the trapezoidal FV (resp. $T-$FV).

3) Information distribution.

4) Building. This method fits only 'precision number' which, in fact, is fuzzy but written down exactly. We build a group of the trapezoidal FV (resp. $T-$FV) with fuzzy-time-series analysis through this 'precision number'.

We shall discuss the determination method on the assumption that the trapezoidal FV (resp. $T-$FV) can be obtained. In fact, we turn trapezoidal FV (resp. $T-$FV) into an ordinary variable depending on platform index $\mathcal{T}$(resp. cone index $\mathcal{T}$) by means of Definitions 6.1.1, 6.1.5, 6.1.6 and Note 6.1 above.

6.2 LP Problem with T−Fuzzy Variables

6.2.1 Advancement of the Problem

Theoretically, we build a new LP model on the basis of T-fuzzy numbers, study its dual form, nonfuzzify it under a cone index $\mathcal{J}$, and turn an LP with $T-$FV into an LP depending on a cone index $\mathcal{J}$. In such a theoretical framework, we can transplant many results of the LP into an LP with $T-$FV [Cao96a].

6.2.2 LP Problem with $T-$FV

Definition 6.2.1 Let the fuzzy LP problem be

$$
\begin{aligned}
(\widetilde{LP}) \qquad & \widetilde{\min}\ c\tilde{x} \\
& \text{s.t. } A\tilde{x} \leqslant \tilde{b} \\
& \qquad \tilde{x} \geqslant \tilde{0}
\end{aligned} \tag{6.4}
$$

where c is a real $1\times m$ matrix, A a real $n\times m$ matrix, $\tilde{x}, \tilde{0}$ two real $m-$dimensional $T-$FV vectors, and $\tilde{b} = (\tilde{b}_1, \tilde{b}_2, \ldots, \tilde{b}_n)^{\mathrm{T}}$ a real $n-$dimensional T-fuzzy vector.

If $\tilde{x}, \tilde{b}$ and $\tilde{0}$ are T-fuzzy data defined as in References [Cao89b,c][Dia87] [DP80], i.e., $\tilde{x} = (\tilde{x}_1, \tilde{x}_2, ..., \tilde{x}_m)^{\mathrm{T}}$; here $\tilde{x}_l = (x_l, \underline{\xi}_l, \overline{\xi}_l)_T, \tilde{1} = (1, \underline{1}, \overline{1})_T, \tilde{0} = (0, \underline{0}, \overline{0})_T$ and (6.4) is called an LP problem with $T-$FV.

Theorem 6.2.1 *Let the LP problem be given from $T-FV$ as $(\widetilde{LP})$. Then $(\widetilde{LP})$ is equivalent to $(\widetilde{LP}(\mathcal{J}))$ for a given cone index $\mathcal{J}$, and $(\widetilde{LP}(\mathcal{J}))$ has an optimal solution depending on a cone index $\mathcal{J}$, which is equivalent to $(\widetilde{LP})$ with a $T-$fuzzy optimal one.* ◊

Proof Let $\tilde{x}_i = (\tilde{x}_{1i}, \tilde{x}_{2i}, ..., \tilde{x}_{mi})^{\mathrm{T}}$ be a column $T-$FV satisfying $(\widetilde{LP})$, where $\tilde{x}_{li} = (x_i, \underline{\xi}_{li}, \overline{\xi}_{li})_T (1 \leqslant l \leqslant m; 1 \leqslant i \leqslant n)$. We classify vectors of the column by subscripts, and might as well let $p = 1, ..., M$ correspond to a smaller fluctuating variable, and the other variables correspond to $l = M + 1, ..., 3M$. Then for $i = 1, ..., N$ and each l,

$$U_{li} = x_i + (\underline{\xi}_{li} + \overline{\xi}_{li})/2;$$

for $i = N + 1, ..., 2N$ and each l,

$$U_{li} = \begin{cases} x_i - \underline{\xi}_{li}, & j_l = 0 \\ x_i + \overline{\xi}_{li}, & j_l = 1; \end{cases}$$

for $i = 2N + 1, ..., 3N$ and each l,

$$U_{li} = \begin{cases} x_l + \overline{\xi}_{li}, & j_l = 0 \\ x_l - \underline{\xi}_{li}, & j_l = 1. \end{cases}$$

Then, under a given cone index $\mathcal{J}$, $(\tilde{P})$ is changed into

$$\widetilde{LP}(\mathcal{J}) \qquad \min \sum_{l=1}^{m} c_l\Big(\sum_{i=1}^{3N} U_{li}/m\Big)$$
$$\text{s.t.} \sum_{l=1}^{m} a_{lk}\Big(\sum_{i=1}^{3N} U_{li}/m\Big) \leqslant b_k(\mathcal{J})(1 \leqslant k \leqslant n)$$
$$U \geqslant 0$$

where $U = (U_{1i}, U_{2i}, \ldots, U_{mi})^{\mathrm{T}}$ is an $m-$dimensional vector and $b_k(\mathcal{J})$ is a number depending on a cone index $\mathcal{J}$.

From the equivalence of $(\widetilde{LP})$ and $(\widetilde{LP}(\mathcal{J}))$, we know that $(\widetilde{LP}(\mathcal{J}))$ has an optimal solution depending on a cone index $\mathcal{J}$, which is equivalent to $(\widetilde{LP})$ with an optimal T-fuzzy solution. Therefore, the theorem holds. □

Theorem 6.2.1 shows us that $(\widetilde{LP})$ can be turned into an ordinary parametric LP $(\widetilde{LP}(\mathcal{J}))$ depending on a cone index $\mathcal{J}$, where $(\widetilde{LP}(\mathcal{J}))$ has many solution methods and an optimal method to it can be found in any work on LP.

6.2.3 Dual Problem

The LP problem with $T-$FV always contains a dual LP problem with $T-$FV corresponding to it. Since a close connection exists between the prime problem and the dual one, we can find an answer to the latter more easily than the former because

$$(\widetilde{LP}(\mathcal{J})) \;\Leftrightarrow\; \begin{array}{ll} \min & \sum\limits_{l=1}^{m} c_l\big(x_l + \sum\limits_{i=1}^{3N} \xi'_{li}/m\big) \\ \text{s.t.} & \sum\limits_{l=1}^{m} a_{lk}\big(x_l + \sum\limits_{i=1}^{3N} \xi'_{li}/m\big) \leqslant b_k(\mathcal{J}) \\ & x_l \geqslant 0\ (1 \leqslant l \leqslant m; 1 \leqslant k \leqslant n) \end{array} \tag{6.5}$$

where ξ'_{li} is $\underline{\xi}'_{li}$ (resp. $-\underline{\xi}'_{li}$) or $\bar{\xi}'_{li}$ (resp. $-\bar{\xi}'_{li}$).

Substitute $x'_l = x_l + \sum\limits_{i=1}^{3N} \xi'_{li}/m$, and then we might as well let $x_l \geqslant \sum\limits_{i=1}^{3N} \xi'_{li}/m$, and turn (6.5) into

$$\begin{array}{ll} \min & \sum\limits_{l=1}^{m} c_l x'_l \\ \text{s.t.} & \sum\limits_{l=1}^{m} a_{lk} x'_l \leqslant b_k(\mathcal{J}) \\ & x'_l \geqslant 0\ (1 \leqslant l \leqslant m; 1 \leqslant k \leqslant n) \end{array} \tag{6.6}$$

$$\Leftrightarrow \begin{array}{l} \min\ cx' \\ \text{s.t. } Ax' \leqslant b(\mathcal{J}) \\ \quad x' \geqslant 0 \end{array}$$

while the dual form of (6.6) is

$$\begin{array}{l} \max\ yb(\mathcal{J}) \\ \text{s.t. } A^{\mathrm{T}}y \geqslant c \\ \quad y \geqslant 0 \end{array} \tag{6.7}$$

Theorem 6.2.2 *Suppose LP problem $(\widetilde{LP})$ is deduced from T−FV. Its dual form is*

$$\begin{array}{l} \widetilde{\max}\ y\tilde{b} \\ \text{s.t. } A^{\mathrm{T}}y \geqslant c \\ \quad y \geqslant 0 \end{array} \tag{6.8}$$

and $(\widetilde{LP})$ has an optimal T-fuzzy solution which is equivalent to (6.8) having an optimal solution, and $(\widetilde{LP})$ has the same optimal T-fuzzy values as (6.8). ◊

Proof As $(\widetilde{LP})$ can be changed into $(\widetilde{LP}(\mathcal{J}))$ under above cone index $\mathcal{J}$, and the dual form of $(\widetilde{LP}(\mathcal{J}))$ is equivalent to (6.7), then (6.8) can be changed into (6.7) under the cone index $\mathcal{J}$ above. Again, $(\widetilde{LP})$ is known to be mutual dual with (6.8) due to the equivalence of $(\widetilde{LP})$ with $(\widetilde{LP}(\mathcal{J}))$, and (6.8) with (6.7), and the mutual duality of $(\widetilde{LP}(\mathcal{J}))$ and (6.7).

Again, $(\widetilde{LP}(\mathcal{J}))$ and (6.7) are, respectively, an ordinary primal LP and a dual LP depending on the same cone index $\mathcal{J}$. As for $(\tilde{P}(\mathcal{J}))$ and (6.7), applying Theorem 2 in Section 4.2 in Reference [GZ83], we know that if one of them has an optimal solution, so has the other. They contain the same optimal values, therefore the theorem holds from the arbitrariness of the cone index $\mathcal{J}$. □

Theorem 6.2.3 *Suppose that LP problem $(\widetilde{LP})$ is deduced from T−FV, then dual program $(\widetilde{LP})$ has optimal T-fuzzy solutions and (6.8) has optimal solutions iff they have T-fuzzy feasible ones and feasible ones, respectively, at the same time.* ◊

Proof Necessity is apparent and sufficiency is proved as follows:

$(\widetilde{LP})$ can be changed into $(\widetilde{LP}(\mathcal{J}))$ and (6.8) into (6.7) under the given cone index $\mathcal{J}$. Meanwhile $(\widetilde{LP}(\mathcal{J}))$ is equivalent to (6.7) under the same cone index $\mathcal{J}$. In a similar way to the proof of Theorem 1 in Section 4.2 in Reference [GZ83], we can prove $(\widetilde{LP}(\mathcal{J}))$ and (6.7) have feasible solutions depending on a cone index $\mathcal{J}$ iff they contain optimal solutions depending on a cone index $\mathcal{J}$. Again, we know the theorem holds because of the equivalence of $(\widetilde{LP})$ and

$(\widetilde{LP}(\mathcal{J}))$, and (6.8) and (6.7), and the duality of $(\widetilde{LP})$ and (6.8). □

Corollary 6.2.1 *If $\tilde{x}^0$ and $\tilde{y}^0$ are feasible T-fuzzy solutions to $(\widetilde{LP})$ and (6.8), respectively, with $\tilde{c}\tilde{x}^0 = \tilde{y}^0\tilde{b}$, then $\tilde{x}^0, \tilde{y}^0$ are optimal T-fuzzy solutions to $(\tilde{LP})$ and (6.8), respectively.* ◊

Proof Straightforward. □

6.2.4 Numerical Example

Example 6.1 Find

$$\begin{aligned} &\widetilde{\max}\ (3\tilde{x}_1 - \tilde{x}_2) \\ &\text{s.t. } 2\tilde{x}_1 - \tilde{x}_2 \leqslant \tilde{2}, \quad \text{where } \tilde{2} = (2,0,0)_T \\ &\quad\ \tilde{x}_1 \leqslant \tilde{4}, \quad \text{where } \tilde{4} = (4,0,0)_T \\ &\quad\ \tilde{x}_1, \tilde{x}_2 \geqslant \tilde{0}, \quad \text{where } \tilde{0} = (0,0,0)_T \end{aligned}$$

and give T-fuzzy data of a column

$$\begin{aligned} &\tilde{x}_1: \ 1.\ (x_1, 0.5, 1.2)_T, \quad 2.\ (x_1, 0.8, 1)_T, \quad 3.\ (x_1, 1, 1.4)_T, \\ &\tilde{x}_2: \ 4.\ (x_2, 0, 0.4)_T, \quad 5.\ (x_2, 0.6, 1)_T, \quad 6.\ (x_2, 1.5, 0.9)_T. \end{aligned}$$

Solution (i) Number the data by means of 1–6.

Group the data into three parts from Definition 6.1.1: I, No. 1,4; II, No. 2,5; $j_2 = 0, j_5 = 1$; and III. No. 3,6; $j_3 = 1, j_6 = 0$, here $j_l = 1$ for odd numbers and $j_l = 0$ for even numbers.

(ii) Nonfuzzification.

Let x_1, x_2 be

$$\begin{aligned} &[(x_1 + 0.85) + (x_1 - 0.8) + (x_1 + 1.4)]/3 = x_1 + 0.483, \\ &[(x_2 + 0.2) + (x_2 + 1) + (x_2 - 1.5)]/3 = x_2 - 0.1. \end{aligned}$$

(iii) Obtain a programming problem corresponding to (6.5) as follows:

$$\begin{aligned} &\max\ (3x_1 - x_2 + 1.55) \\ &\text{s.t. } 2x_1 - x_2 + 1.07 \leqslant 2 \\ &\quad\ x_1 + 0.483 \leqslant 4 \\ &\quad\ x_1, x_2 \geqslant 0 \end{aligned}$$

$\Rightarrow$

$$\begin{aligned} &\max\ (3x_1 - x_2 + 1.55) \\ &\text{s.t. } 2x_1 - x_2 \leqslant 0.93 \\ &\quad\ x_1 \leqslant 3.52 \\ &\quad\ x_1, x_2 \geqslant 0 \end{aligned}$$

The optimal solution depending on a cone index $\mathcal{J}$ is $x_1(\mathcal{J}) = 3.52, x_2(\mathcal{J}) = 6.11$, and the optimal value is 6.00. If $x_1(\mathcal{J})$ stands for an expensive resource, then $x_2(\mathcal{J})$ stands for a cheap resource. Decrease $x_1(\mathcal{J})$ and increase $x_2(\mathcal{J})$ properly and we obtain the same optimal value as in the noncrisp case. Obviously it decreases its cost.

6.3 GP with $T-$Fuzzy Variables

6.3.1 Introduction

First, a GP model with $T-$FV is built after the $T-$FV is introduced to GP in this section. The model is partitioned into an ordinary GP depending on the cone index $\mathcal{J}$ before dual form is obtained with $T-$FV corresponding to a primal PGP. Secondly, a dual problem is discussed on GP with $T-$FV. Thirdly, the relation is developed between optimal solutions of primal PGP and its dual form. Lastly, the effectiveness of this model with algorithm is verified by a numerical example.

6.3.2 Disposal of Nonfuzziness in Primal Problem

Let us consider an ordinary GP problem

$$\begin{aligned} &\min\ g_0(x) \\ &\text{s.t. } g_i(x) \leqslant \sigma_i (1 \leqslant i \leqslant p) \\ &\quad\ x > 0 \end{aligned}$$

Here $x = (x_1, x_2, ..., x_m)^{\mathrm{T}}$ denotes a variable vector, and

$$g_i(x) = \sum_{k=1}^{J_i} v_{ik}(x) = \sum_{k=1}^{J_i} \sigma_{ik} c_{ik} \prod_{l=1}^{m} x_l^{\gamma_{ikl}} (0 \leqslant i \leqslant p)$$

are all polynomials of x; σ_i is a symbol function; it takes $+1$ or -1 in term of an inequality constraint. Now let all or a portion of sign functions σ_i be negative (except σ_0 corresponding to an objective function, stipulating $\sigma_0 = 1$).

Definition 6.3.1 Suppose GP given by T-fuzzy data is said to be a GP with $T-$FV, whose mathematical formula is

$$\begin{aligned} &\widetilde{\min}\ g_0(\tilde{x}) \\ &\text{s.t. } g_i(\tilde{x}) \leqslant \tilde{\sigma}_i \quad (1 \leqslant i \leqslant p) \\ &\quad\ \tilde{x} > \tilde{0} \end{aligned} \tag{6.9}$$

Here, $\tilde{x} = (\tilde{x}_1, \tilde{x}_2, ..., \tilde{x}_m)^{\mathrm{T}}$ stands for an $m-$dimensional $T-$FV vector, $\tilde{x}_i, \tilde{0}, \tilde{\sigma}_i$ for T-fuzzy numbers [Cao89b,c][Cao90a], writing $\tilde{x}_i = (x_i, \underline{\xi}_i, \overline{\xi}_i)_T$;$\tilde{0} = (0, \underline{0}, \overline{0})_T$; $\tilde{\sigma}_i \triangleq \sigma_i \times \tilde{1}$, and $\tilde{1} = (1, \underline{1}, \overline{1})_T$, all of

$$g_i(\tilde{x}) = \sum_{k=1}^{J_i} v_{ik}(\tilde{x}) = \sum_{k=1}^{J_i} \sigma_{ik} c_{ik} \prod_{l=1}^{m} \tilde{x}_l^{\gamma_{ikl}} (0 \leqslant i \leqslant p)$$

are fuzzy posynomials of $\tilde{x}$, and $\widetilde{\min}$ means finding a fuzzy minimum.

Theorem 6.3.1 *Let a GP model given by T-fuzzy data $\tilde{x}$ be denoted as (6.9). Then, for a given cone index $\mathcal{J}$, (6.9) is equivalent to*

$$\begin{aligned} &\min\ g_0(z(\mathcal{J})) \\ &s.t.\ g_i(z(\mathcal{J})) \leqslant \sigma_i \quad (1 \leqslant i \leqslant p) \\ &\qquad z(\mathcal{J}) > 0 \end{aligned} \tag{6.10}$$

and a parameter optimal solution $z(\mathcal{J})$ depending on a cone index $\mathcal{J}$ in (6.10) is equivalent to optimal T-fuzzy solution to (6.9), where $g_i(z(\mathcal{J})) = \sum\limits_{k=1}^{J_i} \sigma_{ik} c_{ik} \prod\limits_{l=1}^{m} (z_l(\mathcal{J}))^{\gamma_{ikl}} (0 \leqslant i \leqslant p)$. ◊

Proof Let $\tilde{x}_l = (\tilde{x}_{l1}, \tilde{x}_{l2}, ..., x_{lp})^{\mathrm{T}}$ be a T-fuzzy data column satisfying (6.9), where $\tilde{x}_{li} = (x_l, \underline{\xi}_{li}, \overline{\xi}_{li})_T (1 \leqslant l \leqslant m, 1 \leqslant i \leqslant p)$, and divide it by subscripts. Now we might as well let $l = 1, ..., M$ corresponding to the smaller fluctuating data and the others be data corresponding to $l = M + 1, ..., 3M$. Then, to $l = 1, ..., M$ and each i, $z_{li} = x_l + (\underline{\xi}_{li} + \overline{\xi}_{li})/2$; and to $l = M + 1, ..., 2M$ and each i,

$$z_{li} = \begin{cases} x_l + \underline{\xi}_{li}; & \text{for } j_l = 0 \\ x_l - \overline{\xi}_{li}; & \text{for } j_l = 1, \end{cases}$$

and to $l = 2M + 1, ..., 3M$ and each i,

$$z_{li} = \begin{cases} x_l - \underline{\xi}_{li}; & \text{for } j_l = 0 \\ x_l + \overline{\xi}_{li}; & \text{for } j_l = 1. \end{cases}$$

Therefore, under the given cone index $\mathcal{J}$, (6.9) is turned into:

$$\begin{aligned} &\min\ \sum_{k=1}^{J_0} c_{0k} \prod_{l=1}^{m} (z_l(\mathcal{J}))^{\gamma_{0kl}} \\ &\text{s.t.}\ \sum_{k=1}^{J_i} \sigma_{ik} c_{ik} \prod_{l=1}^{m} (z_l(\mathcal{J}))^{\gamma_{ikl}} \leqslant \sigma_i \ (1 \leqslant i \leqslant p) \\ &\qquad z_l(\mathcal{J}) > 0 \end{aligned}$$

such that (6.10) can be found out.

Since (6.9) is equivalent to (6.10), a parameter optimal solution to (6.10) depending on a cone index $\mathcal{J}$ is equivalent to an optimal T-fuzzy one to (6.9). □

This shows that the GP with T−FV can be changed into an ordinary GP depending on a cone index $\mathcal{J}$ for solution.

6.3.3 Fuzzy Dual Problem

Just like each fuzzy GP with its dual form, the GP with $T-$FV has a dual program (6.9) with $T-$FV corresponding to it.

Definition 6.3.2 Let us call

$$\begin{aligned} &\max\ \tilde{d}(w)\\ &\text{s.t. } w_{00}=1\\ &\quad \Gamma^{\mathrm{T}}w=0\\ &\quad w\geqslant 0 \end{aligned} \tag{6.11}$$

a dual programming of primal GP (6.9), where

$$\tilde{d}(w)=\sigma\left[\prod_{i=0}^{p}\prod_{k=1}^{J_i}\left(\frac{\tilde{c}_{ik}}{w_{ik}}\sum_{k=1}^{J_i}w_{ik}\right)^{\sigma_i w_{ik}}\right]^{\sigma},$$

here $w=(w_{01},...,w_{0J_0},...,w_{p1},...,w_{pJ_p})^{\mathrm{T}}$ is a $J-$dimensional variable vector $(J=J_0+J_1+\cdots+J_p)$. In order to assume the defined continuity of $\tilde{d}(w)$, we stipulate

$$(w_{ik})^{w_{ik}}|_{w_{ik}=0}=1.$$

Theorem 6.3.2 *If the problem (6.9) is deduced from T-FV $\tilde{x}_l=(\tilde{x}_{l1},\tilde{x}_{l2},...,x_{lp})^T$, the dual form of (6.9) is (6.11).* ◊

Proof As (6.9) is equivalent to (6.10) from Theorem 6.3.1, the dual form of (6.10) is obviously

$$\begin{aligned} &\max\ d(w(\mathcal{J}))\\ &\text{s.t. } w_{00}(\mathcal{J})=1\\ &\quad \Gamma^{\mathrm{T}}w(\mathcal{J})=0\\ &\quad w(\mathcal{J})\geqslant 0 \end{aligned} \tag{6.12}$$

where

$$d(w(\mathcal{J}))=\sigma\left[\prod_{i=0}^{p}\prod_{k=1}^{J_i}\left(\frac{c_{ik}}{w_{ik}(\mathcal{J})}\sum_{k=1}^{J_i}w_{ik}(\mathcal{J})\right)^{\sigma_i w_{ik}(\mathcal{J})}\right]^{\sigma}$$

Here

$$w(\mathcal{J})=(w_{01}(\mathcal{J}),...,w_{0J_0}(\mathcal{J}),...,w_{p1}(\mathcal{J}),...,w_{pJ_p}(\mathcal{J}))^{\mathrm{T}}$$

is a $J-$dimensional variable vector depending on a cone index $\mathcal{J}$, and we stipulate

$$(w_{ik}(\mathcal{J}))^{w_{ik}(\mathcal{J})}|_{w_{ik}(\mathcal{J})=0}=1$$

correspondingly.

We can also prove (6.12) to be equivalent to (6.11) under the above cone index $\mathcal{J}$, while (6.10) and (6.12) are mutually dual under the cone index $\mathcal{J}$. (6.9) and (6.11) are mutually dual problems, so the theorem holds. □

6.3.4 Determination of Optimal Solutions for Primal PGP and Its Dual Form

Definition 6.3.3 If σ is taken to be ± 1 from (6.9)–(6.12), then (6.9) is written as $(\tilde{P})$, (6.10) as $(P(\mathcal{J}))$, (6.11) as $(\tilde{D})$ and (6.12) as $(D(\mathcal{J}))$.

We consider $(\tilde{P})$ and $(\tilde{D})$ to be problems of primal PGP and a dual one with $T-$FV, respectively; and $(P(\mathcal{J}))$ and $(D(\mathcal{J}))$ to be problems of the primal PGP and a dual one depending on a cone index $\mathcal{J}$ as well.

Theorem 6.3.3 (first fuzzy dual theorem) *Let the primal PGP $(\tilde{P})$ be deduced from $T-FV$ $\tilde{x}_l = (\tilde{x}_{l1}, \tilde{x}_{l2}, ..., \tilde{x}_{lp})^T$. If it is fuzzy super-consistent having T-fuzzy optimal solution $\tilde{x}^*$, then there must exist a Lagrange multiplier $\mu^* = (\mu_1^*, \mu_2^*, ..., \mu_p^*)^T \geqslant 0$, such that*

$$\nabla g_0(\tilde{x}^*) + \sum_{i=1}^{p} \mu_i^* \nabla g_i(\tilde{x}^*) = \tilde{0}, \tag{6.13}$$

$$\mu_i^*(g_i(\tilde{x}^*) - 1) = \tilde{0} \quad (1 \leqslant i \leqslant p), \tag{6.14}$$

while w^* defined by

$$w_{ik}^* = \begin{cases} \dfrac{v_{0k}(\tilde{x}^*)}{g_0(\tilde{x}^*)} & (i = 0; 1 \leqslant k \leqslant J_0) \\ \dfrac{\mu_i^* v_{ik}(\tilde{x}^*)}{g_0(\tilde{x}^*)} & (i \neq 0; 1 \leqslant k \leqslant J_p) \end{cases} \tag{6.15}$$

is an optimal one of dual program $(\tilde{D})$, with

$$g_0(\tilde{x}^*) = \tilde{d}(w^*). \tag{6.16}$$

◊

Proof For a given cone index $\mathcal{J}$, it may be proved that, similar to Theorem 6.3.1, the conditions of the theorem are equivalent to $(P(\mathcal{J}))$ $\mathcal{J}-$super consistent containing an optimal solution $z^*(\mathcal{J})$ which depends on a cone index $\mathcal{J}$. But under condition of the $(P(\mathcal{J}))$ $\mathcal{J}-$super consistent having optimal solution $z^*(\mathcal{J})$, it can be proved that, similar to Theorem 1.6.3 in Reference [WY82], there must be a Lagrange multiplier $\mu^* = (\mu_1^*, \mu_2^*, ..., \mu_p^*)^{\mathrm{T}} \geqslant 0$, such that

$$\nabla g_0(z^*(\mathcal{J})) + \sum_{i=1}^{p} \mu_i^* \nabla g_i(z^*(\mathcal{J})) = 0, \tag{6.17}$$

$$\mu_i^*(g_i(z^*(\mathcal{J})) - 1) = 0, \tag{6.18}$$

while $w^*(\mathcal{J})$ defined by

$$w^*_{ik}(\mathcal{J}) = \begin{cases} \dfrac{v_{0k}(z^*(\mathcal{J}))}{g_0(z^*(\mathcal{J}))} & (i = 0; 1 \leqslant k \leqslant J_0) \\ \dfrac{\mu^*_i v_{ik}(z^*(\mathcal{J}))}{g_0(z^*(\mathcal{J}))} & (i \neq 0; 1 \leqslant k \leqslant J_p) \end{cases} \tag{6.19}$$

is an optimal solution depending on the cone index $\mathcal{J}$ of the dual program $(D(\mathcal{J}))$, with

$$g_0(z^*(\mathcal{J})) = d(w^*(\mathcal{J})). \tag{6.20}$$

Under the above cone index, (6.13)–(6.16) are equivalent to (6.17)–(6.20), respectively, therefore the theorem holds. □

Theorem 6.3.4 (second fuzzy dual theorem) *Let the primal PGP $(\tilde{P})$ be deduced from $T-FV$. If $(\tilde{P})$ is fuzzy compatible and a dual problem $(\tilde{D})$ has a feasible solution with components being positive, then $(\tilde{P})$ has an optimal T-fuzzy solution.* ◊

Proof Under a given cone index $\mathcal{J}$, the condition of this theorem is equivalent to the following: If primal problem $(P(\mathcal{J}))$ is $\mathcal{J}-$compatible, its dual problem $(D(\mathcal{J}))$ has a $\mathcal{J}-$feasible solution with components being positive, so, under this condition, it may be proved that, similar to Theorem 1.8.1 in Reference [WY82], $(P(\mathcal{J}))$ has a $\mathcal{J}-$optimal solution. This is equivalent to the truth of the theorem. □

Theorem 6.3.3 shows us that if a primal problem $(\tilde{P})$ with $T-$FV is fuzzy super-consistent, with optimal T-fuzzy solutions $\tilde{x}^*$, the dual problem $(\tilde{D})$ has an optimal solution w^*, but $(\tilde{D})$ and $(\tilde{P})$ has the same optimal T-fuzzy values.

Theorem 6.3.4 further gives us a sufficient condition in which $(\tilde{P})$ has an optimal T-fuzzy solution by determination.

6.3.5 Numerical Example

In practical application, a problem to be solved involves T-fuzzy data to be obtained before finding its solution. The former depends on a practical measure, or we regard obtaining real numbers as "exactness" of fuzzy numbers. The latter means turning the problem into finding an ordinary parametric GP by the method mentioned.

Example 6.2 Find

$$\begin{aligned} &\min\ 2\tilde{x}_1^2\tilde{x}_3 + \tilde{x}_2\tilde{x}_3 \\ &\text{s.t.}\ 2\tilde{x}_1^{-2} + \tilde{x}_2^{-1}\tilde{x}_3^{-1} \leqslant \tilde{2} \\ &\qquad \tilde{x}_1, \tilde{x}_2, \tilde{x}_3 > \tilde{0} \end{aligned}$$

where $\tilde{2} = (2,0,0)_T, \tilde{0} = (0,0,0)_T$, and give T-fuzzy data of a column

$\tilde{x}_1$: 1. $(x_1, 0.4, 0.2)_T$; 2. $(x_1, 1, 0.7)_T$; 3. $(x_1, 1.2, 1.5)_T$;

$\tilde{x}_2$: 4. $(x_2, 1, 0.2)_T$; 5. $(x_2, 1.4, 1.2)_T$; 6. $(x_2, 0.5, 0.2)_T$;

$\tilde{x}_3$: 7. $(x_3, 0.8, 1)_T$; 8. $(x_3, 1.6, 1.2)_T$; 9. $(x_3, 0.2, 0.4)_T$.

1^0 Number the data from 1 to 9. Classify the data under three types from Definition 6.1.5: I. No.: 1,6,9; II. No.: 2,5,8, with $j_2 = 0, j_5 = 1, j_8 = 0$; III. No.: 3,4,7, with $j_3 = 1, j_4 = 0, j_7 = 1$. Here $j_l = 1$ stands for odd numbers and $j_l = 0$ for even ones.

2^0 Nonfuzzify:

$$\tilde{x}_1 \longrightarrow \frac{(x_1+0.2)+(x_1-1)+(x_1+1.5)}{3} = x_1 + 0.27,$$
$$\tilde{x}_2 \longrightarrow \frac{(x_2+0.4)+(x_2+1.2)+(x_2-1)}{3} = x_2 + 0.2,$$
$$\tilde{x}_3 \longrightarrow \frac{(x_3+0.3)+(x_3-1.6)+(x_3+1)}{3} = x_3 - 0.1.$$

3^0 Obtain a programming corresponding to (6.10) as follows:

$$\begin{aligned}&\min\{2(x_1+0.27)^2(x_3-0.1)+(x_2+0.2)(x_3-0.1)^{-1}\}\\ &\text{s.t. } 2(x_1+0.27)^{-2}+(x_2+0.2)^{-1}(x_3-0.1)^{-1} \leqslant 2,\\ &\qquad x_1, x_2, x_3 > 0.\end{aligned}$$

Substituting $u_1 = x_1 + 0.27, u_2 = x_2 + 0.2, u_3 = x_3 - 0.1$ for it, we obtain its dual program

$$\begin{aligned}&\max\Big(\frac{2}{w_{01}}\Big)^{w_{01}}\Big(\frac{1}{w_{02}}\Big)^{w_{02}}\Big(\frac{1}{w_{11}}\Big)^{w_{11}}\Big(\frac{1}{2w_{12}}\Big)^{w_{12}}(w_{11}+w_{12})^{w_{11}+w_{12}}\\ &\text{s.t. } w_{01}+w_{02}=1;\\ &\qquad 2w_{01}-2w_{12}=0;\\ &\qquad w_{01}-w_{12}=0;\\ &\qquad w \geqslant 0.\end{aligned}$$

We obtain a unique dual feasible solution: $w_{01} = 2/3, w_{02} = 1/3, w_{11} = 2/3, w_{12} = 1/3$, which is an optimal solution correspondingly, and the dual optimal value is $M_D = 9/2$. Again from (6.19), combining substitutes, we can obtain a $\mathcal{J}$−optimal solution to a primal problem: $x_1(\mathcal{J}) = \sqrt{\frac{3}{2}} - 0.27, x_2(\mathcal{J}) = \frac{3}{2} - 0.2, x_3(\mathcal{J}) = 1.1$, and the optimal value is $d(x(\mathcal{J})) = 9/2$.

6.4 GP with Trapezoidal Fuzzy Variables

6.4.1 Introduction

Researchers such as H.J.Zimmermann [Zim76], H.Tanaka et al [TOA73] and J. L.Verdegay [Ver84] have contributed much to fuzzy LP with fuzzy variables, while no one had studied a fuzzy PGP until we advanced the problem in 1987 [Cao87a,b][Cao93b]. It has aroused scholars all over the world. Hereafter, we extend the trapezoidal FV into a programming problem. Again we put forward a PGP with the trapezoidal FV [Cao90][Cao92a][Cao93a], which contains more information than a classical model, based on the trapezoidal fuzzy data. Now, having considered sets of the trapezoidal FV to be a fuzzy platform, we build a deterministic model through a platform index $\mathcal{T}$, such that it can be changed into a determined GP depending on the platform index $\mathcal{T}$.

6.4.2 Primal GP with Trapezoidal FV and Its Determined Method

Let the PGP model be

$$\begin{aligned} &\min\ g_0(x) \\ &\text{s.t. }\ g_i(x) \leqslant 1 (1 \leqslant i \leqslant p) \\ &\quad\ \ x > 0 \end{aligned} \tag{6.21}$$

where $x = (x_1, x_2, \cdots, x_m)^{\mathrm{T}}$ denotes a vector with positive variables and all

$$g_i(x) = \sum_{k=1}^{J_i} v_{ik}(x) = \sum_{k=1}^{J_i} c_{ik} \prod_{l=1}^{m} x^{\gamma_{ikl}} (0 \leqslant i \leqslant p)$$

are posynomials of x.

Definition 6.4.1 If (6.21) is given by the trapezoidal FV, i.e.,

$$\begin{aligned} (\tilde{P}) \qquad &\widetilde{\min}\ g_0(\tilde{x}) \\ &\text{s.t. }\ g_i(\tilde{x}) \leqslant \tilde{1} (1 \leqslant i \leqslant p) \\ &\quad\ \ \tilde{x} > \tilde{0} \end{aligned}$$

then it is a PGP with the trapezoidal FV, where $\tilde{x} = (\tilde{x}_1, \tilde{x}_2, \ldots, \tilde{x}_m)^{\mathrm{T}}$ is an $m-$dimensional trapezoidal FV vector, $\tilde{x}_l$, $\tilde{0}$, $\tilde{1}$ all trapezoidal fuzzy numbers like $\tilde{x}_l = (x_l^-, x_l^+, \underline{\xi}_l, \overline{\xi}_l)$, $\tilde{0} = (0^-, 0^+, \underline{0}, \overline{0})$, $\tilde{1} = (1^-, 1^+, \underline{1}, \overline{1})$; all of

$$g_i(\tilde{x}) = \sum_{k=1}^{J_i} v_{ik}(\tilde{x}) = \sum_{k=1}^{J_i} c_{ik} \prod_{l=1}^{m} \tilde{x}_l^{\gamma_{ikl}} (0 \leqslant i \leqslant p)$$

fuzzy posynomials of the trapezoidal FV $\tilde{x}$, and $\widetilde{\min}$ means extracting a fuzzy minimum value.

According to the ideas in References [DP80][Cao93b], if possibility degree of $v(g_i(\tilde{x}), \tilde{1})$ is greater than or equal to 0, $\tilde{x}$ trapezoidal FV, then we call $g_i(\tilde{x}) \leqslant \tilde{1}$ an inequality with trapezoidal FV. In a particular situation when the degree is equal to 0, it is said to be an equality with trapezoidal FV for $g_i(\tilde{x}) = \tilde{1}$.

Theorem 6.4.1 *Suppose that the PGP with the trapezoidal FV is shown as $(\tilde{P})$, for a fixed platform index $\mathcal{T}$. Then $(\tilde{P})$ is equivalent to a PGP depending on platform index $\mathcal{T}$*

$$(P(\mathcal{T})) \qquad \begin{aligned} &\min\ g_0(z(\mathcal{T})) \\ &\text{s.t.}\ \ g_i(z(\mathcal{T})) \leqslant 1 (1 \leqslant i \leqslant p) \\ &\qquad z(\mathcal{T}) > 0 \end{aligned}$$

and the parameter optimum solution with platform index $\mathcal{T}$ in $(P(\mathcal{T}))$ means also a trapezoidal fuzzy one in $(\tilde{P})$, where

$$g_i(z(\mathcal{T})) = \sum_{k=1}^{J_i} c_{ik} \prod_{l=1}^{m} (z_l(\mathcal{T}))^{\gamma_{ikl}} (0 \leqslant i \leqslant p).$$

◇

Proof Let $\tilde{x}_l = (\tilde{x}_{l1}, \tilde{x}_{l2}, \cdots, \tilde{x}_{lp})^{\mathrm{T}}$ be a trapezoidal FV satisfying $(\tilde{P})$, where $\tilde{x}_{li} = (x_{li}^-, x_{li}^+, \underline{\xi}_{li}, \overline{\xi}_{li})$ $(1 \leqslant l \leqslant m; 1 \leqslant i \leqslant p)$. Because x is freely fixed in the closed value interval $[x_{li}^-, x_{li}^+]$, we choose the degree of accomplishment in the light of membership functions

$$\tilde{\phi}_l(x_{li}) = \begin{cases} 0, & \text{if } x_{li} < x_{li}^- \\ \left(\dfrac{x_{li} - x_{li}^-}{x_{li}^+ - x_{li}^-} \right)^n & \text{if } x_{li}^- \leqslant x_{li} \leqslant x_{li}^+ \\ 1, & \text{if } x_{li} > x_{li}^- \end{cases} \tag{6.22}$$

then we deduce

$$x_{li} - x_{li}^- \geqslant \sqrt[n]{\alpha}(x_{li}^+ - x_{li}^-) \Rightarrow x_{li} = x_{li}^- + \sqrt[n]{\alpha}(x_{li}^+ - x_{li}^-)$$

by $\tilde{\phi}_l(x_{li}) \geqslant \alpha$.

We classify variables of the column by subscripts, and might as well let $l = 1, 2, \cdots, M$ correspond to a smaller fluctuating variable while the other variables correspond to $l = M+1, \ldots, 3M$. And then

1^0 As for $l = 1, 2, \ldots, M$ and each i,

$$\tilde{x}_{li} \to x_{li}^- + \sqrt[n]{\alpha}(x_{li}^+ - x_{li}^-) + \frac{\underline{\xi}_{li} + \bar{\xi}_{li}}{2};$$

2^0 As for $l = M+1, \ldots, 2M$ and each i,

$$\tilde{x}_{li} \to \begin{cases} x_{li}^- + \sqrt[n]{\alpha}(x_{li}^+ - x_{li}^-) + \underline{\xi}_{li}, j_l = 0, \\ x_{li}^- + \sqrt[n]{\alpha}(x_{li}^+ - x_{li}^-) - \bar{\xi}_{li}, j_l = 1; \end{cases}$$

3^0 As for $l = 2M+1, \ldots, 3M$ and each i,

$$\tilde{x}_{li} \to \begin{cases} x_{li}^- + \sqrt[n]{\alpha}(x_{li}^+ - x_{li}^-) - \underline{\xi}_{li}, j_l = 0, \\ x_{li}^- + \sqrt[n]{\alpha}(x_{li}^+ - x_{li}^-) + \bar{\xi}_{li}, j_l = 1. \end{cases}$$

Therefore, under the same given platform index $\mathcal{T}$, let

$$z_{li} = x_{li}^- + \sqrt[n]{\alpha}(x_{li}^+ - x_{li}^-) + \xi_{li}^*,$$

where ξ_{li}^* is $\frac{\underline{\xi}_{li} + \bar{\xi}_{li}}{2}$, $\pm\underline{\xi}_{li}$, or $\pm\bar{\xi}_{li}$; then

$$z_l(\mathcal{T}) = \left(\frac{\bar{\xi}_{li} x_{li}^- + \underline{\xi}_{li} x_{li}^+}{\underline{\xi}_{li} + \bar{\xi}_{li}} + \sum_{i=1}^{3M} \xi_{li}^* \frac{3M}{m}\right).$$

Substitute $z_l(\mathcal{T})$ for $\tilde{x}_l$ in $(\tilde{P})$, and we turn $(\tilde{P})$ into $(P(\mathcal{T}))$.

So, under the given platform index $\mathcal{T}$ mentioned above, $(\tilde{P})$ can be turned into $(P(\mathcal{T}))$ and a parameter optimal solution in $(P(\mathcal{T}))$ depending on platform index $\mathcal{T}$ is equivalent to a trapezoidal fuzzy optimal one in $(\tilde{P})$. □

This shows that the PGP with trapezoidal FV can be changed into an ordinary PGP depending on the platform index $\mathcal{T}$ for solution.

6.4.3 Fuzzy Dual Theory

We consider a dual form of PGP with trapezoidal FV as a similar ordinary PGP with dual programming.

Definition 6.4.2 We call

$$\begin{aligned} (\tilde{D}) \qquad & \max\ \tilde{d}(w) \\ & \text{s.t. } w_{00} = 1 \\ & \qquad \Gamma^{\mathrm{T}} w = 0 \\ & \qquad w \geqslant 0 \end{aligned}$$

a dual programming in $(\tilde{P})$, where $\tilde{d}(w) = \prod_{i=0}^{p} \prod_{k=1}^{J_i} \left(\frac{\tilde{c}_{ik}}{w_{ik}} \sum_{k=1}^{J_i} w_{ik}\right)^{w_{ik}}$, $\tilde{c}_{ik} = \tilde{1} \cdot c_{ik}$ are trapezoidal fuzzy numbers, $w = (w_{01}, \cdots, w_{0J_0}, \cdots, w_{p1}, \cdots, w_{pJ_p})^{\mathrm{T}}$ denotes a $J-$dimensional variables vector.

Problem $(\tilde{D})$ is a fuzzy coefficient PGP and when DD $D = 0$ or 1, we have an easy solution to it [Cao87a,b][Cao93a]. Now we only discuss the relation between $(\tilde{P})$ and $(\tilde{D})$, find an answer to a trapezoidal fuzzy optimal value of the primal program $(\tilde{P})$ by means of a dual programming, and determine whether $(\tilde{P})$ contains a trapezoidal fuzzy optimal solution.

Theorem 6.4.2 *For any trapezoidal fuzzy feasible solution $\tilde{x}$ in a primal problem $(\tilde{P})$ and for any feasible solution w in a dual problem $(\tilde{D})$, we have*

$$g_0(\tilde{x}) \gtrsim g_0(\tilde{x}) \prod_{i=1}^{p} (g_i(\tilde{x}))^{w_{i0}} \gtrsim \tilde{d}(w)$$

with $g_0(\tilde{x}) = \tilde{d}(w)$, iff

$$w_{ik} = \begin{cases} v_{0k}(\tilde{x})/g_0(\tilde{x}) & (i = 0, 1 \leqslant k \leqslant J_0) \\ w_{i0} v_{ik}(\tilde{x}) & (i \neq 0, 1 \leqslant k \leqslant J_p) \end{cases}$$

holds; $\tilde{x}$ and w denote a trapezoidal fuzzy optimal solution to $(\tilde{P})$ and an optimal solution to $(\tilde{D})$, respectively. ◇

Proof From Theorem 6.3.1, we know $(\tilde{P}) \Longleftrightarrow (P(\mathcal{T}))$. Similarly, we can confirm $(\tilde{D})$ equivalent to

$$(D(\mathcal{T})) \qquad \begin{aligned} &\max d(w(\mathcal{T})) \\ &\text{s.t. } w_{00}(\mathcal{T}) = 1 \\ &\quad \Gamma^{\mathrm{T}} w(\mathcal{T}) = 0 \\ &\quad w(\mathcal{T}) \geqslant 0 \end{aligned}$$

here, $(D(\mathcal{T}))$ denotes an ordinary programming depending on a platform index $\mathcal{T}$, where

$$d(w(\mathcal{T})) = \prod_{i=0}^{p} \prod_{k=1}^{J_i} (c_{ik}/w_{ik}(\mathcal{T}) \sum_{k=1}^{J_i} w_{ik}(\mathcal{T}))^{w_{ik}(\mathcal{T})},$$

and

$$w(\mathcal{T}) = (w_{01}(\mathcal{T}), \ldots, w_{0J_0}(\mathcal{T}), \ldots, w_{p1}(\mathcal{T}), \ldots, w_{pJ_p}(\mathcal{T}))^{\mathrm{T}}$$

is a $J-$dimensional variable vector depending on a platform index $\mathcal{T}$, and we stipulate $(w_{ik}(\mathcal{T}))^{w_{ik}(\mathcal{T})}|_{w_{ik}(\mathcal{T})=0} = 1$.

Under the same given platform index $\mathcal{T}$, $(P(\mathcal{T}))$ is mutually dual with $(D(\mathcal{T}))$ about the platform index $\mathcal{T}$. From Lemma 1.5.3 in Reference [WY82], any feasible solution of $x(\mathcal{T})$ and $w(\mathcal{T})$ in $(P(\mathcal{T}))$ and $(D(\mathcal{T}))$ contains

$$g_0(x(\mathcal{T})) \geqslant g_0(x(\mathcal{T})) \prod_{i=1}^{p} (g_i(x(\mathcal{T}))^{w_{i0}} \geqslant d((\mathcal{T}))$$

with $g_0(x(\mathcal{T})) = d(w(\mathcal{T})) \Longleftrightarrow$

$$w_{ik}(\mathcal{T}) = \begin{cases} v_{0k}(x(\mathcal{T}))/g_0(x(\mathcal{T})) & (i = 0, 1 \leqslant k \leqslant J_0) \\ w_{i0}(\mathcal{T})v_{ik}(x(\mathcal{T})) & (i \neq 0, 1 \leqslant k \leqslant J_i) \end{cases}$$

holding, and $x(\mathcal{T})$ and $w(\mathcal{T})$ denoting optimal solutions to $(P(\mathcal{T}))$ and $(D(\mathcal{T}))$, respectively.

Again, by the equivalence of $(\tilde{P})$ and $(P(\mathcal{T}))$ as well as $(\tilde{D})$ and $(D(\mathcal{T}))$, we know the equivalence of $(\tilde{P})$ and $(\tilde{D})$. Therefore, the theorem holds. □

We consider the dual problem $(\tilde{D})$ as for program $(\tilde{P})$ with $T-$FV instead of finding an answer to the latter directly, which greatly simplifies the operation.

Obviously, as for $(\tilde{P})$, when the mean value of trapezoidal FV $\tilde{x} = (x^-, x^+, \underline{\xi}, \overline{\xi})$ denotes decision variables and $\underline{\xi}, \overline{\xi}$ denote constants, we can find a solution by the method. But this will increase a difficulty in solution since the number of constraints doubles. In practice, when the number doubles in solution to the programming, we can certainly find their numerical solutions by computer. But we may try regarding them as constants by relaxing the spread condition for the variables (resp. data) in order to decrease the calculation by half.

Theorem 6.4.3 (first fuzzy dual theorem) *Let $(\tilde{P})$ be fuzzy super-consistent having a trapezoidal fuzzy optimal solution $\tilde{x}^*$. Then, there must exist a Lagrange multiplier $\mu^* = (\mu_1^*, \mu_2^*, \cdots, \mu_p^*)^T \geqslant 0$, such that equalities with trapezoidal FV*

$$\nabla g_0(\tilde{x}^*) + \sum_{i=1}^{p} \mu_i^* \nabla g_i(\tilde{x}^*) = \tilde{0}, \tag{6.23}$$

$$\mu_i^*(g_i(\tilde{x}^*) - \tilde{1}) = \tilde{0} \quad (1 \leqslant i \leqslant p) \tag{6.24}$$

hold, where "∇" represents a gradient, while w^* defined by

$$w_{ik} = \begin{cases} v_{0k}(\tilde{x}^*)/g_0(\tilde{x}^*), & (i = 0; 1 \leqslant k \leqslant J_0) \\ \mu_i^* v_{ik}(\tilde{x}^*)/g_0(\tilde{x}^*), & (i \neq 0; 1 \leqslant k \leqslant J_p) \end{cases} \tag{6.25}$$

is an optimal solution to a dual program $(\tilde{D})$, with

$$g_0(\tilde{x}^*) = \tilde{d}(w^*). \tag{6.26}$$

◊

Proof Because $(\tilde{P})$ is fuzzy super-consistent $\Longleftrightarrow$ $(P(\mathcal{T}))$ is super-consistent, (6.23)–(6.26) is equivalent to [Cao93c][WY82]

$$\nabla g_0(z^*(\mathcal{T})) + \sum_{i=1}^{p} \mu_i^* \nabla g_i(z^*(\mathcal{T})) = 0, \tag{6.27}$$

$$\mu_i^*(g_i(z^*(\mathcal{T})) - 1) = 0 \ (1 \leqslant i \leqslant p), \tag{6.28}$$

$$w_{ik}^*(\mathcal{T}) = \begin{cases} v_{0k}(z^*(\mathcal{T}))/g_0(z^*(\mathcal{T})), & (i=0; 1 \leqslant k \leqslant J_0) \\ \mu_i^* v_{ik}(z^*(\mathcal{T}))/g_0(z^*(\mathcal{T})), & (i \neq 0; 1 \leqslant k \leqslant J_p), \end{cases} \tag{6.29}$$

$$g_0(z^*(\mathcal{T})) = d(w^*(\mathcal{T})). \tag{6.30}$$

From Theorem 1.6.3 in Reference [WY82], we know a Lagrange multiplier μ^* must exist when there is an optimal solution, depending on the platform index $\mathcal{T}$, to $(D(\mathcal{T}))$ super-consistence, such that (6.27)–(6.30) hold with $w_{ik}^*(\mathcal{T})$ being an optimal solution dependent upon platform index $\mathcal{T}$. Therefore the theorem holds. □

When does an optimum trapezoidal fuzzy solution exist in $(\tilde{P})$? Now we have the following:

Theorem 6.4.4 (second fuzzy dual theorem) *Let* $(\tilde{P})$ *be fuzzy consistent and* $(\tilde{D})$ *have a feasible solution with components being positive. Then* $(\tilde{P})$ *has an optimal trapezoidal fuzzy solution.* ◊

Proof From the condition of the theorem, we know $(P(\mathcal{T}))$ is consistent and $(D(\mathcal{T}))$ contains a feasible solution with components being positive depending on platform index $\mathcal{T}$. From the second dual theorem (Theorem 1.8.1) in the ordinary PGP [WY82], we know $(P(\mathcal{T}))$ has an optimum solution. For a determinate platform index $\mathcal{T}$, $(\tilde{P})$ can be turned into $(P(\mathcal{T}))$. Conversely, $(P(\mathcal{T}))$ returns to $(\tilde{P})$. Therefore, we know $(\tilde{P})$ has an optimal trapezoidal fuzzy solution. □

6.4.4 Illustrative Examples

Now, variables $\tilde{x}_l = (x_l^-, x_l^+, \underline{\xi}_l, \overline{\xi}_l)$ are divided into two cases:

I. Variable type, where the mean value and the spreads are all decision variables.

II. Datum type, where the mean value and the spreads are all real numbers.

Example 6.3 Find

$$\begin{aligned} \min\ g_0(\tilde{x}) &= \frac{1}{\sqrt{\tilde{x}_1 - 1}} \frac{1}{\sqrt[3]{\tilde{x}_2 - \frac{1}{4}}} \\ \text{s.t.}\ (\tilde{x}_1 - 1)(\tilde{x}_2 - \tfrac{1}{4}) &\leqslant \tilde{1} \\ 0 < \tilde{x}_1 &\leqslant \tilde{2} \\ 0 < \tilde{x}_2 &\leqslant \tilde{\tfrac{5}{4}} \end{aligned} \tag{6.31}$$

where $\tilde{1} = (1,1,0,0), \tilde{2} = (2,2,0,0), \tilde{\frac{5}{4}} = (\frac{5}{4}, \frac{5}{4}, 0, 0)$ are special trapezoidal fuzzy numbers.

i) Take $\tilde{x}_1$ and $\tilde{x}_2$ trapezoidal fuzzy data as:

$\tilde{x}_1$:
1) $(x_1^-, x_1^+, 1, 0)$; 3) $(x_1^-, x_1^+, 2, 1)$; 5) $(x_1^-, x_1^+, 1.5, 1)$;
7) $(x_1^-, x_1^+, \frac{1}{2}, \frac{1}{2})$; 9) $(x_1^-, x_1^+, 2, 2)$; 11) $(x_1^-, x_1^+, 1, 2)$.

$\tilde{x}_2$:
2) $(x_2^-, x_2^+, 1.5, 1.2)$; 4) $(x_2^-, x_2^+, 0, 1)$; 6) $(x_2^-, x_2^+, 2, 1)$;
8) $(x_2^-, x_2^+ 1, 1.5)$; 10) $(x_2^-, x_2^+, 0, 0)$; 12) $(x_2^-, x_2^+, 2, 1)$.

ii) Because $\tilde{x}_1, \tilde{x}_2$ may be freely fixed in a closed value intervals $[1, 2], [\frac{1}{4}, 1\frac{1}{4}]$ respectively, for interval $[m, M]$, from formula (1.14) in Section 1.6.3 in Chapter 1, where x_L and x_U are left and right interval endpoints, respectively.

Let $n = 1$, and then

$$\begin{cases} \psi_1(x_1) = \dfrac{x_1 - 1}{2 - 1} \geqslant \alpha_1 \\ \psi_2(x_2) = \dfrac{x_2 - \frac{1}{4}}{1\frac{1}{4} - \frac{1}{4}} \geqslant \alpha_2 \end{cases} \Rightarrow \begin{cases} x_1 \geqslant \alpha_1 + 1 \\ x_2 \geqslant \alpha_2 + \frac{1}{4} \end{cases}$$

which is equivalent to $\tilde{x}_1 : x_1 = \alpha_1 + 1; \tilde{x}_2 : x_2 = \alpha_2 + \frac{1}{4}, \alpha_1, \alpha_2 \in [0, 1]$. So the data of twelve groups mentioned above are as follows:

1) $(\alpha_1 + 1, 1, 0)$; 3) $(\alpha_1 + 1, 2, 1)$; 5) $(\alpha_1 + 1, 1.5, 1)$;
7) $(\alpha_1 + 1, \frac{1}{2}, \frac{1}{2})$; 9) $(\alpha_1 + 1, 2, 2)$; 11) $(\alpha_1 + 1, 1, 2)$.

2) $(\alpha_2 + \frac{1}{4}, 1.5, 1.2)$; 4) $(\alpha_2 + \frac{1}{4}, 0, 1)$; 6) $(\alpha_2 + \frac{1}{4}, 2, 1)$;
8) $(\alpha_2 + \frac{1}{4}, 1, 1.5)$; 10) $(\alpha_2 + \frac{1}{4}, 0, 0)$; 12) $(\alpha_2 + \frac{1}{4}, 2, 1)$.

iii) Partition them into three groups I, II, and III by applying the proof in Theorem 6.3.1 in Section 6.3.

I. Number 1,4,7 and 10, whose data correspond to

$$\alpha_1 + \frac{3}{2},\ \alpha_2 + \frac{3}{4},\ \alpha_1 + 1\frac{1}{2},\ \alpha_2 + \frac{1}{4}$$

II. Number 2,5,8 and 11, whose data correspond to

$$\alpha_2 - \frac{5}{4},\ \alpha_1 + 2,\ \alpha_2 - \frac{3}{4},\ \alpha_1 + 3$$

III. Number 3,6,9 and 12, whose data correspond to

$$\alpha_1 - 1,\ \alpha_2 + \frac{5}{4},\ \alpha_1 - 1,\ \alpha_2 + \frac{5}{4}$$

Add their factors with $\alpha_j (j = 1, 2)$, such that

$$\alpha_1 + \frac{3}{2} + \alpha_1 + \frac{3}{2} + \alpha_1 + 2 + \alpha_1 + 3 + \alpha_1 - 1 + \alpha_1 - 1 = 6\alpha_1 + 6$$
$$\alpha_2 + \frac{3}{4} + \alpha_2 + \frac{1}{4} + \alpha_2 - \frac{5}{4} + \alpha_2 - \frac{3}{4} + \alpha_2 + \frac{5}{4} + \alpha_2 + \frac{5}{4} = 6\alpha_2 + \frac{6}{4}$$

iv) Substitute $\alpha_1 + 1, \alpha_2 + \frac{1}{4}$ for $\tilde{x}_1$ and $\tilde{x}_2$ in (6.31), respectively, and then change formula (6.31) into an equivalent problem

$$\begin{aligned} \min\ & g_0(\alpha) = \alpha_1^{-\frac{1}{2}} \alpha_2^{-\frac{1}{3}} \\ \text{s.t.}\ & \alpha_1 \alpha_2 \leqslant 1 \\ & 0 < \alpha_1 \leqslant 1 \\ & 0 < \alpha_2 \leqslant 1 \end{aligned} \tag{6.32}$$

An optimal solution in (6.23) is $\alpha = (1, 1)^{\mathrm{T}}$, an optimal value $g_0(\alpha) = 1$, and the dual problem in (6.32) is

$$\begin{aligned} \max\ & (\frac{1}{w_0})^{w_0} (\frac{1}{w_1})^{w_1} (\frac{1}{w_2})^{w_2} (\frac{1}{w_3})^{w_3} w_1^{w_1} w_2^{w_2} w_3^{w_3} \\ \text{s.t.}\ & w_0 = 1 \\ & -\frac{1}{2} w_0 + w_1 + w_2 = 0 \\ & -\frac{1}{3} w_0 + w_1 + w_3 = 0 \\ & w_0, w_1, w_2, w_3 \geqslant 0 \end{aligned}$$

Its optimum solution is $w = (1, \frac{1}{3}, \frac{1}{6}, 0)^{\mathrm{T}}$, optimal value d is also 1, such that we can get an approximately optimal solution of the prime problem which is $x = (2, 1\frac{1}{4})^{\mathrm{T}}$, and its optimal value which is 1.

6.4.5 Extension of Models

Similar to Reference [WY82], it can be proved that any signomial GP with FV can be changed into a reverse PGP with FV, so we only consider the reverse PGP with the trapezoidal FV.

We call

$$\begin{aligned} & \widetilde{\min}\ g_0(\tilde{x}) \\ \text{s.t.}\ & g_i(\tilde{x}) \leqslant \tilde{1} (1 \leqslant i \leqslant p') \\ & g_i(\tilde{x}) \geqslant \tilde{1} (p' + 1 \leqslant i \leqslant p) \\ & \tilde{x} > \tilde{0} \end{aligned}$$

a primal reverse PGP with the trapezoidal FV, where $\tilde{x} = (\tilde{x}_1, \tilde{x}_2, \ldots, \tilde{x}_m)^{\mathrm{T}}$ is an $m-$dimensional trapezoidal vector of FV, all of

$$g_i(\tilde{x}) = \sum_{i=1}^{J_i} v_{ik}(\tilde{x})(0 \leqslant i \leqslant p)$$

is a posynomial with the trapezoidal FV, where

$$v_{ik}(\tilde{x}) = \begin{cases} c_{ik} \prod\limits_{l=1}^{m} \tilde{x}_l^{\gamma_{ikl}}, & (1 \leqslant k \leqslant J_i; 0 \leqslant i \leqslant p'), \\ c_{ik} \prod\limits_{l=1}^{m} \tilde{x}_l^{-\gamma_{ikl}}, & (1 \leqslant k \leqslant J_i; p'+1 \leqslant i \leqslant p). \end{cases}$$

Its dual programming represents

$$\max \tilde{d}(w) = \left(\frac{\tilde{a}_{0k}}{w_{00}}\right)^{w_{00}} \prod_{i=0}^{p'} \prod_{k=1}^{J_i} \left(\frac{\tilde{c}_{ik}}{a_{ik} w_{ik}}\right)^{w_{ik}} \prod_{i=p'+1}^{p} \prod_{k=1}^{J_i} \left(\frac{\tilde{c}_{ik} a_{ik}}{w_{ik}}\right)^{-w_{ik}}$$

$$\prod_{i=1}^{p'} w_{i0}^{w_{i0}} \prod_{i=p'+1}^{p} w_{i0}^{-w_{i0}}$$

$$\text{s.t.} \quad w_{00} = \sum_{k=1}^{J_0} w_{0k} = 1$$

$$\Gamma^{\mathrm{T}} w = 0$$

$$w \geqslant 0$$

where $w = (w_{00}, w_{01}, \ldots, w_{0J_0}, w_{p1}, \ldots, w_{pJ_p})^{\mathrm{T}}$ is a $J'-$dimensional vector ($J' = 1 + J_0 + \cdots + J_p$), with $w_{i0} = w_{i1} + \cdots + w_{iJ_i}$,

$$\Gamma = (\gamma_{ikl})(1 \leqslant k \leqslant J_i; 1 \leqslant i \leqslant p; 1 \leqslant l \leqslant m)$$

an exponent matrix, $-\gamma_{ikl}$ its exponent of each-reversed-direction inequality $g_i(\tilde{x}) \geqslant \tilde{1}$ corresponding to x_l, and $-w_{ik}$ and w_{i0} the inequality exponents corresponding to each factor $(c_{ik}/w_{ik})^{-w_{ik}}$ and $(w_{i0})^{-w_{i0}}$ corresponding to inequality $g_i(\tilde{x}) \geqslant \tilde{1}$.

6.4.6 Concluding Remarks

As for the PGP with type I variables, it is very difficult to find its solution directly. But, it is much easier to find an answer to the PGP with type II variables.

Therefore, the outcome in this section is easily generalized into the reverse PGP with type II trapezoidal fuzzy data in (6.9). Next section will give a survey of the reverse PGP with general fuzzy variables in (6.9).

6.5 Dual Method of Programming with Fuzzy Variables

6.5.1 Introduction

We shall advance a brief solution method to a programming with general fuzzy variables by a fuzzy dual theorem.

Consider a general programming with FV

$$\begin{aligned} &\min\ f_0(\tilde{x}) \\ &\text{s.t. } f_i(\tilde{x}) \leqslant \tilde{b}_i (1 \leqslant i \leqslant p) \\ &\quad \tilde{x} > 0 \end{aligned} \tag{6.33}$$

where $\tilde{x}$ serves as kinds of FV. (6.33) is an LP with FV when $f_i(\tilde{x}) = \sum\limits_{l=1}^{m} a_{il}\tilde{x}_l (0 \leqslant i \leqslant p)$ are linear functions. (6.33) is called a GP with FV when $f_i(\tilde{x})$ denotes polynomials of $\sum\limits_{k=1}^{J_i} c_{ik} \prod\limits_{l=1}^{m} \tilde{x}_l^{r_{ikl}} (0 \leqslant i \leqslant p)$.

We advance general method for such a program (6.33) with various FV, that is a dual method.[Cha83]

Program (6.33) with FV is turned into a dual one with fuzzy coefficients before its solution is found by the method mentioned in Chapter 5. Next follows an introduction for LP and GP with FV, respectively.

6.5.2 LP with FV

Definition 6.5.1 We have the problem

$$\begin{aligned} &\min\ \tilde{z} = c\tilde{x} \\ &\text{s.t. } A\tilde{x} \geqslant \tilde{b} \\ &\quad \tilde{x} \geqslant 0 \end{aligned} \tag{6.34}$$

called an LP problem with FV [AMA93], where $\tilde{x} = (\tilde{x}_1, \tilde{x}_2, \dots, \tilde{x}_m)^{\mathrm{T}}$ is an $m-$dimensional FV vector, $0 \leqslant c \in \Re^m$ and $A \in \Re^{n \times m}$ represent $1 \times m$ and $n \times m$ matrices, respectively, $\tilde{b} \in (\mathcal{F}(\Re))^n$, $\tilde{x} \in (\mathcal{F}(\Re))^m$ is a fuzzy vector.

Consider the problem such as (6.34), then its dual problem is denoted by

$$\begin{aligned} &\max\ \tilde{Z} = y\tilde{b} \\ &\text{s.t. } yA \leqslant c \\ &\quad y \geqslant 0 \end{aligned} \tag{6.35}$$

where $c \in \Re^m$, $A \in \Re^{n \times m}$, $y \in \Re^n$, $\tilde{b} \in (\mathcal{F}(\Re))^n$.

$\tilde{x}$ is said to be a fuzzy feasible solution to (6.34) iff $\tilde{x}$ satisfies the constraints of the problem. By an optimal fuzzy solution to (6.34) we mean a fuzzy feasible solution, say $\tilde{x}^0$, such that we have $c\tilde{x}^0 \leqslant c\tilde{x}$ for all $\tilde{x}$ belonging to the set of all fuzzy feasible solutions to (6.34).

In order to solve problem (6.34), we shall find an optimal solution to problem (6.35). However (6.35) is, in fact, an LP problem with fuzzy coefficient, and we already know how to solve this. It follows that we shall discuss the relationships between the primary and dual problems.

Lemma 6.5.1 *If $\tilde{x}$ is any fuzzy feasible solution to (6.34) and y is any feasible one to (6.35), then $y\tilde{b} \leqslant c\tilde{x}$.* ◊

Proof Straightforward. □

Lemma 6.5.2 *If $\tilde{x}^0$ is a fuzzy feasible solution to (6.34) and y^0 is a feasible one to (6.35), such that $y^0\tilde{b} = c\tilde{x}^0$, then y^0 is an optimal solution to (6.35) and $\tilde{x}^0$ is a fuzzy optimal one to (6.34).* ◊

Proof Straightforward. □

Theorem 6.5.1 *If the dual problem (6.35) has an optimal solution, then problem (6.34) has a fuzzy optimal solution.* ◊

Proof We first transform (6.34) into the form

$$\begin{aligned} \max\ & \tilde{Z} = y\tilde{b} \\ \text{s.t. } & Ay - Iy_s = c \\ & y, y_s \geqslant 0 \end{aligned}$$

where $\tilde{Z} = (\tilde{Z}_1, \tilde{Z}_2, \ldots, \tilde{Z}_n), y \in \Re^n$; A is an $n \times m$ matrix; $\tilde{b} = (\tilde{b}_1, \tilde{b}_2, \ldots, \tilde{b}_n) \in (\mathcal{F}(\Re))^n$; $c \in \Re^m$ and I is a unit matrix.

Let $A' = (A, -I), y' = (y, y_s)^{\mathrm{T}}, c' = (c, 0)^{\mathrm{T}}$. The formula above is simplified as follows:

$$\begin{aligned} \min\ & \tilde{Z}' = y'\tilde{b} \\ \text{s.t. } & A'y' = c' \\ & y' \geqslant 0 \end{aligned} \tag{6.36}$$

Let y'_B be an optimal basic solution to (6.36), such that $\tilde{Z}_j - \tilde{b}_j \geqslant 0$ for all j; thus, $\tilde{b}_B B^{-1} A \geqslant \tilde{b}$, where B is a basic matrix.

If we write $\tilde{x} = \tilde{b}_B B^{-1}$, we can see that $\tilde{x}$ is a fuzzy feasible solution to (6.34). On the other hand, we have

$$\tilde{z} = c\tilde{x} = \tilde{b}_B B^{-1} c = y_B \tilde{b}_B = \tilde{Z}.$$

Hence, $\tilde{x}$ is an optimal solution to (6.33). □

Lemma 6.5.3 *If problem (6.34) has an unbounded solution, then problem (6.33) has no fuzzy feasible solution.* ◊

Proof Straightforward. □

We conclude that, in order to solve an LP problem with FV, it is sufficient to solve its dual problem by the method discussed in Chapter 5. We can then obtain the fuzzy optimal solution to our problem by using the theorem and lemmas of this section.

6.5.3 Dual Method of GP with FV

Let a GP with FV be

$$\begin{aligned} &\min\ g_0(\tilde{x}) \\ &\text{s.t. } g_i(\tilde{x}) \leqslant \tilde{1}(1 \leqslant i \leqslant p) \\ &\quad\ \tilde{x} > 0 \end{aligned} \tag{6.37}$$

where $g_i(\tilde{x})(0 \leqslant i \leqslant p)$ is posynomial with FV, $\tilde{x}$ is an $m-$dimensional FV vector, $\tilde{1}$ is a fuzzy number. Then its dual programming is

$$(\tilde{D}) \qquad \begin{aligned} &\max \prod_{i=0}^{p} \prod_{k=1}^{J_i} \left(\frac{\tilde{c}_{ik}}{w_{ik}}\right)^{w_{ik}} \prod_{i=1}^{p} w_{i0}^{w_{i0}} \\ &\text{s.t. } w_{00} = 1 \\ &\quad\ \Gamma^{\mathrm{T}} w = 0 \\ &\quad\ w \geqslant 0 \end{aligned} \tag{6.38}$$

where $\tilde{c}_{ik} = c_{ik} \cdot \tilde{1}(1 \leqslant i \leqslant p)$ are fuzzy numbers, Γ an exponent matrix; w a $J-$dimensional variables vector.

Obviously, (6.37) can be changed into a dual program (6.38) with fuzzy coefficients, while (6.38) is easier to be found than (6.37) since its variables are nonfuzzy and its optimal solution can be obtained by methods mentioned in the previous chapter.

Example 6.4 Find

$$\begin{aligned} &\min\ g_0(\tilde{x}) = 20\tilde{x}_1\tilde{x}_3 + 40\tilde{x}_2\tilde{x}_3 + 80\tilde{x}_1\tilde{x}_2 \\ &\text{s.t. } g_0(\tilde{x}) = 8\tilde{x}_1^{-1}\tilde{x}_2^{-1}\tilde{x}_3^{-1} \leqslant \tilde{1} \\ &\quad\ \tilde{x} > \tilde{0} \end{aligned}$$

where $\tilde{x}_i$ are special trapezoidal FV and $\tilde{1} = (1, 1, 0, 0)$ is a special trapezoidal fuzzy number.

Its dual programming is

$$\begin{aligned}\max \tilde{d}(w) &= \left(\frac{20}{w_{01}}\right)^{w_{01}} \left(\frac{40}{w_{02}}\right)^{w_{02}} \left(\frac{80}{w_{03}}\right)^{w_{03}} \left(\frac{\tilde{8}}{w_{11}}\right)^{w_{11}} w_{11}^{w_{11}} \\ \text{s.t. } & w_{01} + w_{02} + w_{03} = 1 \\ & w_{01} + w_{03} - w_{11} = 0 \\ & w_{02} + w_{03} - w_{11} = 0 \\ & w_{01} + w_{02} + w_{03} - w_{11} = 0 \\ & w \geqslant 0\end{aligned}$$

where $w = (w_{01}, w_{02}, w_{03}, w_{11})^{\mathrm{T}}$ is a 4–dimensional vector, $\tilde{8} = 8 \cdot \tilde{1}$ is a trapezoidal fuzzy number.

Now, the equivalence of (6.37) and (6.38) is testified. Next we discuss what condition is needed for the existence of a fuzzy optimal solution in (6.37).

Lemma 6.5.4 (basic lemma) *For any fuzzy feasible solution $\tilde{x}$ in a primal PGP (6.37) with FV, and any feasible one w in a dual program (6.38) with fuzzy coefficients, we have*

$$g_0(\tilde{x}) \geqslant g_0(\tilde{x}) \prod_{i=1}^{p} (g_i(\tilde{x}))^{w_{i0}} \geqslant \tilde{d}(w)$$

and $g_0(\tilde{x}) = \tilde{d}(w) \Longleftrightarrow$

$$w_{ik} = \begin{cases} v_{0k}(\tilde{x})/g_0(\tilde{x}), & (i = 0; 1 \leqslant k \leqslant J_0) \\ w_{i0} v_{ik}(\tilde{x}), & (i \neq 0; 1 \leqslant k \leqslant J_i) \end{cases}$$

holds, such that $\tilde{x}$ and w denote a fuzzy optimal solution to (6.37) and an optimal solution to (6.38), respectively. ◊

Theorem 6.5.2 (the first fuzzy dual theorem) *Let the primal PGP (6.37) with FV be fuzzy super-consistent, having fuzzy optimal solution $\tilde{x}^*$. Then there must exist a Lagrange multiplier $\mu^* = (\mu_1^*, \mu_2^*, ..., \mu_p^*)^T \geqslant 0$, such that*

$$\begin{aligned}&\nabla g_0(\tilde{x}^*) + \sum_{i=1}^{p} \mu_i^* \nabla g_i(\tilde{x}^*) = \tilde{0}, \\ &\mu_i^*(g_i(\tilde{x}^*) - \tilde{1}) = \tilde{0} \quad (1 \leqslant i \leqslant p),\end{aligned}$$

while w^* defined by

$$w_{ik}^* = \begin{cases} v_{0k}(\tilde{x}^*)/g_0(\tilde{x}^*) & (i = 0; 1 \leqslant k \leqslant J_0) \\ \mu_i^* v_{ik}(\tilde{x}^*)/g_0(\tilde{x}^*) & (i \neq 0; 1 \leqslant k \leqslant J_i) \end{cases}$$

is an optimal one of a dual program (6.38), with

$$g_0(\tilde{x}^*) = \tilde{d}(w^*).$$

◊

Theorem 6.5.3 (second fuzzy dual theorem) *Let the primal PGP (6.37) be deduced from FV. If (6.37) is fuzzy compatible and dual problem (6.38) has a feasible solution with components being positive, then (6.37) has a fuzzy optimal solution.* ◊

Lemma 6.5.4, Theorem 6.5.2 and Theorem 6.5.3 are proved to be true without difficulty in the same way as Sections 6.3, 6.4 in Chapter 6 and Lemma 4.2.1, Theorem 4.2.4 and 4.2.5 in Chapter 4.

If fuzzy compatible (6.37) in Theorem 6.5.3 is substituted for a finite number $M_{\tilde{D}}$, the result of Theorem 6.5.3 still holds.

Overall, Theorem 6.5.2 gives a method for an optimal solution in (6.37) in a dual way when (6.37) has a fuzzy optimal solution. Theorem 6.5.3 gives a sufficient condition to determine whether there exists a fuzzy optimal solution in (6.37).

6.5.4 Conclusions

This section gives methods to find a programming with fuzzy variables and to determine an optimal solution. We can acquire an analytic solution to the primal problem by finding its dual problem as long as a solution exists in the primal programming problem.

6.6 Genetic Algorithm Based on Fuzzy Imitation

Similar to an algorithm advanced by reference [LZ98], we design a genetic algorithm based on a fuzzy imitation and find an answer to a GP problem (6.33) with FV. Since this algorithm can be programmed, we introduce only algorithm procedures instead of expounding it in detail.

Now we use an $m-$dimensional FV $V = (\tilde{x}_1, \tilde{x}_2, \ldots, \tilde{x}_m)$ as a chromosome, representing a fuzzy solution to (6.33), where component $\tilde{x}_l$ is derived from reference class $X_l(1 \leqslant l \leqslant m)$ composed of fuzzy sets.

6.6.1 Initialization process

As for each fuzzy gene $\tilde{x}_l$, a fuzzy set taken stochastically from X_l forms a fuzzy chromosome $V = (\tilde{x}_1, \tilde{x}_2, \ldots, \tilde{x}_m)$. If the latter proves not to be feasible by fuzzy imitation, another fuzzy set is re-generated till a feasible chromosome appears. Repeat the process *pop-size* times and we can obtain *pop-size* initial feasible chromosomes $V_1, V_2, \ldots, V_{\text{pop-size}}$.

6.6.2 Evaluation function

Let a parameter $a \in (0,1)$ in a genetic system be given. Then we can define the rank-based evaluation function as follows:

$$\text{eval }(V_l) = a(1-a)^{l-1}(l = 1, 2, \ldots, \textit{pop-size})$$

where $l = 1$ means the best chromosome; l=*pop-size* implies the worst one.

6.6.3 Selection process

1^0 Calculate the cumulative probability q_l for each fuzzy chromosome V_l,

$$q_0 = 0,$$
$$q_l = \sum_{j=1}^{l'} \text{eval }(V_j)\ (l = 1, 2, \ldots, \textit{pop-size}).$$

2^0 A random real number r is produced in an interval $(0, q_{\text{pop-size}}]$.
3^0 Choose a chromosome $V_l(1 \leqslant l \leqslant$ *pop-size*), such that

$$q_{l-1} < r \leqslant q_l.$$

4^0 Repeat 2^0 and 3^0 *pop-size* times, and we obtain *pop-size* copies of fuzzy chromosomes.

6.6.4 Crossover operation

A parameter P_c of a genetic system is defined as a probability of a crossover operation. Random real numbers r are produced in interval $[0,1]$ by repeating the process from $l = 1$ to *pop-size*. V_l is selected as a parent if $r < P_c$. The selected parents are denoted as $V_1', V_2', V_3', \cdots$ and divided into the pairs

$$(V_1', V_2'), (V_3', V_4'), (V_5', V_6'), \ldots.$$

Let us generate a random integer l' from 1 to l, as a crossover point before two children are obtained, by exchanging all of the genes after l'th genes of fuzzy chromosomes V_1' and V_2'. If both children are feasible solutions to the GP with FV, then we replace their parents by feasible children. Otherwise, we redo the crossover operation.

6.6.5 Mutation operation

Define parameter P_n of a genetic system as the probability of mutation. Similar to the process of crossover operation, we repeat the following steps from $l = 1$ to *pop-size*: generating a random real number r from the interval $[0,1]$, the chromosome V_l is chosen as a parent if $r < P_n$. For each above selected parent, denoted by $V = (\tilde{x}_1, \tilde{x}_2, \ldots, \tilde{x}_m)$, we mutate it in the following way. Choose a mutation l' between 1 and l, and draw a new fuzzy set from class $x_{l'}$, composed of fuzzy sets, to replace the l'th gene in V, which constitutes a new chromosome V'. If V' is not a feasible solution to the GP with FV, then we redo the mutation until a feasible chromosome is obtained.

6.6.6 Algorithm process:

1^0 Input parameters *pop-size*, a, P_c and P_n.

2^0 Produce *pop-size* chromosomes by initial process and we can test the feasibility of a chromosome by fuzzy simulation.

3^0 Update the chromosomes by crossover and mutation operations, in which the feasibility of offspring may be checked by fuzzy simulation.

4^0 Calculate all the chromosome objective values by fuzzy simulation.

5^0 Compute the fitness of each chromosome by rank-based evaluation functions on the basis of objective values.

6^0 Choose the chromosomes by the spinning roulette wheel.

7^0 Duplicate Steps $3 - 6$ until the given number of cycles ends.

8^0 Find the best chromosome for an optimum solution.

But there exist 3 disadvantages in a genetic algorithm [WX99]:

a. The algorithm can not guarantee convergence to an optimal solution.

b. The coding is irregular and its representation is incorrect as well.

c. The coding in a sole genetic algorithm can not present overall a constraint of an optimal problem with computing time increasing.

Note 6.2 The best chromosome does not necessarily appear in the last generation, so we shall keep the best ones at the beginning. We substitute the better chromosome for the best prime one if we find it in a new group.

Chapter 7

FUZZY MULTI-OBJECTIVE PROGRAMMING

In this chapter, based on the theory of Zadeh fuzzy sets, a problem is raised in three kinds of multi-objective programming as follows

(i) Models with exponents acting as various fuzzy numbers and those with a function equation possessing flexibility.

(ii) Models with various fuzzy coefficients.

(iii) Models with $T-$FV.

Meanwhile, equivalence form and algorithms are discussed for the three models by means of fuzzy-valued sets, the notion of fuzzy numbers, and $\alpha-$cut sets and a threshold value principle.

7.1 Fuzzy-Valued-Set on Multi-Objective GP

A multi-objective GP is advanced in this section. The theory of fuzzy sets is employed to solve the multi-objective GP problem; Zimmermann algorithm is spread and it is nonfuzzified with fuzzy programming before a solution to it is found by means of a classical GP. At the same time, it emphasizes that the product operator reduces the DD. And finally, a multi-objective GP is generalized.

7.1.1 Multi-objective GP Problem

The multi-objective GP problem is defined as follows:

Definition 7.1.1 We call

$$(P_1) \qquad \begin{aligned} &\min\ (g_0^{(1)}(x), ..., g_0^{(n)}(x)) \\ &\text{s.t. } g_i(x) \leqslant 1\ (1 \leqslant i \leqslant p) \\ &\quad x > 0 \end{aligned}$$

a multi-objective GP problem, where

$$g_0^{(j)}(x) = \sum_{k=1}^{J_0^{(j)}} c_{0k}^{(j)} \prod_{l=1}^{m} x_l^{\gamma_{0kl}^{(j)}} (1 \leqslant j \leqslant n), \tag{7.1}$$

$$g_i(x) = \sum_{k=1}^{J_i} c_{ik} \prod_{l=1}^{m} x_l^{\gamma_{ikl}} (1 \leqslant i \leqslant p); \tag{7.2}$$

here $c_{0k}^{(j)}, c_{ik}$ are positive constants; $\gamma_{0kl}^{(j)}, \gamma_{ikl}$ arbitrary real numbers; $J_0^{(j)}$ the number of terms in the $j-$th objective function; J_i the number of terms in the $i-$th constraint function, and $x = (x_1, x_2, ..., x_m)^{\mathrm{T}}$.

The DD of the program (P_1) is

$$D = \left[\sum_{j=1}^{n} J_0^{(j)} + \sum_{i=1}^{p} J_i - m - 1\right]$$

It is not always possible that the weights are known showing the relative importance of each objective function. In such a situation, the fuzzy mathematical programming property of combining all objective functions is exploited. (P_1) is solved via a fuzzy approach. The central idea behind fuzzy mathematical programming is that the problem is defined as a fuzzy model. The equivalent crisp model is then solved by existing techniques.

7.1.2 Crisp Model

Let $x_{\text{opt}}^{(j)}$ be an optimal solution to

$$\begin{aligned} &\min\ g_0^{(j)}(x)\ (1 \leqslant j \leqslant n) \\ &\text{s.t. } g_i(x) \leqslant 1\ (1 \leqslant i \leqslant p) \\ &x > 0 \end{aligned}$$

Let

$$g_0^{(r)} = \max_{\substack{j \neq r \\ (1 \leqslant j \leqslant n)}} [g_0^{(r)}(x_{\text{opt}}^{(j)})] \tag{7.3}$$

be the fuzzy expectation level of the $r-$th objective function. The fuzzified version of program (P_1) is given by

$$(P_2) \qquad \begin{aligned} &\min\ g_0^{(r)}(x) \lesssim g_0^{(r)} \\ &\text{s.t. } g_i(x) \leqslant 1\ (1 \leqslant i \leqslant p) \\ &x > 0 \end{aligned}$$

Now any alternative x^* gives full satisfaction if $g_0^{(r)}(x^*) \leqslant g_0^{(r)}$, partial satisfaction if $g_0^{(r)} < g_0^{(r)}(x^*) \leqslant g_0^{(r)} + d_0^{(r)}$ and no satisfaction if $g_0^{(r)}(x^*) > g_0^{(r)} + d_0^{(r)}$, where $d_0^{(r)}$ denotes a subjectively chosen constant of admissible violation of the $r-$th constraint.

Define the following membership functions as

$$\tilde{B}_r(g_0^{(r)}(x)) = \begin{cases} 1, & \text{for } g_0^{(r)}(x) \leqslant g_0^{(r)} \\ 1 - [g_0^{(r)}(x) - g_0^{(r)}]/d_0^{(r)} & \text{for } g_0^{(r)} < g_0^{(r)}(x) \leqslant g_0^{(r)} + d_0^{(r)} \\ 0, & \text{for } g_0^{(r)} + d_0^{(r)} < g_0^{(r)}(x) \ (1 \leqslant r \leqslant n). \end{cases} \tag{7.4}$$

The maximum value of each of the membership functions lies in the interval [0,1]. The maximization of the $r-$th membership function is equivalent to the minimization of $[g_0^{(r)}(x) - g_0^{(r)}]/d_0^{(r)}$. Choosing $\alpha > 0$ and an infinitesimally small quantity ε, such that $g_0^{(r)} > \varepsilon d_0^{(r)}$ and $\alpha \geqslant \varepsilon + [g_0^{(r)}(x) - g_0^{(r)}]/d_0^{(r)}$, and by using the minimum operator [Zim78], we get the following crisp model:

$$\begin{aligned} (P_3) \qquad & \min \alpha \\ & \text{s.t. } g_0^{(r)}(x)/(\alpha d_0^{(r)} + g_0^{(r)} - \varepsilon d_0^{(r)}) \leqslant 1 (1 \leqslant r \leqslant n) \\ & \quad g_i(x) \leqslant 1 \ (1 \leqslant i \leqslant p), \alpha > 0 \\ & \quad x > 0 \end{aligned}$$

The quantity ε has been chosen just in order to ensure the positivity of the variable α as required by a GP.

The DD of the program (P_3) is equal to that of program (P_1). Hence the minimum operator does not change the DD.

The membership function value will be highest when $g_0^{(r)}(x)$ is less than or equal to $g_0^{(r)}$ at optimality.

Now by using a product operator [Zim78], we get the following crisp model:

$$\begin{aligned} (P_4) \qquad & \min \prod_{r=1}^{n} [g_0^{(r)}(x) - g_0^{(r)}]/d_0^{(r)} \\ & \text{s.t. } g_i(x) \leqslant 1 \ (1 \leqslant i \leqslant p) \\ & \quad x > 0 \end{aligned}$$

Choose the negligibly small quantities ε_r such that $g_0^{(r)} > \varepsilon_r d_0^{(r)} (1 \leqslant r \leqslant n)$, and extra variables $x_0^{(r)} (1 \leqslant r \leqslant n)$, which are constrained to satisfy $x_0^{(r)} > 0$ and

$$x_0^{(r)} \geqslant \varepsilon_r + [g_0^{(r)}(x) - g_0^{(r)}]/d_0^{(r)}. \tag{7.5}$$

Now we obtain the following program formulation:

$$(P_5) \qquad \min \prod_{r=1}^{n} x_0^{(r)}$$
$$\text{s.t. } g_0^{(r)}(x)/(x_0^{(r)} d_0^{(r)} + g_0^{(r)} - \varepsilon_r d_0^{(r)}) \leqslant 1$$
$$g_i(x) \leqslant 1 \ (1 \leqslant i \leqslant p)$$
$$x > 0, x_0^{(r)} > 0 \ (1 \leqslant r \leqslant n)$$

The DD of the program (P_5) is equal to that of program (P_1) minus $(n-1)$. Thus the product operator reduces the DD considerably; the amount by which the DD is reduced denotes [number of objective functions-1].

Program (P_5) is a complementary GP [AW70] and its solution can be obtained by a condensation technique [AW70][BP76].

Since the character of a crisp model does not depend on mathematical structure of the original problem, the above treatment can be analogously made with regard to the multi-objective complementary GP problem and a signomial GP problem.[DP73]

7.1.3 Numerical Example

In this section, we shall solve the following problem:

Example 7.1

$$\min \ (g_1, g_2)$$
$$\text{s.t. } x_1 + x_2 \leqslant 1$$
$$x_1, x_2 > 0$$

where $g_1(x) = x_1^{-1} x_2^{-2}$ and $g_2(x) = 2x_1^{-2} x_2^{-3}$.

In order to solve the problem above, we shall solve subproblems

$$\min \ g_1(x)$$
$$\text{s.t. } x_1 + x_2 \leqslant 1$$
$$x_1, x_2 > 0$$

and

$$\min \ g_2(x)$$
$$\text{s.t. } x_1 + x_2 \leqslant 1$$
$$x_1, x_2 > 0$$

Their respective solutions are given by $(x_1, x_2) = (\frac{1}{3}, \frac{2}{3})$ and $(x_1, x_2) = (\frac{2}{5}, \frac{3}{5})$. The values of g_1 and g_2 at $(\frac{1}{3}, \frac{2}{3})$ are 6.75 and 60.75, respectively, and the values of g_1 and g_2 at $(\frac{2}{5}, \frac{3}{5})$ are 6.9444 and 57.8703, respectively.

Select $g_0^{(1)} = 6.9444, g_0^{(2)} = 60.75, d_1 = 0.1944, d_2 = 2.8897$, and $\varepsilon = 0.0000001$, and we get the following GP by applying the minimum operator, in which it is observed that any alternative gives a value of g_1 and g_2 respectively lying in the interval $(6.75, 6.9444)$ and $(57.8703, 60.75)$:

$$\begin{aligned}
&\min \alpha \\
&\text{s.t. } x_1^{-1}x_2^{-2}/(0.19444\alpha + 6.9444) \leqslant 1 \\
&\quad 2x_1^{-2}x_2^{-3}/(2.8897\alpha + 60.75) \leqslant 1 \\
&\quad x_1 + x_2 \leqslant 1, \alpha \in [0,1] \\
&\quad x_1, x_2
\end{aligned}$$

Its solution is given by $(\alpha, x_1, x_2) = (\varepsilon, 0.35, 0.65)$ where ε serves as a negligibly small positive quantity. The values of g_1 and g_2 are 6.7625 and 59.4504, respectively, and the value of the membership function is 1.

Now take $\varepsilon_r = 10^{-6}(1 \leqslant r \leqslant 2)$ and, by using a product operator, we get the following programming formulation:

$$\begin{aligned}
&\min y_1 y_2 \\
&\text{s.t. } x_1^{-1}x_2^{-2}/(0.1944y_1 + 6.9444) \leqslant 1 \\
&\quad 2x_1^{-2}x_2^{-3}/(2.8897y_2 + 60.75) \leqslant 1 \\
&\quad x_1 + x_2 \leqslant 1 \\
&\quad x_1, x_2, y_1, y_2 > 0
\end{aligned}$$

Its solution is given by $(y_1, y_2, x_1, x_2) = (\varepsilon_1, \varepsilon_2, 0.4, 0.6)$, where ε_1 and ε_2 mean negligibly small quantities. Hence a compromise solution is given by $(x_1, x_2) = (0.4, 0.6)$. The values of g_1 and g_2 at the compromise solution are 6.94444 and 57.87037, respectively.

7.1.4 General Multi-objective GP

We consider GP where objective and constraint functions are flexible and coefficients as well as exponents are fuzzy numbers:

$$\begin{aligned}
&\widetilde{\min} \sum_{k=1}^{J_0^{(j)}} \tilde{c}_{0k}^{(j)} \prod_{l=1}^{m} x_l^{\tilde{\gamma}_{0kl}^{(j)}} (1 \leqslant j \leqslant n) \\
&\text{s.t. } \sum_{k=1}^{J_i} \tilde{c}_{ik} \prod_{l=1}^{m} \tilde{x}_l^{\tilde{\gamma}_{ikl}} \lesssim \tilde{1} (1 \leqslant i \leqslant p) \\
&\quad x > 0
\end{aligned} \tag{7.6}$$

where, coefficients and exponents are defined as (1.14) in Definition 1.6.3.

Theorem 7.1.1 *(7.6)* $\Longleftrightarrow$

$$\begin{aligned} &\min(-\alpha)\\ &\text{s.t. } g_0^{(j)}(\bar{x}) = \sum_{k=1}^{J_0^{(j)}} \tilde{c}_{0k}^{(j)^{-1}}(\beta)/\tilde{B}_0^{(j)^{-1}}(\alpha)\prod_{l=1}^{m} x_l^{\gamma_{0kl}^{-1}(\beta)} \leqslant 1\\ &g_i(\bar{x}) = \sum_{k=1}^{J_i} \tilde{c}_{ik}^{-1}(\beta)/\tilde{B}_i^{-1}(\alpha)\prod_{l=1}^{m} x_l^{\gamma_{ikl}^{-1}(\beta)} \leqslant 1\\ &\alpha,\beta \in [0,1](1\leqslant j\leqslant n, 1\leqslant i\leqslant p)\\ &x>0 \end{aligned} \tag{7.7}$$

◊

where $\bar{x} = (x,\alpha,\beta)$.

Obviously, (7.6) and (7.7) are the same in the DD, but (7.7) means an ordinary GP with parameters α,β, which can be solved in several ways [Biw92][Cao93a] [Cao92a]. Because of (7.6)$\Longleftrightarrow$(7.7), it is as easy to get the most satisfactory fuzzy solution to (7.6) as the most satisfactory parameter one to (7.7).

Definition 7.1.2 Let $\tilde{A}_j(1\leqslant j\leqslant n), \tilde{F}_i(1\leqslant i\leqslant p)$ be defined on $\Re\subset\mathcal{F}(\Re)$, $\tilde{A}_j$ be a fuzzy point set of an objective function $\tilde{g}_0^{(j)}(x)$ and $\tilde{F}_i$ be a fuzzy feasible solution set of constraint functions $\tilde{g}_i(x)$. If objective function $\tilde{g}_0^{(j)}(x)$ denotes an optimal fuzzy point set $\tilde{A}^* = \bigwedge_{j=1}^{n} \tilde{A}_j$, such that

$$\tilde{z}^*(x) = \tilde{A}^*(x)\wedge\tilde{F} \triangleq \min\{\tilde{A}^*(x),\tilde{F}(x)\}$$

where

$$\tilde{F}(x) = \bigwedge_{i=1}^{p} \tilde{F}_i(x),$$

then we call

$$\begin{aligned} &\max\ \tilde{z}^*(x)\\ &\text{s.t. } x\in\tilde{F}(x) \end{aligned} \tag{7.8}$$

a multi-objective GP problem with respect to z^* in $\tilde{g}_0^{(j)}(x)(1\leqslant j\leqslant n)$; here,

$$\tilde{F}(x) = \{x|g_i(x)\lesssim 1(1\leqslant i\leqslant p), x>0\}.$$

Definition 7.1.3 If

$$\begin{aligned} &\tilde{g}_0^{(j)}(x)(1\leqslant j\leqslant n),\\ &\tilde{g}_i'(x) = \tilde{g}_i(x)-1(1\leqslant i\leqslant p) \end{aligned}$$

denote fuzzy convex (resp. strongly fuzzy convex) functions with respect to $\tilde{A}_j$ and $\tilde{F}_i$, respectively, then we call (7.8) a fuzzy convex (resp. strongly fuzzy convex) programming problem with respect to $\tilde{g}_0^{(j)}(x)$.

Theorem 7.1.2 *Suppose (7.8) is a fuzzy convex programming problem with respect to $\tilde{g}_0^{(j)}(x)$; then a strictly fuzzy local maximum solution x^* is its fuzzy global maximum one.* ◊

Theorem 7.1.3 *If we change the condition of Theorem 7.1.2 into a strongly fuzzy convex programming problem, then any fuzzy local maximum solution to (7.8) denotes its unique fuzzy global maximum one.* ◊

Referring to Theorem 3.3 and 3.6 in Reference [Cao93a], we can confirm Theorem 7.1.2 and 7.1.3. Meanwhile, by terms of a product operator [Ver90], model (7.6) can be converted into

$$\begin{aligned} &\min \prod_{j=1}^{n} g_0^{(j)}(\bar{x}') \\ &\text{s.t. } g_i(\bar{x}') \lesssim 1 \; (1 \leqslant i \leqslant p), \alpha, \beta \in [0,1] \\ &\quad\; x > 0 \end{aligned} \tag{7.9}$$

where $\bar{x}' = (x, \beta)$.

Let $x_0^{(j)} > 0$ and $x_0^{(j)} - g_0^{(j)}(\bar{x}') \geqslant 0 (0 \leqslant j \leqslant n)$. Then, by aid of (1.15) in Section 1.6.3 and (7.4), we have

$$\begin{aligned} \tilde{B}_0(g_0^{(j)}(\bar{x}'), x_0^{(j)}) \geqslant \alpha &\Leftrightarrow g_0^{(j)}(\bar{x}') \leqslant x_0^{(j)} + (x_0^{(j)} - \alpha) d_0^{(j)} \\ \tilde{B}_i(g_i(\bar{x}'), 1) \geqslant \alpha &\Leftrightarrow g_i(\bar{x}') \leqslant 1 + (1-\alpha) d_i, \end{aligned}$$

written as

$$x_0^{(j)} + (x_0^{(j)} - \alpha) d_0^{(j)} = \tilde{B}_0^{-1}(\alpha), \; 1 + (1-\alpha) d_i = \tilde{B}_i^{-1}(\alpha),$$

so problem (7.7) can be turned into

$$\begin{aligned} &\min \prod_{j=1}^{n} x_0^{(j)} \\ &\text{s.t. } g_0^{(j)}(\bar{x}')/\tilde{B}_0^{-1}(\alpha) \leqslant 1 (1 \leqslant j \leqslant n) \\ &\quad\; g_i(\bar{x}')/\tilde{B}_i^{-1}(\alpha) \leqslant 1 (1 \leqslant i \leqslant p), \alpha, \beta \in [0,1] \\ &\quad\; x_0^{(j)}, x > 0 \end{aligned} \tag{7.10}$$

(7.8) means a monomial GP with parameters, such that the conclusion of Theorem 7.1.2 and 7.1.3 holds which fits (7.8).

7.1.5 Conclusions

We can draw conclusions as follows [Ver90]:

(1) The multi-objective GP can be solved by a fuzzy approach without difficulty.

(2) The minimum operator does not reduce the DD.

(3) The product operator reduces the DD by an amount equal to [number of objective functions −1].

(4) This section advances the state-of-the-art of both GP and fuzzy mathematical programming.

Meanwhile, it is pointed out that the result in this section can be generalized into a reverse multi-objective GP without difficulty, at which you might have a try.

7.2 Fuzzy Programming Duality Technique to Multi-Objective GP

7.2.1 Introduction

In this section, a fuzzy programming technique is used to solve a multi-objective GP problem as a vector minimum problem. It is nonfuzzified with fuzzy GP before a solution to it is found by means of dual programming of a classical GP. A fuzzy membership function is defined for the multi-objective GP problem. This fuzzy programming is applied to solve some optimal engineering design problems [BP76] where we minimize cost and/or weight, maximize volume and/or efficiency etc. In general, there exist multi-objective functions in an engineering design. In this case, it is not suitable to use any single-objective programming for an optimal compromise solution, so we apply a fuzzy programming [Zim78][Zim91] in this solution.

7.2.2 Dual Algorithm of Multi-objective GP

Definition 7.2.1 A multi-objective GP problem can be expressed to find $x = (x_1, x_2, \ldots, x_m)^{\mathrm{T}}$, such that it satires

$$\min \ g_0^{(j)}(x) = \sum_{k=1}^{J_0^{(j)}} c_{0k}^{(j)} \prod_{l=1}^{m} x_l^{\gamma_{0kl}^{(j)}} \ (1 \leqslant j \leqslant n) \tag{7.11}$$

$$\text{s.t.} \ g_i(x) = \sum_{k=1}^{J_i} c_{ik} \prod_{l=1}^{m} x_l^{\gamma_{ikl}} \leqslant 1 \ (1 \leqslant i \leqslant p) \tag{7.12}$$

$$x_l > 0 \ (1 \leqslant l \leqslant m) \tag{7.13}$$

where $c_{0k} \geqslant 0$, $\forall \ c_{ik} \geqslant 0$, $\forall i, k$. $\gamma_{0kl}^{(j)}, \gamma_{ikl}$ are real numbers. $J_0^{(j)}$ denotes number of terms exhibited in the $j-$th objective function, J_i is the number of terms in the $i-$th constraint.

Definition 7.2.2 We call

$$\begin{aligned} &\max \ d^{(j)}(w) \\ &\text{s.t.} \ w_{00} = 1 \\ &\quad \Gamma^{(j)} w = 0 \\ &\quad w \geqslant 0 \end{aligned} \tag{7.14}$$

a dual programming of (7.11) – (7.13) for $1 \leqslant j \leqslant n$, where

$$d^{(j)} = \prod_{k=0}^{p} \left(\frac{c_{0k}^{(j)}}{w_{0k}} \right)^{w_{0k}} \prod_{k=1}^{J_i} \left(\frac{c_{ik}}{w_{ik}} \right)^{w_{ik}} \prod_{i=1}^{p} w_{i0}^{w_{i0}},$$

$$\Gamma^{(j)} = \begin{pmatrix} \gamma_{011}^{(j)} & \cdots & \gamma_{01l}^{(j)} & \cdots & \gamma_{01m}^{(j)} \\ \cdots & & \cdots & & \cdots \\ \gamma_{0J_01}^{(j)} & \cdots & \gamma_{0J_0l}^{(j)} & \cdots & \gamma_{0J_0m}^{(j)} \\ \vdots & & \vdots & & \vdots \\ \gamma_{p11} & \cdots & \gamma_{p1l} & \cdots & \gamma_{p1m} \\ \vdots & & \vdots & & \vdots \\ \gamma_{pJ_p1} & \cdots & \gamma_{pJ_pl} & \cdots & \gamma_{pJ_pm} \end{pmatrix} \quad (1 \leqslant j \leqslant n).$$

The problem is considered as a vector-minimum one. It is assumed that we have an optimal compromise solution in it, which can solve by the following steps:

1^0 Pick only the first objective function (i.e., $g_0^{(j)}(x), j = 1$) and solve it as a single-objective GP problem subject to all the constraints (7.12) and (7.13) by using a GP algorithm [BP76]. Then pick only the second objective function (i.e., $g_0^{(j)}(x), j = 2$) and solve it as a single-objective GP problem subject to all the constraints (7.12) and (7.13). Continue the process n times for n different objective functions. Let $x^{(1)}, x^{(2)}, \ldots, x^{(n)}$ be the respective optimum solutions for n different GP problems. It is assumed that at least two of these solutions are different, such that the $n-$th objective function (i.e., $g_0^{(j)}(x), 1 \leqslant j \leqslant n$) has different bounded values. If all the solutions are the same (i.e., $x^{(1)} = x^{(2)} = \cdots = x^{(n)}$), then we stop. Now $x^{(1)}$ is the optimal compromise solution, otherwise, go to Step 2.

2^0 Evaluate all these n objective functions over all these n optimal points: $x^{(1)}, x^{(2)}, \cdots, x^{(n)}$.

It is simple to consider using a fuzzy dual algorithm for DD $d = 0$ or 1 in the programming problem, accordingly, we turn (7.11) – (7.13) into a dual problem (7.14) for a determined j.

3^0 For each objective function $g_0^{(j)}(x)$, find the best value (L_j) and the worst value (U_j) corresponding to the set of j optimal solutions. Let $L_j \leqslant g_0^{(j)}(x) \leqslant U_j (1 \leqslant j \leqslant n)$, where L_j is the best value (minimum value) and U_j is the worst value (maximum value) for the $j-$th objective function $g_0^{(j)}(x)$.

4^0 By associating $g_0^{(j)}(x)$ with the $j-$th objective function, we define a fuzzy membership function $\tilde{A}_0^{(j)}(x) = \tilde{B}_0(g_0^{(j)}(x))$ as:

$$\tilde{A}_0^{(j)}(x) = \begin{cases} 1, & \text{if } g_0^{(j)}(x) \leqslant L_j \\ \dfrac{U_j - g_0^{(j)}(x)}{U_j - L_j}, & \text{if } L_j < g_0^{(j)}(x) < U_j \\ 0, & \text{if } g_0^{(j)}(x) \geqslant U_j \end{cases} \tag{7.15}$$

where $L_j \neq U_j$ $(1 \leqslant j \leqslant n)$.

If $L_j = U_j$, then we define

$$\tilde{A}_0^{(j)}(x) = 1, \text{ for any value of } j. \tag{7.16}$$

Then maximize the membership function $\tilde{A}_0^{(j)}(x)(1 \leqslant j \leqslant n)$ subject to the constraints (7.12) and (7.13). Use the max-min operator [Zim91] to find a crisp model.

5^0 Use a dummy variable x_{m+1} and formulate a single-objective fuzzy GP problem as a crisp model and the fuzzy PGP can be formulated as:

$$\max\ x_{m+1} \text{ or } \min\ x_{m+1}^{-1} \tag{7.17}$$

$$\text{s.t. } g_i(x) \leqslant 1\ (1 \leqslant i \leqslant p) \tag{7.18}$$

$$x_{m+1} \leqslant \frac{U_j - g_0^{(j)}(x)}{U_j - L_j}\ (1 \leqslant j \leqslant n) \tag{7.19}$$

$$x_1, x_2, \ldots, x_{m+1} > 0 \tag{7.20}$$

Inequality (7.19) can be rewritten as:

$$g_{l+j}(x) = \frac{1}{U_j} g_0^{(j)}(x) + (1 - \frac{L_j}{U_j}) x_{m+1} \leqslant 1\ (1 \leqslant j \leqslant n)$$

where $g_i(x)(1 \leqslant i \leqslant p + n)$ are posynomials.

6^0 Solve the fuzzy GP problem by using a GP algorithm and we obtain an optimal compromise solution and an optimal compromise value for problem (7.17) – (7.20) by means of dual or other techniques in Chapter 2, such that all the object functions (7.11) are evaluated at this optimal solution.

7.2.3 Numerical Examples

Example 7.2 Find x_1, x_2 and x_3, such that

$$\min\ g_0^{(1)}(x) = 40x_1^{-1}x_2^{-1/2}x_3^{-1} + 20x_1x_3 + 20x_1x_2x_3$$
$$\min\ g_0^{(2)}(x) = 40x_1^{-1}x_2^{-1}x_3^{-1} + 20x_1^{1/3}x_3^{3/4}$$

$$\text{s.t. } g_1(x) = \tfrac{1}{3}x_1^{-2}x_2^{-2} + \tfrac{4}{3}x_2^{1/2}x_3^{-1} \leqslant 1 \tag{7.21}$$

$$x_1, x_2, x_3 > 0. \tag{7.22}$$

We first minimize $g_0^{(1)}(x)$ subject to the constraints (7.21) and (7.22). The DD of this problem is 1. The problem is solved via dual programming [DPZ67]:

$$\begin{aligned}
\max\ d_1(w) = & \left(\frac{40}{w_{01}}\right)^{w_{01}}\left(\frac{20}{w_{02}}\right)^{w_{02}}\left(\frac{20}{w_{03}}\right)^{w_{03}}\left(\frac{\frac{1}{3}}{w_{11}}\right)^{w_{11}}\left(\frac{\frac{4}{3}}{w_{12}}\right)^{w_{12}} \\
& (w_{11}+w_{12})^{w_{11}+w_{12}} \\
\text{s.t. } & w_{01}+w_{02}+w_{03}=1 \\
& -w_{01}+w_{02}+w_{03}-2w_{11}=0 \\
& -\tfrac{1}{2}w_{01}+w_{03}-2w_{11}+\tfrac{1}{2}w_{12}=0 \\
& -w_{01}+w_{02}+w_{03}-w_{12}=0 \\
& w_{01},w_{02},w_{03},w_{11},w_{12}\geqslant 0
\end{aligned}$$

The dual variables are calculated as

$$w_{01}=\frac{1}{5}, w_{02}=\frac{2}{5}, w_{03}=\frac{2}{5}, w_{11}=\frac{3}{10} \quad \text{and} \quad w_{12}=\frac{3}{5}$$

from the maximizing equation. Maximum value of the dual objective function is $d_1(w)=100$.

By using the values of the dual variables, the primal variables are obtained to be $x_1=1, x_2=1$ and $x_3=2$. Then, we minimize $g_0^{(2)}(x)$ subject to the constraints (7.21) and (7.22). The DD of this problem is zero, which is solved via the dual program:

$$\begin{aligned}
\max\ d_2(w) = & \left(\frac{40}{w_{01}}\right)^{w_{01}}\left(\frac{20}{w_{02}}\right)^{w_{02}}\left(\frac{\frac{1}{3}}{w_{11}}\right)^{w_{11}}\left(\frac{\frac{4}{3}}{w_{12}}\right)^{w_{12}} \\
& (w_{11}+w_{12})^{w_{11}+w_{12}} \\
\text{s.t. } & w_{01}+w_{02}=1 \\
& -w_{01}+\tfrac{1}{3}w_{02}-2w_{11}=0 \\
& -w_{01}-2w_{11}+\tfrac{1}{2}w_{12}=0 \\
& -w_{01}+\tfrac{3}{4}w_{02}-w_{12}=0 \\
& w_{01},w_{02},w_{11},w_{12}\geqslant 0
\end{aligned}$$

Dual variables are calculated as

$$w_{01}=\frac{1}{13}, w_{02}=\frac{12}{13}, w_{11}=\frac{3}{26} \quad \text{and} \quad w_{12}=\frac{8}{13}.$$

The maximum value of the dual objective function is $d_2(w)=40.01468$. By using the values of the dual variables, primal variables are obtained as $x_1=8.94345, x_2=31.90009$ and $x_3=0.04554$. By the two different solutions $x^{(1)}=(1,1,2)$ and $x^{(2)}=(8.94345,31.90009,0.04554)$, the values of L_j and U_j are obtained as follows:

$L_1=100, U_1=285.42318, L_2=40.01468, U_2=53.63585.$

By using methods in Steps 3,4,5 and 6, a single-objective fuzzy GP problem is formulated as:

$$
\begin{aligned}
\min\ & x_4^{-1} \\
\text{s.t. } & g_1(x) = \tfrac{1}{3}x_1^{-2}x_2^{-2} + \tfrac{4}{3}x_2^{1/2}x_3^{-1} \leqslant 1 \\
& g_2(x) = 0.003503(40x_1^{-1}x_2^{-1/2}x_3^{-1} \\
& \qquad +20x_1x_3 + 20x_1x_2x_3 + 185.42318x_4) \leqslant 1 \\
& g_3(x) = 0.018644(40x_1^{-1}x_2^{-1}x_3^{-1} + 20x_1^{1/3}x_3^{3/4} \\
& \qquad +13.62117x_4) \leqslant 1 \\
& x_1, x_2, x_3, x_4 > 0
\end{aligned}
$$

The DD of this fuzzy GP problem is equal to 5. Due to the higher DD of the problem, a computer program [KM73] is used to solve the problem as a special purpose algorithm. The values of the decision variables are obtained as $x_1 = 0.31016, x_2 = 4.45459, x_3 = 3.40946$ and $x_4 = 0.82048$. At this optimal compromise solution, objective functions are calculated as $g^{(1)}(x) = 133.28543$ and $g^{(2)}(x) = 42.45969$.

7.2.4 Conclusion

With a fuzzy programming technique, we can solve a multi-objective GP problem as a vector-minimum problem. A vector-maximum problem can be transformed into a vector-minimum problem. If any of the objective function and/or constraint does not satisfy the properties of a posynomial after the transformation, then we use any of the general purpose non-LP (non-linear programming) algorithms to solve the problem. We can also apply this technique to solve a multi-objective signomial GP problem. However, if a GP problem has either a higher or a negative DD, then we use any of the general purpose non-LP algorithms instead of a GP algorithm.

7.3 Multi-Objective GP with Type $(\cdot, c)$ Fuzzy Coefficients

7.3.1 Multi-objective PGP

Definition 7.3.1 We call a model

$$\begin{aligned}
\min\ & \tilde{g}_0^{(j)}(x) = \sum_{k=1}^{J_0^{(j)}} \tilde{c}_{0k}^{(j)} \prod_{l=1}^{m} x_l^{\gamma_{0kl}^{(j)}}\ (1 \leqslant j \leqslant n) \\
\text{s.t. } & \tilde{g}_i(x) = \sum_{k=1}^{J_i} \tilde{c}_{ik} \prod_{l=1}^{m} x_l^{\gamma_{ikl}} \lesssim \tilde{1}\ (1 \leqslant i \leqslant p) \\
& x > 0
\end{aligned} \tag{7.23}$$

a multi-objective PGP with type $(\cdot, c)$ fuzzy coefficients, where

$$\tilde{c}_{0k}^{(0)} = (a_{0k}^{(0)}, c_{0k}^{(0)}), \tilde{c}_{ik} = (a_{ik}, c_{ik}), \tilde{1} = (1, c^1)$$

($c^1 = 1 \pm \delta$, and δ is a sufficiently small positive number) denote type $(\cdot, c)$ fuzzy numbers.

On the basis of the paper [Cao92a], problem (7.23) is converted into finding answers to n PGP with type $(\cdot, c)$ fuzzy coefficients:

$$\begin{aligned}
\min\ & \tilde{g}_0^{(1)}(x) = \sum_{k=1}^{J_0^{(1)}} \tilde{c}_{0k}^{(1)} \prod_{l=1}^{m} x_l^{\gamma_{0kl}^{(1)}} \\
\text{s.t. } & \tilde{g}_i(x) = \sum_{k=1}^{J_i} \tilde{c}_{ik} \prod_{l=1}^{m} x_l^{\gamma_{ikl}} \lesssim \tilde{1}\ (1 \leqslant i \leqslant p) \\
& x > 0 \\
& \quad\vdots \\
\min\ & \tilde{g}_0^{(n)}(x) = \sum_{k=1}^{J_0^{(n)}} \tilde{c}_{0k}^{(n)} \prod_{l=1}^{m} x_l^{\gamma_{0kl}^{(n)}} \\
\text{s.t. } & \tilde{g}_i(x) = \sum_{k=1}^{J_i} \tilde{c}_{ik} \prod_{l=1}^{m} x_l^{\gamma_{ikl}} \lesssim \tilde{1}\ (1 \leqslant i \leqslant p) \\
& x > 0
\end{aligned} \tag{7.24}$$

Obviously, (7.24) denotes n single objective fuzzy PGP such as (7.23) for $j = 1, 2, \cdots, n$, respectively. They are solved respectively by methods mentioned in Section 5.3 before n optimal solution sets are gained as follows:

$$z_0^{(j)*} = \max\{z_j | z_j = g_0^{(j)}(x, h), g_i(x, h) \leqslant 1, h \in [0, 1], x > 0 (1 \leqslant i \leqslant p)\}, \\ (1 \leqslant j \leqslant n).$$

Now, we give flexible indexes $d_0^{(j)}(>0)$ reflecting each objective function of importance and we construct correspondingly a new fuzzy object set $\tilde{F}_j$, whose membership function is stipulated as follows:

$$\mu_{\tilde{F}_j}(x) = B_0^{(j)}(g_0^{(j)}(x,h)) = \begin{cases} 0, & g_0^{(j)}(x,h) \leqslant z_0^{(j)*} - d_0^{(j)} \\ 1 - \dfrac{(z_0^{(j)*} - g_0^{(j)}(x,h))}{d_0^{(j)}}, & z_0^{(j)*} - d_0^{(j)} \leqslant g_0^{(j)}(x,h) < z_0^{(j)*} \\ 1, & z_0^{(j)*} \leqslant g_0^{(j)}(x,h)(1 \leqslant j \leqslant n) \end{cases}$$

written as $\tilde{F} = \tilde{F}_1 \cap \tilde{F}_2 \cap \cdots \cap \tilde{F}_n$, such that a new parameter PGP problem is deduced as follows:

$$\begin{aligned} &\max\ \alpha \\ &\text{s.t. } 1 - \frac{(z_0^{(j)*} - g_0^{(j)}(x,h))}{d_0^{(j)}} \geqslant \alpha(1 \leqslant j \leqslant n) \\ &\quad g_i^*(x,h) \geqslant 0, \alpha \geqslant 0, h > 0(1 \leqslant i \leqslant p) \\ &\quad x > 0 \end{aligned} \tag{7.25}$$

The most satisfactory parameter solution $(x_1^*, \ldots, x_m^*; \alpha^*, h^*)^{\mathrm{T}}$ can be obtained to problem (7.25) by any method mentioned in Section 5.3, such that the most satisfactory fuzzy solution can be gained to the primal problem (7.23).

7.3.2 Multi-objective reversed PGP

When the exponent in (7.23) is a fuzzy number which was related in the Reference [Cao93a], a reversed GP with type $(\cdot, c)$ fuzzy coefficients is denoted by

$$\begin{aligned} &\min\ \sum_{k=1}^{J_0} \tilde{c}_{0k}^{(j)} \prod_{l=1}^{m} x_l^{\tilde{\gamma}_{0kl}^{(j)}}\ (1 \leqslant j \leqslant n) \\ &\text{s.t. } \sum_{k=1}^{J_i} \tilde{c}_{ik} \prod_{l=1}^{m} x_l^{\tilde{\gamma}_{ikl}} \lesssim \tilde{b}_i\ (1 \leqslant i \leqslant p') \\ &\quad \sum_{k=1}^{J_i} \tilde{c}_{ik} \prod_{l=1}^{m} x_l^{\tilde{\gamma}_{ikl}} \gtrsim \tilde{b}_i\ (p'+1 \leqslant i \leqslant p) \\ &\quad x > 0 \end{aligned}$$

Therefore, the problem is turned into

$$\begin{aligned} &\min\ \tilde{g}_0^{(j)}(x,\beta)\ (1 \leqslant j \leqslant n) \\ &\text{s.t. } \tilde{g}_i(x,\beta) \lesssim \tilde{b}_i (1 \leqslant i \leqslant p') \\ &\quad \tilde{g}_i(x,\beta) \gtrsim \tilde{b}_i\ (p'+1 \leqslant i \leqslant p), \beta \in [0,1] \\ &\quad x > 0 \end{aligned}$$

by the method mentioned in Reference [Cao93a] before we solve it by a mentioned one in this section.

We can transplant the outcome in this chapter into the fuzzy reversed PGP since fuzzy reversed PGP means only turning parts of constraints "$\leqslant$" for fuzzy PGP into "$\geqslant$".

A series of results can be gained on the basis of the discussion above.

i) Any GP with type $(\cdot, c)$ fuzzy coefficients

$$\begin{aligned} \min\ \tilde{g}_0^{(j)}(x) &= \sum_{k=1}^{J_0^{(j)}} \tilde{c}_{0k}^{(j)} \prod_{l=1}^{m} x_l^{\gamma_{0kl}^{(j)}}\ (1 \leqslant j \leqslant n) \\ \text{s.t. } \tilde{g}_i(x) &= \sum_{k=1}^{J_i} \tilde{c}_{ik} \prod_{l=1}^{m} x_l^{\gamma_{ikl}} \lesssim \tilde{1}\ (1 \leqslant i \leqslant p') \\ \tilde{g}_i(x) &= \sum_{k=1}^{J_i} \tilde{c}_{ik} \prod_{l=1}^{m} x_l^{\gamma_{ikl}} \gtrsim \tilde{1}\ (p'+1 \leqslant 1 \leqslant p) \\ x &> 0 \end{aligned} \tag{7.26}$$

can be turned into an n GP with type $(\cdot, c)$ fuzzy coefficients

$$\begin{aligned} \min\ \tilde{g}_0^{(1)}(x) &= \sum_{k=1}^{J_0^{(1)}} \tilde{c}_{0k}^{(1)} \prod_{l=1}^{m} x_l^{\gamma_{0kl}^{(1)}} \\ \text{s.t. } \tilde{g}_i(x) &= \sum_{k=1}^{J_i} \tilde{c}_{ik} \prod_{l=1}^{m} x_l^{\gamma_{ikl}} \lesssim \tilde{1}\ (1 \leqslant i \leqslant p') \\ \tilde{g}_i(x) &= \sum_{k=1}^{J_i} \tilde{c}_{ik} \prod_{l=1}^{m} x_l^{\gamma_{ikl}} \gtrsim \tilde{1}\ (p'+1 \leqslant i \leqslant p) \\ x &> 0 \\ &\vdots \\ \min\ \tilde{g}_0^{(n)}(x) &= \sum_{k=1}^{J_0^{(n)}} \tilde{c}_{0k}^{(n)} \prod_{l=1}^{m} x_l^{\gamma_{0kl}^{(n)}} \\ \text{s.t. } \tilde{g}_i(x) &= \sum_{k=1}^{J_i} \tilde{c}_{ik} \prod_{l=1}^{m} x_l^{\gamma_{ikl}} \lesssim \tilde{1}\ (1 \leqslant i \leqslant p') \\ \tilde{g}_i(x) &= \sum_{k=1}^{J_i} \tilde{c}_{ik} \prod_{l=1}^{m} x_l^{\gamma_{ikl}} \gtrsim \tilde{1}\ (p'+1 \leqslant i \leqslant p) \\ x &> 0 \end{aligned} \tag{7.27}$$

ii) Since problems (7.26) and (7.27) have the same DD, a solution to (7.26) is turned into finding an answer to (7.27) without increasing DD in it. We can obtain n objective value $\tilde{F}_1, \tilde{F}_2, \cdots, \tilde{F}_n$ after we use techniques in Chapter 4 to solve n reverse PGP in (7.27). Again we get the most satisfactory fuzzy solution to (7.26) by means of (7.25) and of the method similar to 1 in this section.

7.3.3 Another Algorithm

According to [TA84],we assume programming problem (7.23):

1) Constraints and goals which are given by sets.

2) Constraints and goals are just identical concepts in the sense that the decision x must be given so as to satisfy both sets.

Without distinction of constraints and goals, we can write a fuzzy GP problem (7.23) in the following form:

$$\begin{array}{ll}
\text{(goal)} & \tilde{g}_1^{(1)} = \tilde{b}_1 G_{00} - \sum\limits_{k=1}^{J_0^{(1)}} \tilde{c}_{0k}^{(1)} G_{0k}^{(1)}(x) \gtrsim 0 \\
\vdots & \quad \vdots \\
\text{(goal)} & \tilde{g}_n^{(n)} = \tilde{b}_n G_{00} - \sum\limits_{k=1}^{J_0^{(n)}} \tilde{c}_{0k}^{(n)} G_{0k}^{(n)}(x) \gtrsim 0 \\
\text{(constraint)} & \tilde{g}_1 = \tilde{1} G_{00} - \sum\limits_{k=1}^{J_i} \tilde{c}_{ik} G_{ik}(x) \gtrsim 0 \; (1 \leqslant i \leqslant p') \\
\vdots & \quad \vdots \\
\text{(constraint)} & \tilde{g}_p = -\hat{1} G_{00} + \sum\limits_{k=1}^{J_i} \tilde{c}_{ik} G_{ik}(x) \gtrsim 0 \; (p'+1 \leqslant i \leqslant p)
\end{array} \tag{7.28}$$

where $G_{00} = 1, G_{0k}^{(j)}(x) = \prod\limits_{l=1}^{m} x_l^{\gamma_{0kl}^{(j)}}, G_{ik}(x) = \prod\limits_{l=1}^{m} x_l^{\gamma_{ikl}}, \tilde{b}_j (1 \leqslant j \leqslant n, 0 \leqslant i \leqslant p)$ denote objective expectation value.

From the result in Section 5.3, we know (7.28) become simply

$$\begin{array}{ll}
(\alpha_{0k}^{*(j)} - hc_{0k}^{*(j)})^{\mathrm{T}} G_0^{(j)}(x) \geqslant 0 & (1 \leqslant j \leqslant n) \\
(\alpha_{ik}^{*} - hc_{ik}^{*})^{\mathrm{T}} G_i(x) \geqslant 0 & (1 \leqslant k \leqslant J_i, 1 \leqslant i \leqslant p)
\end{array} \tag{7.29}$$

if they are written as $G_0^{(j)}(x) = (G_{01}^{(j)}(x), G_{02}^{(j)}(x), \ldots, G_{0J_0}^{(j)}(x)), G_i(x) = (G_{i1}(x), G_{i2}(x), \ldots, G_{iJ_i}(x))(1 \leqslant j \leqslant n, 1 \leqslant k \leqslant J_i, 1 \leqslant i \leqslant p)$, such that (7.28) is equivalent to

$$\begin{array}{l}
\max\ h \\
\text{s.t. } (\alpha_{0k}^{*(j)} - hc_{0k}^{*(j)})^{\mathrm{T}} G_0^{(j)}(x) \geqslant 0 \quad (1 \leqslant j \leqslant n) \\
\qquad (\alpha_{ik}^{*} - hc_{ik}^{*})^{\mathrm{T}} G_i(x) \geqslant 0 \quad (1 \leqslant k \leqslant J_i, 1 \leqslant i \leqslant p) \\
\qquad x > 0,\ h \in [0,1].
\end{array} \tag{7.30}$$

Algorithm

1^0 Determine a small intimal value of h so that there exists an admissible set satisfying (7.29).

2^0 Let $\Delta > 0$ be small increment. Find $h + (k+1)\Delta$ which is the smallest value not compatible with (7.29), in the process of increasing h by Δ, 2Δ, ...

3^0 Since there exists an admissible set in the level of $h + k\Delta$, regard this set as the constraint set, and choose the most interesting inequality from (7.29). If it is the rth inequality in (7.29), take the left side of this inequality as an objective function

$$Z = \alpha_{r_1} G_1 + \cdots + \alpha_{r_p} G_p.$$

Then, obtain an x^* that maximizes Z subject to

$$\begin{array}{ll}(\alpha_{0k}^{(j)} - (h + \Delta k)c_{0k}^{(j)})^{\mathrm{T}} G_0^{(j)} \geqslant 0 & (1 \leqslant j \leqslant n) \\ (\alpha_{ik} - (h + \Delta k)c_{ik})^{\mathrm{T}} G_i(x) \geqslant 0 & (1 \leqslant k \leqslant J_i,\ 1 \leqslant i \leqslant p)\end{array}$$

This x^* is an approximate solution of (7.29).

7.4 Multi-Objective GP with Type L-R Fuzzy Coefficients

In this section, the multi-objective GP with Type L-R fuzzy coefficients are introduced and also corresponding algorithm is raised.

As for the definition and operative properties of type L-R fuzzy coefficients, we refer to Section 5.1.

Definition 7.4.1 Model:

$$\begin{aligned}\widetilde{\min} \ & \sum_{k=1}^{J_0^{(j)}} \tilde{c}_{0k}^{(j)} \prod_{l=1}^{m} x_l^{\gamma_{0kl}^{(j)}} \ (1 \leqslant j \leqslant n) \\ \text{s.t.} \ & \sum_{k=1}^{J_i} \tilde{c}_{ik} \prod_{l=1}^{m} x_l^{\gamma_{ikl}} \leqslant \tilde{1}(1 \leqslant i \leqslant p) \\ & x > 0\end{aligned} \tag{7.31}$$

calling (7.31) a multi-objective PGP with a type L-R fuzzy coefficients, where

$$\tilde{c}_{0k}^{(j)} = (c_{0k}^{(j)}, \underline{c}_{0k}^{(j)}, \overline{c}_{0k}^{(j)})_{LR}, \tilde{c}_{ik} = (c_{ik}, \underline{c}_{ik}, \overline{c}_{ik})_{LR}, \tilde{1} = (1, \underline{1}, \overline{1})_{LR}$$

denote a type L-R fuzzy number.

As for multi-objective reversed PGP with type L-R fuzzy coefficients, we can define and discuss it, similarly.

7.4.1 PGP of Objective Functions with L-R Fuzzy Coefficients

Consider program

$$\begin{aligned}\widetilde{\min} \ & \sum_{k=1}^{J_0} \tilde{c}_{0k} \prod_{l=1}^{m} x_l^{\gamma_{0kl}} \\ \text{s.t.} \ & g_i(x) = \sum_{k=1}^{J_i} c_{ik} \prod_{l=1}^{m} x_l^{\gamma_{ikl}} \leqslant 1 \ (1 \leqslant i \leqslant p) \\ & x > 0\end{aligned}$$

where $\tilde{c}_{0k}$ denote L-R fuzzy coefficients. We can turn it into program (5.26), such that we only consider the state of a monomial PGP with L-R fuzzy coefficients, i.e.,

$$\begin{aligned}\widetilde{\min} \ & \tilde{c}_0 \prod_{l=1}^{m} x_l^{\gamma_{0l}} \\ \text{s.t.} \ & g_i(x) = c_i \prod_{l=1}^{m} x_l^{\gamma_{il}} \leqslant 1 \ (1 \leqslant i \leqslant p) \\ & x > 0\end{aligned} \tag{7.32}$$

Theorem 7.4.1 *Program (7.32)* $\Longleftrightarrow$

$$\begin{aligned}&\widetilde{\min}\sum_{l=1}^{m}\gamma_{0l}z_l+\widetilde{\log c_0}\\&\text{s.t. }\sum_{l=1}^{m}\gamma_{il}z_l+\log c_i\leqslant 0\ (1\leqslant i\leqslant p)\end{aligned}\tag{7.33}$$

An approximate fuzzy optimal solution in (7.33) also is that in (7.32). ◊

Proof

$$(7.32)\Longleftrightarrow\quad\begin{aligned}&\min\ c_0\prod_{l=1}^{m}x_l^{\gamma_{0l}}\\&\max\ \underline{c}_0\prod_{l=1}^{m}x_l^{\gamma_{0l}}\\&\min\ \bar{c}_0\prod_{l=1}^{m}x_l^{\gamma_{0l}}\\&\text{s.t. }c_i\prod_{l=1}^{m}x_l^{\gamma_{il}}\leqslant 1\ (1\leqslant i\leqslant p)\\&\qquad x>0\end{aligned}\tag{7.34}$$

$$\Longleftrightarrow\quad\begin{aligned}&\min\ (\sum_{l=1}^{m}\gamma_{0l}z_l+\log c_0)\\&\max\ (\sum_{l=1}^{m}\gamma_{0l}z_l+\log \underline{c}_0)\\&\min\ (\sum_{l=1}^{m}\gamma_{0l}z_l+\log \bar{c}_0)\\&\text{s.t. }\sum_{l=1}^{m}\gamma_{il}z_l+\log c_i\leqslant 0\ (1\leqslant i\leqslant p)\end{aligned}\tag{7.35}$$

$$\Longleftrightarrow\ (7.33)$$

Therefore the theorem holds. □

(7.34) is an ordinary multi-objective PGP and (7.35) is an ordinary multi-objective LP. Obviously, (7.34) and (7.35) are mutually changeable through the replacement of $z_l=\log x_l (1\leqslant l\leqslant m)$, such that an optimal solution of (7.35) is that of (7.34). Therefore, an approximately fuzzy optimal solution of (7.33) is also that of (7.32) and the theorem holds.

7.4.2 Mixed with L-R Fuzzy Coefficients

Theorem 7.4.2 *(7.31)* $\Longleftrightarrow$

$$\min\ g_0^{(j)}=\sum_{k=1}^{J_0^{(j)}}c_{0k}^{(j)}\prod_{l=1}^{m}x_l^{\gamma_{0kl}^{(j)}}\tag{7.36}$$

$$\max \ \underline{g}_0^{(j)} = \sum_{k=1}^{J_0^{(j)}} \underline{c}_{0k}^{(j)} \prod_{l=1}^{m} x_l^{\gamma_{0kl}^{(j)}} \tag{7.37}$$

$$\min \ \overline{g}_0^{(j)} = \sum_{k=1}^{J_0^{(j)}} \overline{c}_{0k}^{(j)} \prod_{l=1}^{m} x_l^{\gamma_{0kl}^{(j)}} \tag{7.38}$$

$$(1 \leqslant j \leqslant n)$$

$$\begin{aligned} s.t. \quad & \sum_{k=1}^{J_i} c_{ik} \prod_{l=1}^{m} x_l^{\gamma_{ikl}} \leqslant 1 \\ & \sum_{k=1}^{J_i} \underline{c}_{ik} \prod_{l=1}^{m} x_l^{\gamma_{ikl}} \geqslant \underline{1} (1 \leqslant i \leqslant p) \\ & \sum_{k=1}^{J_i} \overline{c}_{iK} \prod_{l=1}^{m} x_l^{\gamma_{ikl}} \leqslant \overline{1} \\ & x > 0 \end{aligned} \tag{7.39}$$

◇

Proof According to approximate formula of $\widetilde{\min}$ and the operative properties of type L-R fuzzy numbers, the theorem can be confirmed in combination with a number-vector multiplication [DP80]. □

Now, we present the importance of a flexible index to each objective function $d_s^{(j)}(> 0)(1 \leqslant j \leqslant n; s = 1, 2, 3)$. The smaller $d_s^{(j)}$ is, the more important $g_s^{(j)}$ is. Correspondingly, we construct a fuzzy object set $\tilde{F}_i$, whose membership functions stipulate

$$\mu_{\tilde{F}_1^{(j)}}(x) = g_1\Big(\sum_{k=1}^{J_0^{(j)}} c_{0k}^{(j)} \prod_{l=1}^{m} x_l^{\gamma_{0kl}^{(j)}}\Big)$$
$$= \begin{cases} 0, & \text{for } g_0^{(j)}(x) < g_0^{(j)} - d_1^{(j)} \\ 1 - \frac{1}{d_1^{(j)}}\Big(g_0^{(j)} - \sum_{k=1}^{J_0^{(j)}} c_{0k}^{(j)} \prod_{l=1}^{m} x_l^{\gamma_{0kl}^{(j)}}\Big), & \text{for } g_0^{(j)} - d_1^{(j)} \leqslant g_0^{(j)}(x) < g_0^{(j)} \\ 1, & \text{for } g_0^{(j)} \leqslant g_0^{(j)}(x), \end{cases}$$

$$\mu_{\tilde{F}_2^{(j)}}(x) = g_2\Big(\sum_{k=1}^{J_0^{(j)}} \underline{c}_{0k}^{(j)} \prod_{l=1}^{m} x_l^{\gamma_{0kl}^{(j)}}\Big)$$
$$= \begin{cases} 0, & \text{for } g_0^{(j)}(x) < \underline{g}_0^{(j)} - d_2^{(j)} \\ 1 - \frac{1}{d_2^{(j)}}\Big(\underline{g}_0^{(j)} - \sum_{k=1}^{J_0^{(j)}} \underline{c}_{0k}^{(j)} \prod_{l=1}^{m} x_l^{\gamma_{0kl}^{(j)}}\Big), & \text{for } \underline{g}_0^{(j)} - d_2^{(j)} \leqslant \underline{g}_0^{(j)}(x) < \underline{g}_0^{(j)} \\ 1, & \text{for } \underline{g}_0^{(j)} \leqslant \underline{g}_0^{(j)}(x), \end{cases}$$

$$\mu_{\tilde{F}_3^{(j)}}(x) = g_3\Big(\sum_{k=1}^{J_0^{(j)}} \bar{c}_{0k}^{(j)} \prod_{l=1}^{m} x_l^{\gamma_{0kl}^{(j)}}\Big)$$
$$= \begin{cases} 0, & \text{for } \bar{g}_0^{(j)}(x) < \bar{g}_0^{(j)} - d_3^{(j)} \\ 1 - \frac{1}{d_3^{(j)}}\Big(\bar{g}_0^{(j)} - \sum_{k=1}^{J_0^{(j)}} \bar{c}_{0k}^{(j)} \prod_{l=1}^{m} x_l^{\gamma_{0kl}^{(j)}}\Big), & \text{for } \bar{g}_0^{(j)} - d_3^{(j)} \leqslant \bar{g}_0^{(j)}(x) < \bar{g}_0^{(j)} \\ 1, & \text{for } \bar{g}_0^{(j)} \leqslant \bar{g}_0^{(j)}(x), \end{cases}$$

where $g_0^{(j)*}, \underline{g}_0^{(j)*}$ and $\bar{g}_0^{(j)*}$ are optimal values for each objective function obtained under the constraint of (7.39).

Therefore, solutions to problem (7.31) are concluded as follows:

1^0 $\forall j \in [1, n]$, we obtain optimal values $g_0^{(j)}, \underline{g}_0^{(j)}\ \bar{g}_0^{(j)}$ for (7.36) with (7.39), (7.37) with (7.39) and (7.38) with (7.39), respectively.

2^0 If $\tilde{F} = \bigcap_{j=1}^{n}(\tilde{F}_1^{(j)} \cap \tilde{F}_2^{(j)} \cap \tilde{F}_3^{(j)})$, we find hgt$\tilde{F}$ by means of $\mu_{\tilde{F}_1}^{(j)}, \mu_{\tilde{F}_2}^{(j)}$ and $\mu_{\tilde{F}_3}^{(j)}$.

3^0 Find an ordinary parameter GP [Ver84]

$$\begin{aligned} &\max \alpha \\ &\text{s.t. } \mu_{\tilde{F}_s}^{(j)}(x) \leqslant \alpha (S = 1, 2, 3) \\ &\quad (7.39), (1 \leqslant j \leqslant n) \\ &\quad \alpha \in [0, 1] \\ &\quad x > 0 \end{aligned}$$

The most satisfactory parameter solution obtained to the ordinary parameter GP corresponds with the most satisfactory one obtained to problem (7.37)–(7.39): $g_0^{(j)^*}, \underline{g}_0^{(j)^*}, \bar{g}_0^{(j)^*}, \alpha^*$, such that we have the most satisfactory one of fuzziness to Model (7.31):

$$z_0^{(j)^*} = (g_0^{(j)^*}, \underline{g}_0^{(j)^*}, \bar{g}_0^{(j)^*})_{LR}.$$

7.5 Multi-Objective GP with Flat Fuzzy Coefficients

Definition 7.5.1 Call model

$$\begin{aligned} \min & \sum_{k=1}^{J_0^{(j)}} \tilde{c}_{0k}^{(j)} \prod_{l=1}^{m} x_l^{\gamma_{0kl}^{(j)}} (1 \leqslant j \leqslant n) \\ \text{s.t.} & \sum_{k=1}^{J_i} \tilde{c}_{ik} \prod_{l=1}^{m} x_l^{\gamma_{ikl}} \leqslant \tilde{1}\ (1 \leqslant i \leqslant p) \\ & x > 0 \end{aligned} \tag{7.40}$$

a multi-objective GP with flat fuzzy coefficients, where

$$\tilde{c}_{0k}^{(j)} = (c_{0k}^{-(j)}, c_{0k}^{+(j)}, \sigma^-_{c_{0k}^{(j)}}, \sigma^+_{c_{0k}^{(j)}}), \tilde{c}_{ik} = (c_{ik}^-, c_{ik}^+, \sigma^-_{c_{ik}}, \sigma^+_{c_{ik}}),$$
$$\tilde{1} = (1^-, 1^+, \sigma_1^-, \sigma_1^+).$$

Theorem 7.5.1 *Let* $\tilde{g}_0^{(j)}(x) =$

$$(\sum_{k=1}^{J_0} c_{0k}^{(j)-} \prod_{l=1}^{m} x_l^{\gamma_{0kl}}, \sum_{k=1}^{J_0} c_{0k}^{(j)+} \prod_{l=1}^{m} x_l^{\gamma_{0kl}}, \sum_{k=1}^{J_0} \sigma^-_{c_{0k}^{(j)}} \prod_{l=1}^{m} x_l^{\gamma_{0kl}}, \sum_{k=1}^{J_0} \sigma^+_{c_{0k}^{(j)}} \prod_{l=1}^{m} x_l^{\gamma_{0kl}}),$$
$$F_0^{(j)} = (\overline{F}_0^{(j)}, \overline{F}_0^{(j)}, 0, \sigma^+_{\overline{F}_0^{(j)}})\ (1 \le j \le n).$$

Then

$$\min\ \tilde{g}_0^{(j)}(x) \Longleftrightarrow \min \frac{\sum_{k=1}^{J_0} c_{0k}^{(j)-} \prod_{l=1}^{m} x_l^{\gamma_{0kl}} - \overline{F}_0^{(j)}}{\sum_{k=1}^{J_0} \sigma^-_{c_{0k}^{(j)}} \prod_{l=1}^{m} x_l^{\gamma_{0kl}} + \sigma^+_{\overline{F}_0^{(j)}}}\ (1 \le j \le n).$$

◇

Proof It is proved similar to Theorem 5.5.1. □

Theorem 7.5.2 *(7.40)*$\Longleftrightarrow$

$$\begin{aligned} \max\ & \theta \\ \text{s.t.} & \sum_{k=1}^{J_0^{(j)}} [c_{0k}^{-(j)} - (1-\theta)\sigma^-_{c_{0k}^{(j)}}] \prod_{l=1}^{m} x_l^{\gamma_{0kl}^{(j)}} \leqslant \bar{F}_0^{(j)} + (1-\theta)\sigma^+_{\bar{F}_0^{(j)}} \\ & \qquad (1 \leqslant j \leqslant n) \\ & \sum_{k=1}^{J_i} [c_{ik}^- - (1-\theta)\sigma^-_{c_{ik}}] \prod_{l=1}^{m} x_l^{\gamma_{ikl}} \leqslant 1^+ + (1-\theta)\sigma_1^+ \\ & \theta \in [0,1], (1 \leqslant i \leqslant p) \\ & x > 0 \end{aligned} \tag{7.41}$$

Here $\tilde{F}_0^{(j)} = (\bar{F}_0^{(j)}, \bar{F}_0^{(j)}, 0, \sigma^+_{\bar{F}_0^{(j)}})$ *represents the* j*-th expectation object in (7.40).* ◊

Proof From the definition of $\mathrm{PD}(\tilde{g}_i, \tilde{g}_j)$ in Section 5.5.1 [RT91], (7.41) is equivalent to

$$\begin{aligned} &\min \mathrm{PD}(\tilde{F}_0^{(j)}, \tilde{g}_0^{(j)}(x))\ (1 \leqslant j \leqslant n) \\ &\text{s.t. } \mathrm{PD}(\tilde{g}_i(x), \tilde{1}) \leqslant \theta \\ &\qquad \theta \in [0,1], (1 \leqslant i \leqslant p) \\ &\qquad x > 0 \end{aligned}$$

$$\Longleftrightarrow$$

$$\min\left\{ g_0^{(j)} = \frac{\sum\limits_{k=1}^{J_0^{(j)}} c_{0k}^{-(j)} \prod\limits_{l=1}^{m} x_l^{\gamma_{0kl}^{(j)}} - \bar{F}_0^{(j)}}{\sum\limits_{k=1}^{J_0^{(j)}} \sigma^-_{c_{0k}^{(j)}} \prod\limits_{l=1}^{m} x_l^{\gamma_{0kl}^{(j)}} + \sigma^+_{\bar{F}_0^{(j)}}} \right\} (1 \leqslant j \leqslant n) \tag{7.42}$$

$$\left.\begin{aligned} \text{s.t. } &\sum_{k=1}^{J_i} [c_{ik}^- - (1-\theta)\sigma^-_{c_{ik}}] \prod_{l=1}^{m} x_l^{\gamma_{ikl}} \leqslant 1^+ + (1-\theta)\sigma_1^+ \\ &\theta \in [0,1], (1 \leqslant i \leqslant p) \\ &x > 0 \end{aligned}\right\} \tag{7.43}$$

But

$$(7.42) \Leftrightarrow \max\{1 - g_0^{(j)}\}.$$

Make substitution $\theta = 1 - \tilde{g}_0^{(j)}(x)$. Let

$$\mathrm{PD}(\tilde{F}_0^{(j)}, \tilde{g}_0^{(j)}(x)) \geqslant \theta (0 \leqslant j \leqslant n).$$

i.e,

$$\sum_{k=1}^{J_0^{(j)}} [c_{0k}^{-(j)} - (1-\theta)\sigma^-_{c_{0k}^{(j)}}] \prod_{l=1}^{m} x_l^{\gamma_{0kl}^{(j)}} \leqslant \bar{F}_0^{(j)} + (1-\theta)\sigma^+_{\bar{F}_0^{(j)}} (1 \leqslant j \leqslant n). \tag{7.44}$$

Then

$$\begin{aligned} &\max\ \theta \\ &\text{s.t. } (7.44) \\ &\qquad (7.43) \end{aligned}$$

is problem (7.41).

To turn (7.40) into (7.41) does not involve changing DD, and (7.41) serves as an ordinary parameter GP, whose satisfactory parameter solution is easy to get, so that it is easy to obtain a fuzzy satisfactory solution to (7.40). □

If constraint condition in (7.40) is denoted by $\tilde{g}_i(x) \approx \tilde{1}$, then by means of

$$\tilde{g}_i(x) \approx \tilde{1}\ (1 \leqslant i \leqslant p)$$

$$\Longleftrightarrow \begin{cases} \sum\limits_{k=1}^{J_i} [c_{ik}^- - (1-\theta)\sigma_{c_{ik}}^-] \prod\limits_{l=1}^{m} x_l^{\gamma_{ikl}} \leqslant 1^+ + (1-\theta)\sigma_1^+ \\ \sum\limits_{k=1}^{J_i} [c_{ik}^+ + (1-\theta)\sigma_{c_{ik}}^+] \prod\limits_{l=1}^{m} x_l^{\gamma_{ikl}} \geqslant 1^- - (1-\theta)\sigma_1^- \end{cases} \quad (1 \leqslant i \leqslant p). \tag{7.45}$$

from References [RT91] [Cao92a][Cao95b][Cao00], (7.40) is converted into

$$\begin{aligned} &\min \tilde{g}_0^{(j)}(x)\ (1 \leqslant j \leqslant n) \\ &\text{s.t. } \tilde{g}_i \approx \tilde{1}\ (1 \leqslant i \leqslant p) \\ &\quad\ \ x > 0 \end{aligned}$$

which is equivalent to a reversed PGP with parameter vectors

$$\begin{aligned} &\max\ \theta \\ &\text{s.t. } (7.44) \\ &\quad\ \ (7.45) \\ &\quad\ \ \theta \in [0,1] \\ &\quad\ \ x > 0 \end{aligned}$$

and its solution can be obtained by a dual or a direct algorithm without difficulty [Cao87a,b][Cao93a].

Now, we are going to study another method to (7.40), i.e., (7.40) is turned into a monomial PGP before changing it into an LP with parameter vectors.

Take advantage of inequality [Cao87a,b][Cao93a][WY82]

$$\sum_{k=1}^{J_i} c_{ik} \prod_{l=1}^{m} x_l^{\gamma_{ikl}} \geqslant \prod_{k=1}^{J_i} \left(\frac{c_{ik}}{\varepsilon_{ik}}\right)^{\varepsilon_{ik}} \prod_{l=1}^{m} x_l^{\overline{\gamma}_{ik}}.$$

Here,

$$\overline{\gamma}_{il} = \sum_{k=1}^{J_i} \gamma_{ikl}\varepsilon_{ik}$$

and

$$\sum_{k=1}^{J_i} \varepsilon_{ik} = 1, c_{ik} > 0, x_l > 0 (1 \leqslant k \leqslant J_i, 0 \leqslant i \leqslant p, 1 \leqslant l \leqslant m),$$

such that we turn (7.40) into

$$\begin{aligned}
&\min\ (-\theta)\\
&\text{s.t. } \frac{1}{\bar{F}_0^{(j)}+(1-\theta)\sigma^+_{\bar{F}_0^{(j)}}}\prod_{k=1}^{J_0^{(j)}}\left(\frac{\bar{c}_{0k}^{-(j)}-(1-\theta)\sigma^-_{c_{0k}^{(j)}}}{\varepsilon_{0k}^{(j)}}\right)^{\varepsilon_{0k}^{(j)}}\prod_{l=1}^{m}x_l^{\overline{\gamma}_{0l}^{(j)}}\leqslant 1\\
&(1\leqslant j\leqslant n)\\
&\frac{1}{1^+ +(1-\theta)\sigma_1^+}\prod_{k=1}^{J_i}\left(\frac{c_{ik}^- -(1-\theta)\sigma_{c_{ik}^-}}{\varepsilon_{ik}}\right)^{\varepsilon_{ik}}\prod_{l=1}^{m}x_l^{\overline{\gamma}_{il}}\leqslant 1\\
&\theta\in[0,1],(1\leqslant i\leqslant p)\\
&x_l>0
\end{aligned}\tag{7.46}$$

Suppose $z_l=\log x_l$ and we set up

$$\begin{aligned}
a_0^{(j)}&=\log[\bar{F}_0^{(j)}+(1-\theta)\sigma^-_{\bar{F}_0^{(j)}}]-\sum_{k=1}^{J_0^{(j)}}\varepsilon_{0k}^{(j)}\log\frac{c_{0k}^{-(j)}-(1-\theta)\sigma^-_{c_{0k}^{(j)}}}{\varepsilon_{0k}^{(j)}},\\
a_i&=\log[1^+ +(1-\theta)\sigma_1^+]-\sum_{k=1}^{J_i}\varepsilon_{ik}\log\frac{c_{ik}^- -(1-\theta)\sigma_{c_{ik}^-}}{\varepsilon_{ik}}\\
&(1\leqslant j\leqslant n),(0\leqslant i\leqslant p)
\end{aligned}$$

for $a_0^{(j)}>0, a_i>0$ with (7.46) converted into

$$\begin{aligned}
&\min\ (-\theta)\\
&\text{s.t. } G_0^{(j)}(z)=\frac{1}{a_0^{(j)}}\sum_{l=1}^{m}\overline{\gamma}_{0l}^{(j)}z_l\leqslant 1(1\leqslant j\leqslant n)\\
&G_i(z)=\frac{1}{a_i}\sum_{l=1}^{m}\overline{\gamma}_{il}z_l\leqslant 1(1\leqslant i\leqslant p)\\
&\theta\in[0,1]
\end{aligned}\tag{7.47}$$

Design algorithm steps to problem (7.47) as follows:

1^0 Define lower and upper bounds of θ, and we might as well let $\theta_0^-=0,\theta_0^+=1$ for $l=0$.

2^0 Fix θ_{l+1} and let θ_{l+1} =small +(big− small)×0.618 (where, small and big refer to a left endpoint and a right one at the interval).

3^0 Turn to Step 4^0, if a feasible solution set $\mathcal{Z}$ exists in (7.47) for $\theta=\theta_{l+1}$. Otherwise turn back to step 2^0, supposing $\theta_{l+1}^+=\theta_l,\theta_{l+1}^-=\theta_l^-$.

4^0 Suppose $z^*\in\mathcal{Z}$, then we define $\theta^h=\min\left\{\min\limits_{1\leqslant j\leqslant n}G_0^{(j)}(z),\ \min\limits_{1\leqslant i\leqslant p}G_i(z)\right\}$. Select $\theta_{l+1}^-=\theta^h,\theta_{l+1}^+=\theta_l^+$ before turning back to 2^0.

Continue like this till the most satisfactory solution is found to (7.47).

7.6 Multi-Objective LP with $T-$Fuzzy Variables

There are a lot of fuzzy and undetermined phenomena in the real world. If such phenomena are described with T-fuzzy numbers, more information can be obtained. We once developed a regression and self-regression forecasting model with T-fuzzy data [Cao89b,c], an input and output one [Cao93b] and an LP one with $T-$FV [Cao96a]. Here we shall generalize a model in Reference [Cao96][Caoupd] into a multi-objective LP with $T-$FV and discuss its algorithm.

7.6.1 Building the Model

Consider an ordinary multi-objective LP to be

$$\begin{aligned} V - \max\ & c^{(j)}x \ (1 \leqslant j \leqslant r) \\ \text{s.t.}\ & Ax \leqslant b \\ & x \geqslant 0 \end{aligned} \tag{7.48}$$

where, $x = (x_1, x_2, \ldots, x_m)^{\mathrm{T}}$ is an $m-$dimensional vector, x_j a variable, $b = (b_1, b_2, \cdots, b_n)^{\mathrm{T}}$ an $n-$dimensional constant vector; $0 = (0, 0, \cdots, 0)^{\mathrm{T}}$ an $m-$dimensional zero vector, $c^{(j)}$ and A denote $r \times m$ and $n \times m$ matrices, respectively.

Because of practical problems, we generalize (7.48) into the problem of an LP with $T-$FV:

$$\begin{aligned} V - \max\ & c^{(j)}\tilde{x} \ (1 \leqslant j \leqslant r) \\ \text{s.t.}\ & A\tilde{x} \lesssim \tilde{b} \\ & \tilde{x} \geqslant \tilde{0} \end{aligned} \tag{7.49}$$

where, $\tilde{x} = (\tilde{x}_1, \tilde{x}_2, \ldots, \tilde{x}_m)^{\mathrm{T}}$ is an $m-$dimension $T-$FV vector, $\tilde{x}_l = (x_l, \underline{\xi}_l, \overline{\xi}_l)_T$ is a $T-$FV; $\tilde{b} = (\tilde{b}_1, \tilde{b}_2, \ldots, \tilde{b}_n)^{\mathrm{T}}$ an $n-$dimension T-fuzzy constant vector, $\tilde{0} = (\tilde{0}, \tilde{0}, \ldots, \tilde{0})^{\mathrm{T}}$ an $n-$dimension T-fuzzy zero vector, $\tilde{b}_i = (b_i, \underline{b}_i, \overline{b}_i)_T$, $\tilde{0} = (0, \underline{0}, \overline{0})_T$ T-fuzzy numbers; $c^{(j)}$ and A denote $r \times m$ and $n \times m$ constant matrices, respectively.

We call (7.48) a multi-objective LP model with $T-$FV.

7.6.2 Non-fuzziness of Model

Theorem 7.6.1 *If (7.49) is given by $T-$FV, then, to the given cone index $\mathcal{J}$, (7.49) can be turned into*

$$\begin{aligned} V - \max\ & c^{(j)}U(\mathcal{J})(1 \leqslant j \leqslant r) \\ \text{s.t.}\ & AU(\mathcal{J}) \leqslant b(\mathcal{J}) \\ & U(\mathcal{J}) \geqslant 0 \end{aligned} \tag{7.50}$$

where

$$c^{(j)}U(\mathcal{J}) = \sum_{l=1}^{m} c_l^{(j)}(\sum_{i=1}^{3N} U_{li}(\mathcal{J}))\ (1 \leqslant j \leqslant r),$$
$$AU(\mathcal{J}) = \sum_{l=1}^{m} a_{kl}(\sum_{i=1}^{3N} U_{li}(\mathcal{J}))\ (1 \leqslant k \leqslant n).$$

$U(\mathcal{J}) = (U_{1i}(\mathcal{J}), U_{2i}(\mathcal{J}), \ldots, U_{mi}(\mathcal{J}))^{\mathrm{T}}$ and $U_{li}(\mathcal{J})$ are a vector and a variable with a cone index $\mathcal{J}$, respectively. $b(\mathcal{J}) = (b_1(\mathcal{J}), b_2(\mathcal{J}), \ldots, b_n(\mathcal{J}))^{\mathrm{T}}$ and $b_k(\mathcal{J})$ are constant vectors and constants with a cone index $\mathcal{J}$, respectively; $c^{(j)}$ and A denote $r \times m$ and $n \times m$ matrices, respectively. And (7.50) contains a satisfactory solution depending on a cone index $\mathcal{J}$, which is equivalent to (7.49) having a fuzzy satisfactory solution. ◊

Proof Let $\tilde{x}_l = (\tilde{x}_{l1}, \tilde{x}_{l2}, \ldots, \tilde{x}_{ln})^{\mathrm{T}}$ be a column T−FV satisfying (7.49), where $\tilde{x}_{li} = (x_{li}, \underline{\xi}_{li}, \overline{\xi}_{li})_T (1 \leqslant l \leqslant m; 1 \leqslant i \leqslant 3N)$. We classify the T−FV of a column by subscripts, and might as well make $1 \leqslant l \leqslant M$ correspond to smaller fluctuating variables, while the other variables correspond to $M + 1 \leqslant l \leqslant m$. Then to $1 \leqslant i \leqslant N$ and each l,

$$U_{li} = x_{li} + (\underline{\xi}_{li} + \overline{\xi}_{li})/2,$$

to $N + 1 \leqslant i \leqslant 2N$ and each l,

$$U_{li} = \begin{cases} x_{li} + \underline{\underline{\xi}}_{li}, & \text{if } j_l = 0 \\ x_{li} - \overline{\xi}_{li}, & \text{if } j_l = 1, \end{cases}$$

to $2N + 1 \leqslant i \leqslant 3N$ and each l,

$$U_{li} = \begin{cases} x_{li} - \underline{\underline{\xi}}_{li}, & \text{if } j_l = 0 \\ x_{li} + \overline{\xi}_{li}, & \text{if } j_l = 1. \end{cases}$$

Then, under the given cone index $\mathcal{J}$, (7.49) is turned into (7.50), such that (7.50) can be found.

Since (7.49) is equivalent to (7.50), (7.50) with a parameter optimal solution depending on a cone index $\mathcal{J}$ is equivalent to (7.49) having an optimal T-fuzzy solution, the theorem holds. □

From such a theorem, we know that we can turn (7.49) into an ordinary multi-objective parameter LP (7.50) depending on a cone index $\mathcal{J}$. And for (7.50), we adopt such methods to multi-objective programming as changing a multi-objective majority problem into a single one or a series of single ones.

We conclude the solutions to Model (7.49) as follows:

1^0 To the given T−FV $\tilde{x}_l$, we partition natural number sets $\{1, 2, \ldots, n\}$ into three parts by subscripts:

$$\text{I:} \qquad U_{li} = x_{li} + (\underline{\xi}_{li} + \overline{\xi}_{li})/2, (1 \leqslant i \leqslant N) \text{ and each } l;$$

$$\text{II:} \quad U_{li} = \begin{cases} x_{li} + \underline{\xi}_{li}, \text{ if } j_l = 0, \\ x_{li} - \overline{\xi}_{li}, \text{ if } j_l = 1, \end{cases} \quad N + 1 \leqslant i \leqslant 2N \text{ and each } l;$$

$$\text{III:} \quad U_{li} = \begin{cases} x_{li} - \underline{\xi}_{li}, \text{ if } j_l = 0, \\ x_{li} + \overline{\xi}_{li}, \text{ if } j_l = 1, \end{cases} \quad 2N + 1 \leqslant i \leqslant 3N \text{ and each } l.$$

2^0 Nonfuzzify $\tilde{x}_l$.
We select

$$U_{li} = x_{li} + \sum_{i=1}^{3N} \xi_{li}^* / 3N$$

where ξ_{li}^* is $(\underline{\xi}_{li} + \overline{\xi}_{li})/2$, or $\pm\, \underline{\xi}_{li}$, or $\pm\overline{\xi}_{li}$.

3^0 Substitute U_{li} for $\tilde{x}_l$ in (7.49) and we can get (7.50).

4^0 Determine a satisfactory (resp. effective) solution to problem (7.50) by means of solution of an ordinary multi-objective LP and we can get a fuzzy satisfactory solution to (7.49).

There are a lot of methods in finding satisfactory (resp. effective) solutions to problem (7.50). Here, we advance two ways to nonfuzzification (7.49):

1) Non-fuzziness before we use a weighted method

Nonfuzzify (7.49) into an LP (7.50) with a cone index $\mathcal{J}$.

Give weight to r objective functions

$$f_j(U(\mathcal{J})) = \sum_{l=1}^{m} c_l^{(j)} \sum_{i=1}^{3N} U_{li}(\mathcal{J}) \ (1 \leqslant j \leqslant r)$$

and we have

$$f(U(\mathcal{J})) = \alpha_1 f_1(U(\mathcal{J})) + \alpha_2 f_2(U(\mathcal{J})) + \cdots + \alpha_r f_r(U(\mathcal{J}))$$

where $\alpha_j (1 \leqslant j \leqslant r)$ is a factor of weight satisfying $0 \leqslant \alpha_j \leqslant 1$, and

$$\alpha_1 + \alpha_2 + \cdots + \alpha_r = 1.$$

Turn (7.50) into a single object LP

$$\begin{aligned} &\max \ \ f(U(\mathcal{J})) \\ &\text{s.t.} \ \ AU(\mathcal{J}) \leqslant b(\mathcal{J}) \\ &\qquad U(\mathcal{J}) \geqslant 0 \end{aligned} \tag{7.51}$$

2) The weighted method before nonfuzzification

Consider (7.49) and we make r fuzzy objective functions weighted in (7.49):

$$f(\tilde{x}) = \alpha_1 f_1(\tilde{x}) + \alpha_2 f_2(\tilde{x}) + \cdots + \alpha_r f_r(\tilde{x})$$

and then change problem (7.49) into

$$\begin{aligned} &\min \; f(\tilde{x}) \\ &\text{s.t.} \; A\tilde{x} \lesssim \tilde{b} \\ &\qquad \tilde{x} \gtrsim \tilde{0} \end{aligned} \tag{7.52}$$

Now, (7.52) is nonfuzzified by the method mentioned in Section 4 in this chapter, so (7.50) can be obtained also.

Note 7.1: To determine the weighted vectors mentioned above, we can use an Analytic Hierarchy Process or expert experience.

7.6.3 Finding Optimum Solutions

We have many algorithms such as genetic and simulated annealing algorithms [XX99] by which, to single objective LP problem (7.51) or (7.52), we can finally get a satisfactory solution with practical value. Now, we are searching for a better algorithm since the overall optimum constraint by a single algorithm coding fails to show or to ensure its convergence for the optimum solution.

Suppose there are computer programs for solution to (7.49) or (7.50) in order to discuss the search strategy theoretically. Now we consider the following:

Example 7.3 Find

$$\begin{aligned} \max \; & (\tilde{z}_1, \tilde{z}_2) \\ & \tilde{z}_1 = 5\tilde{x}_1 + \tilde{x}_2 \\ & \tilde{z}_2 = \tilde{x}_1 + 4\tilde{x}_2 - 0.5 \\ \text{s.t.} \; & \tilde{x}_1 + \tilde{x}_2 \lesssim \tilde{6} \\ & \tilde{0} \lesssim \tilde{x}_1 \lesssim \tilde{5} \\ & \tilde{x}_2 \gtrsim \tilde{0} \end{aligned}$$

where $\tilde{0} = (0,0,0)_T, \tilde{5} = (5,0,0)_T, \tilde{6} = (6,0,0)_T$.

We take $T-$ fuzzy data as follows:

$\tilde{x}_1$: 1. $(x_1,1,0)_T$, 2. $(x_1,0,1)_T$, 3. $(x_1,2,1)_T$,

$\tilde{x}_2$: 4. $(x_2,0,1)_T$; 5. $(x_2,1,0)_T$; 6. $(x_2,2,2)_T$.

Now, divide the data into three groups including: No. 1,4; No. 2,5 and No. 3,6.

For data No. 1,4, we get a value by formula I. For the rest, we use the formulas corresponding to $j_l = 1$ and $j_l = 0$ in formulas II and III when odd numbers and even numbers appear, respectively.

So, we can nonfuzzify $\tilde{x}_1, \tilde{x}_2$,

$$\begin{aligned}\tilde{x}_1: &\ [(x_1+0.5)+(x_1-0)+(x_1+1)]/3\\ &= x_1+0.5,\\ \tilde{x}_2: &\ [(x_2+0.5)+(x_2-0)+(x_2-2)]/3\\ &= x_2-0.5,\end{aligned}$$

such that

$$\begin{aligned}f(\tilde{x}) &= \alpha_1\tilde{z}_1+\alpha_2\tilde{z}_2 : f(U(\mathcal{J}))\\ &= 6x_1+5x_2\end{aligned}$$

for $\alpha_1 = \alpha_2 = 1$, hence an LP corresponding to (7.50) is denoted by

$$\begin{aligned}\max\ \ & z = 6x_1+5x_2\\ \text{s.t.}\ \ & x_1+x_2 \leqslant 6\\ & 0 \leqslant x_1 \leqslant 5\\ & x_2 \geqslant 0\end{aligned}$$

whose superior corresponding solution is

$$x_1 = 5, x_2 = 1, z_1 = 26, z_2 = 8.5.$$

7.6.4 Conclusion

First, we turn a multi-objective LP with $T-$FV into a multi-objective ordinary LP depending on a cone index $\mathcal{J}$. Besides we can transplant many results of multi-objective LP into the multi-objective LP with $T-$FV. And last in a similar method to the Reference [DP80], we can extend this method into multi-objective non-LP with $T-$FV, which, obviously, contains more information than a classical one.

7.7 Multi-Objective GP with $T-$Fuzzy Variables

7.7.1 Introduction

In this section, we present a multi-objective GP with $T-$FV, by which the result in the previous section can be completely transplanted. But we shall lay particular emphasis on presenting algorithms such as a direct primal one and a dual one instead of introducing how to transplant them.

7.7.2 Model

Definition 7.7.1 The programming problem

$$\begin{aligned} &\min\ g_0^{(j)}(\tilde{x})\ (1 \leqslant j \leqslant n) \\ &\text{s.t. } g_i(\tilde{x}) \leqslant \tilde{\sigma}_i\ (1 \leqslant i \leqslant p) \\ &\qquad \tilde{x} > 0 \end{aligned} \tag{7.53}$$

is said to be a multi-objective GP model with $T-$FV, where

$$\begin{aligned} g_0^{(j)}(\tilde{x}) &= \sum_{k=1}^{J_0^{(j)}} \sigma_{0k} c_{0k}^{(j)} \prod_{l=1}^{m} \tilde{x}_l^{\gamma_{0kl}^{(j)}}, \\ g_i(\tilde{x}) &= \sum_{k=1}^{J_i} \sigma_{ik} c_{ik} \prod_{l=1}^{m} \tilde{x}_l^{\gamma_{ikl}}, \end{aligned}$$

$\tilde{\sigma}_i = \sigma_i \cdot \tilde{1}; \sigma_i, \sigma_{ik} = \pm 1, \tilde{x}_l$ are $T-$FV, $\tilde{1}$ is a T-fuzzy number.

Theorem 7.7.1 *Let the multi-objective GP model with $T-FV$ be as in (7.53). Then (7.53) can be converted into a multi-objective GP with a cone index $\mathcal{J}$*

$$\begin{aligned} &\min\ g_0^{(j)}(z(\mathcal{J}))\ (1 \leqslant j \leqslant n) \\ &\text{s.t. } g_i(z(\mathcal{J})) \leqslant \sigma_i(\mathcal{J})\ (1 \leqslant i \leqslant p) \\ &\qquad z(\mathcal{J}) > 0 \end{aligned} \tag{7.54}$$

where

$$\begin{aligned} g_0^{(j)}(z(\mathcal{J})) &= \sum_{k=1}^{J_0^{(j)}} \sigma_{0k} c_{0k}^{(j)} \prod_{l=1}^{m} z_l^{\gamma_{0kl}^{(j)}}(\mathcal{J}), \\ g_i(z(\mathcal{J})) &= \sum_{k=1}^{J_i} \sigma_{ik} c_{ik} \prod_{l=1}^{m} z_l^{\gamma_{ikl}}(\mathcal{J}), \end{aligned}$$

$\sigma_{ik}, \sigma_i(\mathcal{J}) = \pm 1$, and (7.54) contains an optimal solution with $T-FV$, which is equivalent to (7.54) containing an optimal solution depending on a cone index $\mathcal{J}$. ◊

Proof The theorem is easily proved to be true in a similar way to Theorem 6.3.1. □

Corollary 7.7.1 *$\tilde{\sigma}_i$ is taken to be $\tilde{1}$ when σ_i is chosen as 1 in (7.53) and (7.54), so the conclusion corresponding to Theorem 7.7.1 still holds.* ◊

Definition 7.7.2 We call

$$\begin{aligned} \max\ & \tilde{d}^{(j)}(w)\ (1 \leqslant j \leqslant n) \\ \text{s.t.}\ & w_{00} = 1 \\ & \Gamma^{(j)\mathrm{T}} w = 0 \\ & w \geqslant 0 \end{aligned} \tag{7.55}$$

a multi-objective dual GP with $T-$FV corresponding to (7.53), where

$$\tilde{d}^{(j)}(w) = \sigma\left[\prod_{k=1}^{J_0}(\tilde{c}_{0k}^{(j)}/w_{0k}\sum_{k=1}^{J_0}w_{0k})^{\sigma_{0k}w_{0k}}\prod_{i=1}^{p}\prod_{k=1}^{J_i}(\tilde{c}_{ik}/w_{ik}\sum_{k=1}^{J_i}w_{ik})^{\sigma_{ik}w_{ik}}\right]^{\sigma},$$

$$\Gamma^{(j)} = \begin{pmatrix} \sigma_{01}\gamma_{011}^{(j)} & \cdots & \sigma_{01}\gamma_{01n}^{(j)} & \cdots & \sigma_{01}\gamma_{01m}^{(j)} \\ \vdots & & \vdots & & \vdots \\ \sigma_{0J_0}\gamma_{0J_01}^{(j)} & \cdots & \sigma_{0J_0}\gamma_{0J_0n}^{(j)} & \cdots & \sigma_{0J_0}\gamma_{0J_0m}^{(j)} \\ \sigma_{11}\gamma_{111} & \cdots & \sigma_{11}\gamma_{11n} & \cdots & \sigma_{11}\gamma_{11m} \\ \vdots & & \vdots & & \vdots \\ \sigma_{pJ_p}\gamma_{pJ_p1} & \cdots & \sigma_{pJ_p}\gamma_{pJ_pn} & \cdots & \sigma_{pJ_p}\gamma_{pJ_pm} \end{pmatrix} \quad (1 \leqslant j \leqslant n)$$

denote an exponent matrix. Its l-th column is composed of exponents from a variable x_l at each item in objective functions $g_0^{(j)}(\tilde{x})(1 \leqslant j \leqslant n)$ and constraint ones $g_i(\tilde{x})(1 \leqslant i \leqslant p)$; w represents a J-dimension variables vector.

7.7.3 Algorithm

For a multi-objective GP (7.53) with $T-$FV, we can either weight the objective functions before nonfuzzification or nonfuzzify it before its being weighted. Similarly to Section 7.2 and 7.4, two algorithms can be advanced on the assumption that (7.53) has a solution and it is nonfuzzified into (7.54) for our discussion.

I. Direct primal algorithm

Now weight several objective functions in (7.54) and we have

$$g_0^*(z(\mathcal{J})) = \alpha_1 g_0^{(1)}(z(\mathcal{J})) + \alpha_2 g_0^{(2)}(z(\mathcal{J})) + \cdots + \alpha_n g_0^{(n)}(z(\mathcal{J}));$$

thereafter we calculate

$$\begin{aligned}&\min\ g_0^*(z(\mathcal{J}))\\&\text{s.t. } g_i(z(\mathcal{J}))\leqslant \sigma_i(\mathcal{J})\ (1\leqslant i\leqslant p)\\&\qquad z(\mathcal{J})>0\end{aligned}\tag{7.56}$$

which behaves as a single objective GP problem with respect to a cone index $\mathcal{J}$. It is obvious that we have a variety of direct solutions with respect to (7.56) [Cao93a][Eck80] [Pet78] [WB67].

II. Dual algorithm

A single objective GP is obtained, for a certain j, over (7.54), and its corresponding dual programming is (7.55). n groups of optimal solutions and optimal value U_j, L_j are obtained by solving n dual GP in terms of $j(1\leqslant j\leqslant n)$ respectively. Thereafter a single objective fuzzy GP like (7.56) is also obtained in the light of steps 3^0 and 4^0 in Section 7.2.

Finally, the optimal compromise solution to problem (7.54) is acquired respectively by adopting a primal algorithm [Biw92][Cao93a] [WY82], such that a fuzzy optimal compromise solution comes out.

7.7.4 Conclusion

Obviously, we turn multi-objective GP with $T-$FV into a dual problem for solution (except that exponent is fuzzy), whereas variables and constraints in a dual problem are fuzzy, so it is convenient for us to do the research. This result has been used a lot in this book. Meanwhile it is clear that this solution method is easily generalized for models with a variety of FV.

Chapter 8

APPLICATION OF FUZZY GEOMETRIC PROGRAMMING

Having been introduced only a short time ago, and with their initial application only in power engineering and postal services, a fuzzy GP and a multi-objective fuzzy GP have great vitality and potential for function applications.

In this chapter, we introduce application of a fuzzy GP for seeking schemes for waste-water disposal in a power plant [Cao95b][Cao97b], and determining the power supply radius in a transformer substation (TS) [YCL95][Cao99] [Cao01b,c], and of multi-objective GP in postal services [Biw92].

8.1 Fuzzy GP for Seeking Schemes in Waste Water Disposal in Power Plants

We apply a fuzzy sets theory to building a fuzzy GP model in seeking a waste-water disposal scheme in a power plant, use a fuzzy dual algorithm to find a fuzzy annual total cost as well as an optimum fuzzy technological parameter for various schemes from waste-water disposal stations, and determine an optimum scheme by using a minimum value in the annual total expense.

8.1.1 Technological Scheme Design

A thermal power plant is required to design a station which is mainly used to dispose of water-waste and polluted water. The station is made up of continuous 3-level sections with 10 different technological process schedules (see Table 8.1)

The quality of purified water is required to amount to the national standard (removal of the biochemistry oxygen BOD content must be 97.1%) and the annual cost in operation for each disposal process is exhibited in Table 8.2.

Table 8.1 Technological scheme design in waste-water-disposal stations

sec. x sche.	lev I	lev II	lev III	Note
1	PC	$TF + AS$	CA	PC— primary settler
2	PC	$TF + AL$	CSF	TF— drip filter
3	PC	AS	CA	AS— stirring-dreg and
4	PC	AL	CSF	secondary settiring
5	PC	$TF + AS$	CSF	AL— charge air pool
6	PC	AS	CSF	CA— active carbon
7	PC	AS	/	absorber
8	PC	$TF + AS$	/	CSF— condenser and
9	PC	AL	/	filter
10	PC	$TF + AL$	/	

Table 8.2 Annual cost for different disposal processes

proc.	content	ann. charge(unit$ thousand)	
1	PC	$19.4x_1^{-1.47}$	
2	TF	$16.8x_2^{-1.66}$	
3	$TF \to AS$	$91.5x_3^{-0.30}$	where x_l means
4	$PC \to AS$	$86.0x_4^{-0.38}$	remaining amount of
5	$PC \to AL$	$45.9x_5^{-0.45}$	BOD (percentage) in
6	$TF \to AL$	$27.4x_6^{-0.63}$	water in l−th process
7	$AS \to CSF$	$152x_7^{-0.27}$	
8	$AS \to CA$	$120x_8^{-0.33}$	
9	$AL \to CSF$	$179x_9^{-0.37}$	

There are different annual costs for disposal processes in Table 8.2 including ones for capital construction and equipment in depreciation (usage deadline is 18 years with 6% annual depreciation), insurance and operation. The annual cost of the l-th disposal process is denoted by $y_l = c_l x_l^{\gamma_l} (1 \leqslant l \leqslant 9)$.

8.1.2 Building of A Fuzzy GP Model

Generally, after waste-water is purified through m disposal processes, the remaining amount of the BOD (percentage) in water is limited in permissible value $b(b > 0)$. Hence the idea that the annual total cost is made as small as possible involves finding an answer to the fuzzy GP

$$\begin{aligned} &\widetilde{\min}\, g_0(x) = c_1 x_1^{\gamma_1} + c_2 x_2^{\gamma_2} + \cdots + c_m x_m^{\gamma_m} \\ &\text{s.t. } x_1 x_2 \cdots x_m \lesssim b \\ &\quad x_1, x_2, \ldots, x_m > 0, \end{aligned} \tag{8.1}$$

where, γ_l and $c_l (1 \leqslant l \leqslant m)$ are constants, $\widetilde{\min}$ means solving fuzzy minimal value (mentioned in Chapter 2) and "$\lesssim$" denotes "approximate $\leqslant$".

A standard form in (8.1) is

$$\begin{aligned} &\widetilde{\min}\, g_0(x) = c_1 x_1^{\gamma_1} + c_2 x_2^{\gamma_2} + \cdots + c_m x_m^{\gamma_m} \\ &\text{s.t. } b^{-1} x_1 x_2 \cdots x_m \lesssim 1 \\ &\quad x_1, x_2, \ldots, x_m > 0 \end{aligned} \tag{8.2}$$

Give flexible indexes d_0, d_1 to (8.1) (as for (8.2), $d_0' = d_0, d_1' = b^{-1} d_1$ is selected) and we turn (8.1) into a parameter GP:

$$\begin{aligned} &\max\ \alpha \\ &\text{s.t. } c_1 x_1^{\gamma_1} + c_2 x_2^{\gamma_2} + \cdots + c_m x_m^{\gamma_m} - \alpha d_0 \leqslant z_0 \\ &\quad d_1(\alpha - 1) + b^{-1} x_1 x_2 \cdots x_m \leqslant 0, \alpha \in [0, 1] \\ &\quad x_1, x_2, \ldots, x_m > 0 \end{aligned} \tag{8.3}$$

According to the discussion [Cao89a], the steps for problem (8.1) are concluded as follows:

1^0 Solve an ordinary GP

$$\begin{aligned} &\min\ g_0(x) = c_1 x_1^{\gamma_1} + c_2 x_2^{\gamma_2} + \cdots + c_m x_m^{\gamma_m} \\ &\text{s.t. } x_1 x_2 \cdots x_m \leqslant b \\ &\quad x_1, x_2, \ldots, x_m > 0 \end{aligned} \tag{8.4}$$

Find an optimal solution $x^{*1} = (x_1^{*1}, x_2^{*1}, \cdots, x_m^{*1})^{\mathrm{T}}$ and an optimal value $g_0(x^{*1})$.

2^0 Solve an ordinary GP problem

$$\begin{aligned} &\min\ g_0(x) \\ &\text{s.t. } x_1 x_2 \cdots x_m \leqslant b + d_1 \\ &\quad x_1, x_2, \ldots, x_m > 0 \end{aligned} \tag{8.5}$$

Find an optimal solution $x^{*2} = (x_1^{*2}, x_2^{*2}, \ldots, x_m^{*2})^{\mathrm{T}}$ and an optimal value $g_0(x^{*2}) = g_0(x^{*1}) + d_0$.

3^0 Solve the ordinary GP problem

$$\begin{aligned} &\max\ \alpha = \min(-\alpha) \\ &\text{s.t. } x_1 x_2 \cdots x_m \leqslant b + d_1 \\ &\quad g_0(x) \leqslant z_0 + d_0\alpha \\ &\quad 0 \leqslant \alpha \leqslant 1 \\ &\quad x_1, x_2, \ldots, x_m > 0 \end{aligned} \tag{8.6}$$

If we gain the optimal solution to (8.6) $(x_1^*, x_2^*, \ldots, x_m^*; \alpha^*)^{\mathrm{T}}$, then the optimal point of problem (8.1) is $x^* = (x_1^*, x_2^*, \ldots, x_m^*)^{\mathrm{T}}$ and its correspondent objective function value is $\tilde{G}^* = \tilde{g}_0(x^*)$, with its optimal level $\tilde{D}(x^*) = \alpha^*$.

There exist m variables and $m+1$ terms in problem (8.4) and (8.5), all of their DD is $d = (m+1) - m - 1 = 0$, with its dual constraint below:

$$\begin{aligned} &w_{01} + w_{02} + \cdots + w_{0m} = 1, \\ &\gamma_l w_{0l} + w_{11} = 0\ (1 \leqslant l \leqslant m). \end{aligned}$$

A unique dual solution to it means a dual optimal one

$$\begin{aligned} &w_{0l}^* = -w_{11}/\gamma_l (1 \leqslant l \leqslant m), \\ &w_{11} = -1\Big/\sum_{l=1}^{m} \frac{1}{\gamma_l}, \end{aligned} \tag{8.7}$$

which is substituted for a dual objective function

$$\max \left(\frac{-\alpha}{w_{00}}\right)^{w_{00}} \prod_{l=1}^{m} \left(\frac{c_l/w_{0l}}{z_0 + d_0\alpha}\right)^{w_{0l}} \left(\frac{1/w_{11}}{b + d_1}\right)^{w_{11}} w_{11}^{w_{11}}.$$

We can obtain a dual optimal value for (8.6), such that it is a fuzzy optimal one for (8.1).

8.1.3 Numerical Example

According to Table 8.1 and 8.2, we can calculate 10 different schemes corresponding to dual optimal solutions and optimal value $g_0(x^{*1})$ and $g_0(x^{*2})$ of (8.4) and (8.5), respectively. Now 10 different schemes obtained by (8.4) (the upper left-corner numbers) and (8.5) (the lower right-corner numbers) are listed in Table 8.3, showing an expenditures schedule in which the total annual cost is required to be expended differently for the BOD content in the removed water.

In Table 8.3, the removed BOD content in water of schemes 7,8,9 and 10 shown in "×" amounts to over 90% at least, but no more than 95%, let alone the required 97.1%, so that they are not feasible according to the national standard.

Table 8.3 Annual total cost for disposal schemes by each requirement

sche \ requ	97.1%	98%	95%	95.9%	90%	90.9%	85%	85.9%	80%	80.9%
1	388	409	361	370	330	334	312	315	301	303
2	391	418	354	367	312	313	292	294	277	278
3	392	415	359	371	322	327	302	305	289	291
4	395	422	358	371	317	322	296	298	282	282
5	419	438	393	402	362	366	344	347	331	334
6	423	446	392	403	355	359	335	338	321	323
7	×	×	×	×	211	217	187	190	171	174
8	×	×	×	×	211	215	193	195	181	183
9	×	×	×	×	143	148	124	127	114	115
10	×	×	×	×	138	142	120	122	108	110

Now, let us have a look at scheme 1, it consists of 4 series of disposal processes. In order to find an answer to the optimal technological parameter x_l^* to the process, we have

Example 8.1 Find

$$\begin{aligned}\min\ & g_0(x) = 19.4x_1^{-1.47} + 16.8x_2^{-1.66} + 91.5x_3^{-0.3} + 120x_8^{-0.33}\\ \text{s.t.}\ & x_1x_2x_3x_8 \leqslant b\\ & x_1, x_2, x_3, x_8 > 0.\end{aligned}$$

When we take

$$b = 1 - 0.971 = 0.029,\ \text{i.e.,}\ b^{-1} = 34.5,$$

the problem is equivalent to finding a solution to

$$\begin{aligned}\min\ & g_0(x) = 19.4x_1^{-1.47} + 16.8x_2^{-1.66} + 91.5x_3^{-0.3} + 120x_8^{-0.33} \\ \text{s.t.}\ & 34.5x_1x_2x_3x_8 \leqslant 1 \\ & x_1, x_2, x_3, x_8 > 0,\end{aligned} \tag{8.8}$$

whose objective function of the dual programming behaves as

$$\max \left(\frac{19.4}{w_{01}}\right)^{w_{01}} \left(\frac{16.8}{w_{02}}\right)^{w_{02}} \left(\frac{91.5}{w_{03}}\right)^{w_{03}} \left(\frac{120}{w_{04}}\right)^{w_{04}} \left(\frac{34.5}{w_{11}}\right)^{w_{11}} w_{11}^{w_{11}}.$$

From (8.7), we can obtain optimal solution

$$w_{01}^* = 0.089, w_{02}^* = 0.081, w_{03}^* = 0.436, w_{04}^* = 0.394, w_{11}^* = 0.131$$

in the dual programming, with optimal value $g_0^* = 388$.

Give a flexible index $d_1 = 0.09$ and we have $b' = b - d_1 = 0.029 - 0.009 = 0.02, b'^{-1} = 50$. The above optimal solution can be obtained in the same way, with optimal value $g_0^{*1} = 409$, such that $d_0 = g_0^{*1} - g_0^* = 409 - 388 = 21$.

The problem (8.8) is relevant to the equivalent form of (8.6) as follows:

$$\begin{aligned}& \max \alpha = \min(-\alpha) \\ \text{s.t.}\ & g_0(x) = 19.4x_1^{-1.47} + 16.8x_2^{-1.66} + 91.5x_3^{-0.3} + 120x_8^{-0.33} \\ & \qquad \leqslant 388 + 21\alpha \\ & x_1x_2x_3x_8 \leqslant 0.02 \\ & 0 \leqslant \alpha \leqslant 1 \\ & x_1, x_2, x_3, x_8 > 0\end{aligned}$$

and its dual form

$$\begin{aligned}\max\ & \left(\frac{-\alpha}{w_{00}}\right)^{w_{00}} \left(\frac{19.4/w_{01}}{388+21\alpha}\right)^{w_{01}} \left(\frac{16.8/w_{02}}{388+21\alpha}\right)^{w_{02}} \\ & \left(\frac{91.5/w_{03}}{388+21\alpha}\right)^{w_{03}} \left(\frac{120/w_{04}}{388+21\alpha}\right)^{w_{04}} \left(\frac{1}{0.02w_{11}}\right)^{w_{11}} (w_{11})^{w_{11}} \\ \text{s.t.}\ & w_{00} = w_{01} + w_{02} + w_{03} + w_{04} = 1 \\ & \gamma_l w_{0l} + w_{11} = 0, 0 \leqslant \alpha \leqslant 1\ (1 \leqslant l \leqslant 4) \\ & w \geqslant 0\end{aligned} \tag{8.9}$$

Solve

$$w_{00} = 1, w_{01} = 0.089, w_{02} = 0.081, w_{03} = 0.436, w_{04} = 0.394, w_{11} = 0.31,$$

which is put into (8.9) and then

$$
\begin{aligned}
&406.846/(388+21\alpha)=1\Rightarrow\alpha=0.8974,\\
&d(w^*)=(19.4/0.089)^{0.089}(16.8/0.081)^{0.081}(91.5/0.436)^{0.436}\\
&\qquad(120/0.394)^{0.394}\left(\frac{1}{0.02\times 0.131}\right)^{0.131}(0.131)^{0.131}\\
&\quad=406.846.
\end{aligned}
$$

Again

$$
\begin{aligned}
&19.4x_1^{-1.47}=406.846\times 0.089,\quad 16.8x_2^{-1.66}=406.846\times 0.81,\\
&91.5x_3^{-0.3}=406.846\times 0.436,\quad 120x_8^{-0.33}=406.846\times 0.394,\\
&50x_1x_2x_3x_8=1,
\end{aligned}
$$

and we obtain

$$x_1^*=0.654, x_2^*=0.666, x_3^*=0.11, x_8^*=0.416.$$

Obviously, scheme 1 is the best one, where 4 series of composite disposal processes are $PC \to TF \to AS \to CA$, with the optimal technological parameter of each process being

$$x_1^*=0.654,\ x_2^*=0.666,\ x_3^*=0.11,\ x_8^*=0.416.$$

The removal of the BOD in water is 34.6% from PC, 33.4% from TF, 89% from AS and 58.4% from CA.

Through four steps of the disposal process, BOD is removed from waste water by 98%, which is lower than the set quota 97.1% by the national standard. And its annual total cost is minimum, with only 406.85 (thousand$), which means the content of BOD in waste water will decrease approximately by 1% if we add $18,000 to the initial investment.

The method mentioned here can be applied to optimal disposal of other pollution, and it is of practical use in research for environment protection and pollution disposal as well.

8.2 Application of GP and GP with Fuzzy Coefficients in Supply Radius Option in A City TS

8.2.1 Introduction

With the fast development of China's power industry, it is of practical value for us to choose power supply radiuses in the TS optimally. So far, optimal mathematical models have been built in static and dynamic states [Obr86][YZY87] [YWY91], but the search for a power supply radius is difficult in a complex practical environment. In order to make our built models more practical with more information, we apply the fuzzy GP advanced in 1987 [Cao87a,b] into the search for an economic supply radius in TS by means of a fuzzy set theory [Cao95b][Cao01][Zad65a]. This result has more correct and practical characteristics than a classical model.

8.2.2 Building A GP Model

Consider $110kV$-TS power supply to its consumers in a city by a direct-step-down method of $10kV$ and we shall build a model in the annual-cost way.

Suppose [YWY91]:

1) The power supply range centers at $110kV$-TS, with radius r and 10kV middle voltage distribution net-work being the radiation structure.

2) The cost of investment is annual operational cost μ under a certain load level, without consideration of an investment process.

3) An investment-recovery deadline is determined to be redeemable years N(8—10 years), i.e., the total investment is returned within redeemable years.

4) The load density is even over the whole electrified wire netting cover.

Therefore, a static model is built by means of an annual cost as follows:

$$F = Z/N + \mu$$

where Z denotes a cost of total investment.

According to Reference [YWY91], the annual cost function in a unit capacity is denoted by

$$F_0 = \frac{F}{S} = \frac{(Z_b + Z_l)/N + \mu_1 + \mu_2 + \mu_3}{S} \tag{8.10}$$

where

$$Z_b = a_1 + b_1 S \text{ (yuan)} \tag{8.11}$$

is the investment in the construction of $110kV$-TS [YZY87].

$$Z_l = L(Ma_2 + b_2 S_l) \text{ (yuan)} \tag{8.12}$$

denotes the construction investment in the main-supply lines of $10kV$ middle voltage distribution net-work, where $M = S\cos\phi/P_{av}$ is a circle line of $10kV$ middle voltage distribution net-work, $L = Er(\text{km})$ is each circle-line length of it, and $S_l = \frac{S}{\sqrt{3}U_{Nj}}(mm^2)$ denotes the wire total selection in the main-supply lines of all $10kV$ middle voltage distribution net-work in $110kV$-TS; S is capacity in the TS (unit, kVA).

$$\mu_1 = H(Z_b + Z_l) \tag{8.13}$$

behaves as a direct proportion function of unchangeable part in operations cost (large repair, small repair and depreciation charge) and in the total investment of $110kV$-TS and $10kV$ middle voltage distribution net line.

$$\begin{aligned}\mu_2 &= \Delta P\tau C_0\\ &= 7.26\frac{E\rho j\tau C_0}{U_N\cos^2\phi}\frac{1}{\sqrt{\sigma}}S^{3/2}10^{-5}\end{aligned} \tag{8.14}$$

stands for the depreciation charge of $10kV$ line in a year, while the depreciation charge to transformers of $110kV$-TS is [Obr86]

$$\mu_3 = (C_0K_{Fe}8760 + C_0\tau_bK_{cu}) \times \gamma S_{NL}^{3/4} \tag{8.15}$$

where γ denotes the number of transformers. When the rated capacity as S_N hours for a chosen transformer is $10kV$-TS, γ is taken to denote an average number of transformers S/S_N in the $10kV$-TS.

Substitute (8.11)—(8.15) for (8.10), then

$$\begin{aligned}F_0 =&(\frac{1}{N} + H)(\frac{a_1}{S} + b_1) + E[(\frac{1}{N} + H)\sqrt{\frac{1}{\pi\sigma K_c}}(\frac{a_2\cos\phi}{P_{av}} + \frac{b_2}{\sqrt{3}U_{Nj}})\\ &+ 7.26\frac{\rho j\tau C_0}{U_N\cos^2\phi}\frac{10^{-5}}{\sqrt{\sigma}}]S^{\frac{1}{2}} + (8760C_0K_{Fe} + C_0\tau_bK_{cu})\frac{S_{NL}^{\frac{3}{4}}}{S_N}\end{aligned} \tag{8.16}$$

where

H– annually extracted coefficient is a yearly operation cost in total extracted investment.

a_1– the unconnected part of capacity for TS in the investment (unit, $yuan$);

a_2–the investment of each kilometer unconnected in wire section in $10kV$ line investment (unit, $yuan/km$);

b_1–the coefficient of the part of connected with capacity of the TS in investment (unit, $yuan/kVA$).

b_2–the coefficient of the part of $10kV$ line investment connected with wire section (unit, $yuan/km\cdot mm^2$);

E–the terrain correction coefficient;

σ–the mean load density (unit, kW/km^2);
K_c–the hold ratio (the $110kV$-TS takes $2.2 \sim 2.5$);
P_{av}–line mean load each time (unit, kW);
U_N–the rated voltage of middle voltage distribution net-work ($10kV$);
j–the wire in economical current density (unit, A/mm^2);
ρ–the wire in resistivity (unit, $\Omega/km \cdot mm^2$);
τ–wasted hours of mean maximum load per year;
C_0–cost price of wasted watt-hour (unit, $yuan/kW \cdot h$);
K_{Fe}–transformer iron loss coefficient ($\approx 0.0085kW/(kVA)^{3/4}$);
τ_b–equivalent hours of transformer copper loss;
K_{cu}–transformer copper loss coefficient ($\approx 0.055kW/(kVA)^{3/4}$);
S_{NL}–rated load of transformer (unit, kVA);
S_N–rated capacity of transformer (unit, kVA);
Q–voltage range in electrically-distributed cities (unit, km^2).

The determination of objective function of a static model aims at making unit capacity annual cost minimum, i.e., $\min F_0$, with the limits to constraint $S > 0$, such that we have a model as follows:

$$\begin{aligned} &\min \ F_0 \\ &\text{s.t.} \ \ S > 0 \end{aligned} \tag{8.17}$$

where F_0 is illustrated as (8.16), and (8.17) is called a GP model, written as Model I.

As far as problem (8.17) is concerned, we have various solutions[Cao93a] [Cao95b][Cao01b,c]. Now, of them one is introduced by a numerical example.

Example 8.2 Select a group of parameter value as follows:
N=10, $\cos\phi$=0.9, τ=2400h, E=1.3, C_0=0.06yuan/$kW \cdot h$, H=8.0%, ρ=31.5 $\Omega/km \cdot mm^2$, U_N=10kV, j=1.15A/mm^2, a_1=510000yuan, b_1=50yuan/kVA, a_2=18640yuan/km, b_2=89yuan/$km \cdot mm^2$, P_{av}=4000kW, K_c=2.2, Q=49.03 km^2, K_{Fe}=0.0085$kW/(kVA)^{3/4}$, K_{cu}=0.055$kW/(kVA)^{3/4}$.

Put them into (8.16), such that

$$F_0 = 91800S^{-1} + 0.77S^{\frac{1}{2}}\sigma^{-\frac{1}{2}} + 9.63,$$

thus (8.17) is simplified into solving the following

Model I

$$\begin{aligned} &\min \ \{91800S^{-1} + 0.77S^{\frac{1}{2}}\sigma^{-\frac{1}{2}}\} \\ &\text{s.t.} \ S > 0 \end{aligned} \tag{8.18}$$

That is a GP model we obtained.

8.2.3 Optimal Solution to Model

There exist varieties of solutions to GP (8.18). If adopting a dual algorithm for its optimal solution [Cao89a], we turn (8.18) into a dual form first

$$\begin{aligned}
&\max\ d = \max\Big(\frac{91800}{w_1}\Big)^{w_1}\Big(\frac{0.77}{\sqrt{\sigma}w_2}\Big)^{w_2}\\
&\text{s.t. } w_1 + w_2 = 1\\
&\quad - w_1 + \frac{1}{2}w_2 = 0\\
&\quad w \geqslant 0
\end{aligned}$$

whose optimal solution is denoted by $w_1 = \frac{1}{3}, w_2 = \frac{2}{3}$, and its optimal value is

$$d = 65.061 \times \Big(\frac{2.31}{2\sqrt{\sigma}}\Big)^{2/3}.$$

Select $\sigma = 5915, d \approx 3.9571$. Again from

$$\begin{cases} 91800S^{-1} = \frac{1}{3} \times 3.9571 \\ 0.01S^{\frac{1}{2}} = \frac{2}{3} \times 3.9571 \end{cases} \tag{8.19}$$

$$\xrightarrow{\text{take an average value}} \quad S \approx 69595$$

and from calculating formula r of the mean radius in supply area, average item N_b and numbers n_b in unit area in the TS as follows:

$$r = \sqrt{\frac{S}{\pi\sigma K_c}} \tag{8.20}$$

$$N_b = \frac{\sigma Q K_c}{S} = \frac{Q}{\pi r^2} \tag{8.21}$$

$$n_b = \frac{N_b}{Q} = \frac{\sigma K_c}{S}\frac{1}{\pi r^2} \tag{8.22}$$

we get $r \approx 1.3$.

Calculate the economic capacity S_l and economic supply radius r_l of $110kV$-TS for different values of σ_l, and we have the result in Table 8.4 If the mean load density of a certain urban district shall have gone up to $\sigma = 5915kW/km^2$ by the year 2000, we can determine the economic capacity of the TS $S = 69595$, a mean supply radius $r \approx 1.3km$. Thus we find the answer to be $N_b \approx 9.2$, i.e., about 9.2 $110kV$-TS needs to be built up in this district, which conforms

Table 8.4 Model I for S_l and r_l

$\sigma_l(kW/km^2)$	1566.3	3234.9	4877.6	5915	8300	11044
$S_l(kVA)$	44656	56869	65211	69595	77854	85631
$r_l(km)$	2	1.59	1.39	1.3	1.17	1.06

with the practical case.

8.2.4 GP with Fuzzy Coefficients and Its Application.

We fuzzify an objective function F_0, so that an annual cost in unit capacity is the fuzzy minimum, i.e., $\widetilde{\min} F_0$, because there exist some uncertainty factors in load, investment process and electricity prices, where they contain a lot of stochastic and fuzzy phenomena, and because classical mathematical models neglect such important phenomena which not only distorts the models but also severely limits their adaptation.

We know that the capacity in the TS is nonnegative, i.e., $S > 0$, and F_0 is an exponential polynomial function with fuzzy coefficients with respect to S [Cao01b]. Therefore the problem can be changed into finding an answer to the fuzzy minimum of an annual cost under the capacity $S > 0$ in constraint, such that GP with fuzzy coefficients can be obtained, written as model II.

Example 8.3 Find a solution to

$$\begin{aligned} &\min\ \tilde{F}_0 \\ &\text{s.t. } S > 0 \end{aligned} \tag{8.23}$$

where objective function is

$$\tilde{F}_0 = \widetilde{91800} S^{-1} + \widetilde{0.7\sigma}^{-\frac{1}{2}} S^{\frac{1}{2}}$$

and its coefficient not only an ordinary real number but also a fuzzy number varying in a certain interval.

When a function's coefficient denotes a fuzzy number varying in the interval, we might as well suppose the second coefficient of objective function in (8.18) is $\tilde{c}_2$, and then it may be freely fixed in the value interval $[0.008, 0.03]$, with its membership function being [Ver84]

$$\tilde{\phi}(c_2) = \begin{cases} 0, & c_2 \leqslant 0.008 \\ \dfrac{(c_2 - 0.008)^2}{0.000484}, & 0.008 < c_2 < 0.03 \\ 1, & c_2 \geqslant 0.03. \end{cases} \tag{8.24}$$

Then $c_2 = 0.008 + 0.022\sqrt{1-\alpha}$ is obtained, such that, program (8.18) is turned into

$$\begin{array}{l} \max\ \{91800S^{-1} + (0.008 + 0.022\sqrt{1-\alpha})S^{\frac{1}{2}}\} \\ \text{s.t.}\ \ S > 0 \end{array} \tag{8.25}$$

which denotes a PGP with parameters. Because of the DD $D = 0$, it is easy to find a solution by means of dual programming, and its dual form is denoted by

$$\begin{aligned} &\max\left(\frac{91800}{w_0}\right)^{w_0}\left(\frac{0.008 + 0.022\sqrt{1-\alpha}}{w_1}\right)^{w_1} \\ &\text{s.t.}\ \ w_0 + w_1 = 1 \\ &\qquad - w_0 + \frac{1}{2}w_1 = 0, \alpha \in [0, 1] \\ &\qquad w_0, w_1 > 0 \end{aligned}$$

Find $w_0 = \dfrac{1}{3}, w_1 = \dfrac{2}{3}$. When we take $\alpha = 0.8$, then

$$d = \left(\frac{91800}{1/3}\right)^{1/3}\left(\frac{0.009}{2/3}\right)^{2/3} \approx 3.6887.$$

In terms of formula (8.19), an optimal solution obtained for (8.25) is $S = 74660.45$, S and d are approximate optimal solutions and optimal values for (8.23). When we take $\alpha = 0.8$ and use formula (8.20)–(8.22), we have $r = 1.35$ for economical power supply radius as well as number $N_b \approx 8.56$ for the TS.

8.2.5 Conclusion

In this section, we tried initially applying PGP and PGP with a fuzzy coefficient in power systems, the effect of which is wonderful. As for application of the fuzzy GP and multi-objective fuzzy GP, we shall have further discussion in 8.3 and 8.4, respectively.

8.3 Further Study of Fuzzy GP in Optimum Option Supply Radius

8.3.1 Model Building

An objective function possesses a character of some flexibility because of errors such as fluctuations in the TS and investment cost coefficient of middle voltage distribution lines as well as influence on electricity prices varying in time. Since σ is a parameter with its unit being kW/km^2, it is proper to take 1566.3, 3234.9, 4877.6, 5915, 8300, 11044 for σ through determination, and we fuzzify an objective function F_0, which enables the annual cost in unit capacity to be fuzzy minimum, i.e.,

$$\widetilde{\min}\, F_0.$$

But capacity in the TS is nonnegative, i.e., $S > 0$, and F_0 denotes an exponent polynomial function with respect to S, which is still like (8.16). Therefore, the problem can be converted into finding a fuzzy minumum for the annual expense subject to capacity $S > 0$, such that a model can be obtained as follows:

$$\begin{aligned} &\widetilde{\min}\, F_0 \\ &\text{s.t. } S > 0 \end{aligned} \tag{8.26}$$

written as Model III, which is a fuzzy GP model, where the objective function is $F_0 \lesssim z_0$ [Cao93a][Cao95c][Cao01b,c].

Find the answer to (8.26) and we obtain its fuzzy optimal solution and then we get mean-electric-supplying radius r, mean numbers N_b and mean numbers n_b per unit area in the TS by means of formula (8.20)–(8.22).

Example 8.4 There exist various solutions in fuzzy GP model (8.26)[Cao95c] [Cao01b][TA84]. We are to introduce two of them with numerical examples. For the sake of simple calculation, we still use the same group of parameter values from Example 8.2, which are put into (8.26), such that

$$\begin{aligned} &\widetilde{\min}\{F_0 = 91800S^{-1} + 0.77S^{\frac{1}{2}}\sigma^{-\frac{1}{2}}\} \\ &\text{s.t. } S > 0 \end{aligned} \tag{8.27}$$

Now, we use parameter $\sigma = 5915$ for solution to (8.27) in this section. The programming model is turned into

$$\begin{aligned} &\widetilde{\min}\{91800S^{-1} + 0.01S^{\frac{1}{2}}\} \\ &\text{s.t. } S > 0 \end{aligned} \tag{8.28}$$

8.3.2 Solution

The solution to (8.28) is equivalent to finding three determinacy GP, whose steps to solution [Cao93a][Cao95b] are concluded as follows:

1) Find a solution to a determinacy GP

$$\begin{aligned}&\min\{91800S^{-1}+0.01S^{\frac{1}{2}}\}\\&\text{s.t. } S>0\end{aligned} \tag{8.29}$$

whose dual programming means

$$\begin{aligned}&\max\ d=\max\left(\frac{91800}{w_1}\right)^{w_1}\left(\frac{0.01}{w_2}\right)^{w_2}\\&\text{s.t. } w_1+w_2=1\\&\quad -w_1+\frac{1}{2}w_2=0\\&\quad w_1,w_2\geqslant 0\end{aligned}$$

and its optimal solution and objective optimal value denote $w_1=\frac{1}{3}, w_2=\frac{2}{3}$ and $d_0=3.9571$, respectively, such that we can have an approximately optimal solution $S_0\approx 69595$ to the determinacy GP (8.29).

2) Find a solution to another determinacy GP

$$\begin{aligned}&\min\{91800S^{-1}+0.01S^{\frac{1}{2}}\}\\&\text{s.t. } 0<S\leqslant(S_0+t_1)=71320\end{aligned} \tag{8.30}$$

where $t_1=1725$ is determined by testing and by the formula

$$\tilde{\psi}_i(t_1)=\begin{cases}1-\alpha t_1, & 0\leqslant t_1<t_0\\0, & t_1\geqslant t_0.\end{cases}$$

Here $t_0=1/\alpha$, α presents a properly chosen number in the interval [0,1].

Then we have $d_1=3.9609$ after the solution to (8.30).

3) Find a solution to the third determinacy GP

$$\begin{aligned}&\max\ \alpha\\&\text{s.t. } S+1725\alpha\leqslant 71320\\&\quad -91800S^{-1}-0.01S^{\frac{1}{2}}-z_0\alpha\leqslant -d_1, \alpha\in[0,1]\\&\quad S>0\end{aligned} \tag{8.31}$$

where $z_0=d_1-d_0=0.0038$.

Turn (8.31) into

$$\begin{aligned}&\max\ \alpha = \min(-\alpha)\\ &\text{s.t.}\ \frac{-91800S^{-1} - 0.01S^{\frac{1}{2}}}{-3.9609 + 0.0038\alpha} \leqslant 1\\ &\qquad \frac{1}{71320 - 1725\alpha}S \leqslant 1, \alpha \in [0,1]\\ &\qquad S > 0\end{aligned}$$

whose term number $J = 4$, variable number $m = 2$, its DD $D = 4-2-1 = 1$, and its dual programming is

$$\max\left(\frac{-\alpha}{w_{00}}\right)^{w_{00}}\left(\frac{-91800/w_{01}}{-3.9609 + 0.0038\alpha}\right)^{w_{01}}\left(\frac{-0.001/w_{02}}{-3.9609 + 0.0038\alpha}\right)^{w_{02}}$$
$$\left(\frac{1/w_{11}}{71320 - 1725\alpha}\right)^{w_{11}}(w_{01} + w_{02})^{w_{01}+w_{02}}w_{11}^{w_{11}} \tag{8.32}$$

$$\begin{aligned}\text{s.t.}\ & w_{00} = 1\\ & w_{01} + w_{02} = 1\\ & -w_{01} + \tfrac{1}{2}w_{02} + w_{11} = 0\\ & \alpha \in [0,1]\\ & w_{00}, w_{01}, w_{02}, w_{11} \geqslant 0\end{aligned} \tag{8.33}$$

A general solution to (8.33) behaves as

$$\begin{cases} w_{01} = 1 - w_{02}\\ w_{11} = 1 - \dfrac{3}{2}w_{02}\end{cases}$$

$$\Rightarrow \quad \begin{cases} w = \left(w_{00}, 1 - w_{02}, w_{02}, 1 - \dfrac{3}{2}w_{02}\right)^{\mathrm{T}}\\ 0 \leqslant w_{02} \leqslant \dfrac{2}{3},\end{cases}$$

which is a 1–dimensional linear constraint optimization problem. If we take $w_{02} = \frac{2}{3}$, then

$$w = (w_{00}, w_{01}, w_{02}, w_{11})^{\mathrm{T}} = (1, \frac{1}{3}, \frac{2}{3}, 0)^{\mathrm{T}}$$

and its objective function of dual programming is

$$\bar{d} = (-\alpha)\frac{(-275400)^{1/3}(-0.015015)^{2/3}}{-3.9609 + 0.0038\alpha}.$$

By the formula

$$
\begin{aligned}
&-\alpha = \bar{d} \\
\Rightarrow\ &-3.9609 + 0.0038\alpha = -3.95967,
\end{aligned}
$$

i.e., $\alpha = 0.337$, such that

$$
\begin{aligned}
d &= \frac{-3.9609 + 0.0038\alpha}{-\alpha}\bar{d} \\
&= -3.9609.
\end{aligned}
$$

Now, the minimum value gained from (8.28) is 3.9609.

Again, from the formula [Cao93a],

$$
\begin{cases}
91800S^{-1} = \frac{1}{3} \times 3.9609 \\
0.01S^{\frac{1}{2}} = \frac{2}{3} \times 3.9609
\end{cases}
$$

$$
\Rightarrow \quad \begin{cases}
S_1 \approx 69530 \\
S_2 \approx 69728,
\end{cases}
$$

therefore

$$
S = \frac{S_1 + S_2}{2} \approx 69629.
$$

And $S = 69629$ is an approximate optimal solution to (8.28), and approximately optimal value is 3.9609. From formulas (8.20)–(8.22), we have

$$
r \approx 1.31, \ N_b = \frac{49.03}{\pi r^2} \approx 9.09
$$

so that the economical power supply radius is $r \approx 1.31km$ and 9.09 $110kV$-TS need building in the city when σ is chosen as 5915 and economic capacity as $S = 69629 \in [69530, 69728]$ in $110kV$-TS.

8.3.3 Comparison Analysis in Results

We list the results in Table 8.5 for static, dynamic models, Model I, II and III [Cao01b,c].

As far as different σ_l is concerned, we can calculate the economic capacity S_l, the economical supply radius r_l and the numbers of N_b in the TS, respectively.

In fact, 6~10 $110kV$-TS shall have been planned to be built by the end of 2000 in the city, which coincides with the results mentioned above. Therefore, the models can be applied to a city-electric-net planning.

From Table 8.5, we can know that a dynamic model is superior to a static one, also fuzzy GP model II and III are obviously better than a static one

Table 8.5 Comparison of S, r and N_b

σ=5915(kW/km^2)	S/kVA	r/km	N_b
static	66100	1.27	9.68
dynamic	71320	1.32	8.96
I	69595	1.3	9.2
II	74660.45	1.35	8.56
III	69629	1.31	9.09

and model I, even much better than a dynamic one. Meanwhile, as for the high load level, the chosen capacity of the TS in Table 8.5 is little smaller, yet numbers of TS are more. This lies in failure to consider the process of load growth and of the construction investment in the TS such as increase of land requisition and distribution line construction cost, so that a classical model and its method operate with more difficulty in urban electrified wire netting construction. In consideration of objective flexibility and coefficient fluctuation in Model II and III (of course, the flexibility of constraint and its coefficient fluctuation can be discussed in a similar way, which is omitted here), they contain more information than classical ones, leaving plenty of room for decisions. Therefore decision makers can adjust themselves to their practical requirement. For example, we shall achieve good results when we choose the optimal schedule in the choice of larger economic capacity with fewer numbers of TS.

The effect of the obtained schedule will be better if we consider other factors again, such as investment-cost coefficients both in the TS and in the middle voltage distribution lines as well as the impact on electricity prices varying in time.

8.3.4 Conclusion

The model built in this section contributes much to quick-finding of a mean power supply radius r_l for the determined threshold value α and an urban different mean load density σ_l, under economical capacity S_l permitted by the TS. Especially, when the DD of fuzzy GP is $D > 0$, a corresponding optimal solution with parameter GP like (8.32) and (8.33) can be obtained by a search. This inspires us with an idea that a systematic control model can be set up by means of fuzzy GP, so that the most economic power supply radius with smallest annual cost can be obtained under the maximal economic capacity in the TS after parameters α, σ_l and $S_l(1 \leqslant l \leqslant m)$ are adjusted to practical requirements.

8.4 Application of Multi-Objective Fuzzy GP in Postal services

Postal regulations are placed on the size of parcels. For example, the length of a parcel is restricted to no more than three and a half feet while its length plus girth is restricted to no more than six feet. Let us assume that a rectangular box-shape parcel is to be sent by mail, where we wish to maximize the volume as well as the surface area of the parcel, satisfying the postal regulations. This problem can be formulated as a vector-maximum/vector-minimum problem.

Let x_1, x_2 and x_3 feet be the length, width and height of the box, respectively. Its volume and surface area are denoted by v and s, respectively.

The problem is stated as:

$$\max\ v = x_1 x_2 x_3$$

or

$$\min\ v^{-1} = g_0^{(1)}(x) = x_1^{-1} x_2^{-1} x_3^{-1} \tag{8.34}$$

$$\max\ s = 2(x_1 x_2 + x_2 x_3 + x_3 x_1)$$

or

$$\min\ s^{-1} = g_0^{(2)}(x) = \frac{1}{2}(x_1 x_2 + x_2 x_3 + x_3 x_1)^{-1} \tag{8.35}$$

$$\begin{aligned} \text{s.t. } g_1(x) &= \frac{2}{7}x_1 \leqslant 1 \\ g_2(x) &= \frac{1}{6}x_1 + \frac{1}{3}x_2 + \frac{1}{3}x_3 \leqslant 1 \\ x_1, x_2, x_3 &> 0 \end{aligned} \tag{8.36}$$

The above vector-maximum/vector-minimum problem may be considered as a multi-objective non-LP problem.

Example 8.5 Find (8.34), (8.35) and (8.36).

The second objective function (8.35) is not a posynomial for the vector-minimum problem. So, we use a general purpose non-LP algorithm, namely the ellipsoid algorithm [EK87] to solve the problem instead of a GP algorithm. The maximum volume is $v = 2$ cubic feet at $x_1 = 2, x_2 = 1$ and $x_3 = 1$ and the maximum surface area $s = \dfrac{504}{49}$ square feet at $x_1 = \dfrac{18}{7}, x_2 = \dfrac{6}{7}$ and $x_3 = \dfrac{6}{7}$. For the vector-minimum problem $g_0^{(1)}(x) = \dfrac{1}{2}$ and $g_0^{(2)}(x) = \dfrac{49}{504}$, by using the two different solutions $x^{(1)} = (2, 1, 1)$ and $x^{(2)} = (\frac{18}{7}, \frac{6}{7}, \frac{6}{7})$, we obtain the values of L_j and U_j as follows:

$$L_1 = \frac{1}{2}, U_1 = \frac{343}{648}, L_2 = \frac{49}{504} \text{ and } U_2 = \frac{1}{10}.$$

By Steps 3, 4, 5 and 6 of the method in Section 7.2, a single-objective fuzzy non-LP problem is formulated as:

$$\begin{aligned}
&\min\ x_4^{-1} \\
&\text{s.t. } g_1(x) = \frac{2}{7}x_1 \leqslant 1 \\
&\quad g_2(x) = \frac{1}{6}x_1 + \frac{1}{3}x_2 + \frac{1}{3}x_3 \leqslant 1 \\
&\quad g_3(x) = \frac{648}{343}x_1^{-1}x_2^{-1}x_3^{-1} + \frac{19}{343}x_4 \leqslant 1 \\
&\quad g_4(x) = 5(x_1x_2 + x_2x_3 + x_3x_1)^{-1} + \frac{1}{36}x_4 \leqslant 1 \\
&\quad x_1, x_2, x_3, x_4 > 0
\end{aligned}$$

In this problem the constraint $g_4(x)$ is not a posynomial, so the problem is solved by using a general purpose non-LP algorithm, namely the ellipsoid algorithm. The values of the decision variables are obtained as $x_1 = 2.28200, x_2 = 0.92949, x_3 = 0.92949$ and $x_4 = 0.74375$. At this optimal compromise solution, volume is $v = 1.97157$ cubic feet and surface area is $s = 10.21241$ square feet. However, for the single objective non-LP problem, maximum volume is 2 cubic feet and maximum surface area is 10.28571 square feet.

Chapter 9

ANTINOMY AND FUZZY GP RESEARCH DIRECTIONS

In this chapter, we first research antinomy in LP [Ck71][CDR87]by means of fuzzy sets. Second, we deduce antinomy of fuzzy PGP, advance a method for determining antinomy of a fuzzy PGP by use of a fuzzy LP and acquire a necessary and sufficient condition for antinomy resulting from the fuzzy PGP. According to the optimal solution of close relationships between fuzzy LP and LP, fuzzy GP and GP, we solve antinomy in LP or fuzzy LP, GP or fuzzy GP by fuzziness, which can be better used to reveal the system potentiality, so that we can be effective in business management by antinomy. And finally, we introduce an idea of extensible GP for solution to antinomy, and point out some research directions to readers in fuzzy GP.

9.1 Research for Antinomy in Fuzzy LP

9.1.1 Introduction

Let an LP problem like (the primal problem)

$$\begin{aligned} &\min z = cx \\ &\text{s.t. } Ax = b \\ &\quad\ x \geqslant 0 \end{aligned} \tag{9.1}$$

with dual form [Dan63]

$$\begin{aligned} &\max\ yb \\ &\text{s.t. } yA \leqslant c \\ &\quad\ y \geqslant 0 \end{aligned} \tag{9.2}$$

where $x = (x_1, x_2, ..., x_m)^{\mathrm{T}}, c = (c_1, c_2, ..., c_m), y = (y_1, y_2, ..., y_n), b = (b_1, b_2, ..., b_n)^{\mathrm{T}}$, $A = (a_{ij})_{n\times m}$ is an $n \times m$ matrix.

If a constant b in (9.1) increases by $d(> 0)$, then an objective value z decreases instead. If constant c in (9.2) decreases by $d(> 0)$, then an objective value yb increases instead. Such a strange phenomenon is called antinomy in mathematics, to which we give an accurate description by a mathematical model—fuzzy LP, and present a fuzzy set method for its investigation.

9.1.2 Reason for Antinomy Emergence

Example 9.1 Consider finding

$$\begin{aligned}
\min\ \ & z = 2x_1 + 3x_2 + x_3 + 2x_4 \\
\text{s.t.}\ \ & x_1 + x_2 + x_3 + x_4 = b_1' \\
& 4x_2 + 2x_3 + 6x_4 = b_2' \\
& 5x_1 + 6x_2 + 5x_3 + 4x_4 = b_3' \\
& x_1, x_2, x_3, x_4 \geqslant 0
\end{aligned}$$

where $b_i' = b_i + d_i (i = 1, 2, 3)$.

If we suppose x_1, x_2, x_3 to be basis variables, accordingly, a basis matrix B as well as an inverse matrix B^{-1} is denoted by

$$B = \begin{pmatrix} 1 & 1 & 1 \\ 0 & 4 & 2 \\ 5 & 6 & 5 \end{pmatrix},\ B^{-1} = \begin{pmatrix} -4 & -1/2 & 1 \\ -5 & 0 & 1 \\ 10 & 1/2 & -2 \end{pmatrix}.$$

When assignment volume of three products $b = (9, 12, 46)^{\mathrm{T}}$ is added to $b' = (10, 18, 50)^{\mathrm{T}}$, respectively, the minimum cost $z = c_B x_B$ decreases from $z = 15$ to $z = 11$. Why? If the problem nondegenerates and when a negative component exists in $y = c_B B^{-1}$ or in a certain evaluation coefficient $z_s < 0$, then the objective function is

$$c_B B^{-1} b' = yb' = yb + yd < yb = cx^*$$

or

$$c_B x_B = c_B B^{-1} b' = c_B B^{-1} b + c_B B^{-1} P_s = cx^* + \delta z_s < cx^*$$

so that antinomy comes out. Therefore, we have a discussion as follows [CDR87][Lin86]:

Proposition 9.1.1 *Let a basic solution $x^* = (x_B, x_N)$ be a nondegeneration optimum solution. Antinomy arises iff a negative component exists in $y = c_B B^{-1}$.* ◊

Corollary 9.1.1 *Under the condition of Proposition 9.1.1, if $\exists j_0 : z_{j_0} < 0$, then antinomy takes shape in (9.1).* ◊

Does the above conclusion hold if problem (9.1) degenerates?

Proposition 9.1.2 *If definition (9.1) denotes a degeneration LP, then*

$$\min\{cx|Ax = b + \sum_{j=1}^{n} \varepsilon^j P_j = b(\varepsilon), x \geqslant 0, \varepsilon > 0 \text{ sufficiently small}\} \quad (9.3)$$

is an LP problem of nondegeneration. ◊

Theorem 9.1.1 *If any basic feasible solution is $\varepsilon = 0$ in (9.3) when ε is sufficiently small, a basis feasible solution can be obtained to a degenerated LP (9.1).* ◊

Proposition 9.1.3 *If a basis solution to LP (9.1) $x^*(0) = (x_B(0), x_N)$ denotes a degeneration optimum solution, then there exhibits antinomy iff a negative component exists in $y = c_B B^{-1}$.* ◊

Proof Take a basis solution $x^*(\varepsilon) = (x_B(\varepsilon), x_N)$ in (9.3) into consideration, where $x_B(\varepsilon) = B^{-1}b_0 + \varepsilon_B + B^{-1}N\varepsilon_N - B^{-1}Nx_N = B^{-1}b(\varepsilon) - B^{-1}Nx_N$ is nondegeneration. According to Proposition 9.1.1, if $x_B(\varepsilon)$ is a nondegeneration optimum solution, then there appears antinomy iff a negative component exists in $y = c_B B^{-1}$. When ε is sufficiently small and if we suppose $\varepsilon = 0$ in any basis feasible solution to (9.3), we can obtain a basis feasible solution to (9.1). If (9.3) is solved when ε is sufficiently small, we can get a list of basis feasible solutions $x(\varepsilon) = \{x^0(\varepsilon), x^1(\varepsilon), \ldots\}$ till we have an optimal solution $x^*(\varepsilon)$. If $\varepsilon = 0$, we also have a list of basis feasible solutions $x(0) = \{x^0(0), x^1(0), \ldots\}$ to (9.1). Since the coefficient matrices and the objective functions are all equal in (9.3) and (9.1), accordingly, the test numbers are identical in basis feasible solutions $x^i(\varepsilon)$ and $x^i(0)$. Therefore, $x^*(0)$ is also an optimal solution to (9.1). This demonstrates that $x^*(\varepsilon)$ serving as nondegeneration optimum solution to (9.3) is equivalent to $x^*(0)$ serving as a degeneration optimum one to (9.1). At this time, the objective function denoted by

$$\begin{aligned} c_B B^{-1} b'(\varepsilon) &= yb'(\varepsilon) - yNx_N \\ &= yb_0(\varepsilon) + \delta yT - yNx_N \\ &< yb_0(\varepsilon) = cx^*(\varepsilon). \end{aligned}$$

holds when a negative component exists in y and there exists $c_B B^{-1} b'(0) < cx^*(0)$ for $\varepsilon = 0$. Therefore the proposition holds. □

Corollary 9.1.2 *Antinomy arises in (9.1) under the condition of Proposition 9.1.3 and in the event of $\exists j_0 : z_{j_0} < 0$.* ◊

Proof Because they have identical coefficient matrices and objective functions, with (9.1) and (9.3) having the same test numbers in their basis feasible solutions $x^i(\varepsilon)$ as well as $x^i(0)$, we know a negative component must exist in

y, in the event of $z_{j0} = c_B B^{-1} P_j = y P_j < 0$, with

$$\begin{aligned} c_B x_B(\varepsilon) &= c_B B^{-1} b_0(\varepsilon) + \delta c_B B^{-1} P_s - c_B B^{-1} N x_N \\ &= cx^*(\varepsilon) + \delta z_s < cx^*(\varepsilon) \end{aligned}$$

($\delta > 0$ sufficiently small), so, $c_B x_B(0) < cx^*(0)$ for $\varepsilon = 0$. Therefore, the corollary holds.

In conclusion, whether a classical LP problem degenerates or not results in the fact that antinomy comes into being. If we try to keep antinomy from being contrary, we only change the equal-sign into an inequality sign in constraint condition.

Consider fuzzy LP problem

$$\begin{aligned} &\min \tilde{z} = \tilde{c}x \\ &\text{s.t. } Ax = b \\ &\quad x \geqslant 0 \end{aligned} \tag{9.4}$$

□

Lemma 9.1.1 *The dual form of (9.4) is*

$$\begin{aligned} &\max \tilde{y}b \\ &s.t. \ \tilde{y}A = \tilde{c} \\ &\quad \tilde{y} \geqslant \tilde{0} \end{aligned} \tag{9.5}$$

where $\tilde{c}$ denotes a fuzzy vector, $\tilde{y}$ a fuzzy variable. Therefore, there exists the same fuzzy optimum value in (9.4) and (9.5) for CSM function $\tilde{\phi}$. ◊

Proof According to formula (1.14) in Section 1.6.3, (9.4) is turned into the following problem for solution:

$$\begin{aligned} &\min cx, \\ &\text{s.t. } \ \tilde{\phi}(c) \geqslant 1-\alpha, \alpha \in [0,1] \\ &\qquad Ax = b, c \in \Re^n \\ &\qquad x \geqslant 0 \end{aligned}$$

We define [Ver84]

$$\forall c \in \Re^n: \ \tilde{\phi}(c) = \text{Inf}_i \tilde{\phi}_i(c_i)(1 \leqslant i \leqslant l, l \leqslant n), c = (c_1, c_2, \ldots, c_n).$$

But if $\tilde{\phi}(c) \geqslant 1-\alpha$, then

$$\begin{aligned} \text{Inf}_i \tilde{\phi}_i(c_i) \geqslant 1-\alpha &\Longleftrightarrow \tilde{\phi}_i(c_i) \geqslant 1-\alpha (1 \leqslant i \leqslant l) \\ &\Longleftrightarrow c_i \geqslant \tilde{\phi}_i^{-1}(1-\alpha). \end{aligned}$$

Therefore, we have

$$\begin{aligned}
&\min \sum_{i=1}^{n} c_i x_i \\
&\text{s.t. } c_i \geqslant \tilde{\phi}_i^{-1}(1-\alpha)\ (1 \leqslant i \leqslant n) \\
&\quad\ \ Ax = b, \alpha \in [0,1] \\
&\quad\ \ x \geqslant 0
\end{aligned}$$

This problem is equivalent to

$$\begin{aligned}
&\min \sum_{i=1}^{n} c_i x_i \\
&\text{s.t. } c_i = \tilde{\phi}_i^{-1}(1-\alpha) \\
&\quad\ \ Ax = b, \alpha \in [0,1] \\
&\quad\ \ x \geqslant 0 \\
\Longleftrightarrow\ &\min \tilde{\phi}^{-1}(\beta) x \\
&\text{s.t. } Ax = b, \beta \in [0,1] \\
&\quad\ \ x \geqslant 0
\end{aligned} \tag{9.6}$$

where $\beta = 1 - \alpha$, whose dual form is

$$\begin{aligned}
&\max yb \\
&\text{s.t. } yA = \tilde{\phi}^{-1}(\beta), \beta \in [0,1] \\
&\quad\ \ y \geqslant 0
\end{aligned} \tag{9.7}$$

$$\begin{aligned}
\Longleftrightarrow\ &\max yb \\
&\text{s.t. } yA = c, \tilde{\phi}(c) \geqslant \beta, \beta \in [0,1] \\
&\quad\ \ y \geqslant 0 \\
\Longleftrightarrow\ &(9.5)
\end{aligned}$$

We can see that the lemma holds because of the same parameter solutions to (9.5) as well as to (9.7), by the equivalence of (9.6) with (9.4), and (9.7) with (9.5), and the mutual dual problems of (9.4) and (9.5). □

Proposition 9.1.4 *Let $\tilde{\phi}$ be a CSM function. If a basis solution $x^* = (x_B, x_N)^T$ nondegenerates in (9.4), then antinomy arises iff a negative component exists in a fuzzy shadow price $\tilde{y} = \tilde{c}_B B^{-1}$.* ◊

Proof *Necessity.* Since (9.4)$\Longleftrightarrow$(9.6) and (9.5)$\Longleftrightarrow$(9.7), if $\tilde{y} = \tilde{c}_B B^{-1} \geqslant \tilde{0}$ (fuzzy number $\tilde{0}$ is an extension of number 0) $\Longleftrightarrow y(\beta) = (\tilde{\phi}^{-1}(\beta))_B B^{-1} \geqslant 0$, and then to $\forall T \geqslant 0$, when $b \to b+T$, for any feasible solution in this problem with a soft constraint, we know

$$\begin{aligned}
\mu_{\tilde{D}}(x) &= \tilde{\phi}^{-1}(\beta) x \geqslant y(b+T) \\
&\geqslant yb = \tilde{\phi}^{-1}(\beta) x^* = \mu_{\tilde{D}}(x^*)
\end{aligned}$$

from a dual theorem of ordinary parameter LP. i.e., $\tilde{c}x \geqslant \tilde{y}b = \tilde{c}x^*$.

Sufficiency. If there exists a negative component in $\tilde{y}$, then $\exists : T \geqslant 0, \sum_{i=1}^{m} t_i > 0$, such that $\tilde{y}T < \tilde{0} \Longleftrightarrow \tilde{\phi}^{-1}(\beta)B^{-1}T \leqslant 0$. Let $b' = b + \delta T(\delta > 0)$. Then the problem with a soft constraint concerning a basis solution in basis B is

$$x_B = B^{-1}b' = B^{-1}b + \delta B^{-1}T, \ x_N = 0.$$

x^* nondegenerates on the proposition assumption. $x_B \geqslant 0$ means a basis feasible solution having test numbers unchangeable when $\delta > 0$ is sufficiently small. Therefore it also belongs to an optimal solution with soft constraints and $\tilde{y} = \tilde{c}_B B^{-1}$ is still a fuzzy optimal solution to the dual problem. But the objective value $\forall \beta \in [0,1]$ is denoted by

$$\begin{aligned} &(\tilde{\phi}^{-1}(\beta))_B B^{-1}b' = y(\beta)b' \\ &\quad = y(\beta)b + \delta y(\beta)T < y(\beta)b = cx^* \\ &\Longleftrightarrow \tilde{c}_B B^{-1}b' = \tilde{y}b' \\ &\quad = \tilde{y}b + \delta\tilde{y}T < \tilde{y}b = \tilde{c}x^* \end{aligned}$$

such that antinomy arises. □

Corollary 9.1.3 *Let $\tilde{\phi}$ be a CSM function. If a basis solution $x^* = (x_B, x_N)$ to (9.4) denotes a nondegeneration optimum solution, then the condition where antinomy arises is $\exists : j_0$, such that $\tilde{z}_{j_0} < \tilde{0}$.* ◊

Proof From the proof of Proposition 9.1.4, we know

$$\begin{aligned} \sup_{Ax=b} \mu_0(x) &= \sup \mu_0(x_B) \ (x_B = B^{-1}b_0 + \delta B^{-1}T, x_N = 0) \\ &= \sup_{\beta\in[0,1]} (\tilde{\phi}^{-1}(\beta))_B x_B \\ &= \sup_{\beta\in[0,1]} \{(\tilde{\phi}^{-1}(\beta))_B B^{-1}b + P_j\delta(\tilde{\phi}^{-1}(\beta))_B B^{-1}P_jT\} \\ &< \sup_{\beta\in[0,1]} (\tilde{\phi}^{-1}(\beta))_B B^{-1}bP_j. \end{aligned}$$

(Because $z_j = (\tilde{\phi}^{-1}(\beta))_B B^{-1}P_j < 0$, we know there must exist a negative component in $y(\beta) = (\tilde{\phi}^{-1}(\beta))_B B^{-1}$.)

It is equivalent that there must be a fuzzy negative component in $\tilde{y}$ from the knowledge of $\tilde{z}_j = \tilde{c}_B B^{-1}P_j = \tilde{y}P_j < \tilde{0}$, such that we have

$$\tilde{c}_B x_B = \tilde{c}x^* + \delta\tilde{z}_j < \tilde{c}x^*.$$

□

Proposition 9.1.5 *If definition (9.4) serves as a degeneration fuzzy LP problem, then*

$$\begin{aligned} &\min \tilde{z} = \tilde{c}x \\ &\text{s.t. } Ax = b + \textstyle\sum_{j=1}^{n} \varepsilon^j P_j = b(\varepsilon) \\ &\qquad x \geqslant 0 \end{aligned} \tag{9.8}$$

serves as a nondegeneration fuzzy LP one, where ε^j is a sufficiently small positive number. ◊

Proposition 9.1.6 *Let $\tilde{\phi}$ be a CSM function. If a basis solution $x^*(0) = (x_B(0), x_N)$ to (9.4) denotes a degeneration optimum solution, then antinomy appears iff a fuzzy negative component exists in $\tilde{y} = \tilde{c}_B B^{-1}$.* ◊

Corollary 9.1.4 *Let $\tilde{\phi}$ be a CSM function. Suppose that a basis solution $x^*(0) = (x_B(0), x_N)$ to (9.4) is a degeneration optimum solution, then the condition where antinomy arises is $\exists : j_0$, such that $\tilde{z}_{j_0} < \tilde{0}$.* ◊

In fact, because

$$\begin{aligned} &\min\ \tilde{c}x \\ &\text{s.t. } Ax = b(\varepsilon) \\ &\quad\ \ x \geqslant 0 \end{aligned}$$

$$\Longleftrightarrow \begin{aligned} &\min\ \tilde{\phi}^{-1}(\beta)x \\ &\text{s.t. } Ax = b(\varepsilon), \beta \in [0,1] \\ &\quad\ \ x \geqslant 0 \end{aligned} \tag{9.9}$$

the dual form of (9.9) is

$$\begin{aligned} &\max\ yb(\varepsilon) \\ &\text{s.t. } yA = \tilde{\phi}^{-1}(\beta), \beta \in [0,1] \\ &\quad\ \ y \geqslant 0 \end{aligned}$$

$$\Longleftrightarrow \quad \begin{aligned} &\max\ \tilde{y}b(\varepsilon) \\ &\text{s.t. } \tilde{y}A = \tilde{c} \\ &\quad\ \ \tilde{y} \geqslant \tilde{0} \end{aligned}$$

From Proposition 9.1.3 and Corollary 9.1.2, we know properties as follows:

a. If there exists a degeneration optimum basis solution $x^*(0, \beta)$ in the classical LP (9.9) with parameter variable β, then the antinomy appears iff a negative component exists in $\tilde{y} = \tilde{c}_B B^{-1}$.

b. Under the condition of a, if $\exists : j_0 < 0$, then antinomy appears in (9.9).

Therefore, a) and b) are equivalent to Proposition 9.1.6, and the Corollary 9.1.4 holds.

On the whole, no matter whether (9.4) is a generation or nongeneration fuzzy LP problem, antinomy appears in both of them. If we prevent antinomy in fuzzy LP from being contrary, the constraint equal-sign can be turned into a soft constraint.

9.1.3 Connection of Optimal Solutions Between LP and Fuzzy LP

Let $\tilde{\phi}$ be (1.14) in Section 1.6.3. If a fuzzified form of (9.1) is (9.4), its crisp primal problem is

$$\begin{aligned} \min\ & (m+\beta dn^{-1})x \\ \text{s.t.}\ & Ax = b, \beta \in [0,1] \\ & x \geqslant 0 \end{aligned} \tag{9.10}$$

where m_i, n_i are real numbers, c_i is freely fixed in the value interval $[m_i, n_i]$ with $c_i \geqslant m_i + \beta dn_i^{-1} \Longleftrightarrow c_i = m_i + \beta dn_i^{-1}$, d denoting a flexible index, $\beta = 1 - \alpha$, while the dual problem in (9.10) is

$$\begin{aligned} \max\ & yb \\ \text{s.t.}\ & yA = m + \beta dn^{-1}, \beta \in [0,1] \\ & y \geqslant 0 \end{aligned}$$

Theorem 9.1.2 *Let $\tilde{\phi} : \Re \to [0,1]$ be a CSM membership function. x_0 is a unique solution to (9.1) iff x_0 remains a parameter solution to (9.6) $\forall \beta \in [0,1] (\beta = 1 - \alpha)$.* $\Diamond$

Proof Similar to the proof of Reference [Man79], we have
x_0 is a unique solution to (9.1)

$$\begin{aligned}
\Longleftrightarrow\ & \forall d/n \in \Re^n, \exists x, y, \lambda \in \Re^{n+m+1} \\
& Ax - b\lambda \geqslant 0, -yA + c\lambda = 0, y \geqslant 0 \\
& yb - cx \geqslant 0, -dn^{-1}x + dn^{-1}x_0\lambda > 0, \lambda > 0 \\
\Longleftrightarrow\ & \forall dn^{-1} \in \Re^n, \exists x, u, y, \kappa, r \in \Re^{n+m+k+2} \\
& -Ax + b\kappa + u = 0 \\
& yA - (\kappa c + \beta dn^{-1}) = 0, -yb + cx + dn^{-1}x_0\beta + r = 0 \\
& u, \kappa, r \geqslant 0, \beta \in [0,1], \beta + r > 0 \\
\Longleftrightarrow\ & \forall dn^{-1} \in \Re^n, \exists x, y, \kappa \in \Re^{n+m+1} \\
& Ax \geqslant b\kappa, cx = \kappa c x_0, \kappa \geqslant 0 \\
& yA = \kappa c + \beta dn^{-1}, yb \geqslant (\kappa c + \beta dn^{-1})x_0, \beta \in [0,1] \\
\Longleftrightarrow\ & \forall dn^{-1} \in \Re^n, \exists y, \kappa \in \Re^{m+1} \\
& yA = \kappa c + \beta dn^{-1}, yb = (\kappa c + \beta dn^{-1})x_0 \\
& \kappa \geqslant 0, \beta \in [0,1] \\
& (cx + \beta dn^{-1}x_0 \leqslant yb \leqslant yAx_0 = (\kappa c + \beta dn^{-1})x_0) \\
\Longleftrightarrow\ & \forall dn^{-1} \in \Re^n, \exists y \in \Re^m \\
& yA = c + \beta dn^{-1}, yb = (c + \beta dn^{-1})x_0 \\
& \beta \in [0,1], \text{ and let } \kappa = 1 \\
\Longleftrightarrow\ & \forall dn^{-1} \in \Re^n, \exists \beta \in [0,1]
\end{aligned}$$

so a solution to (9.6) is found to be x_0.

Because x_0 denotes a feasible solution to (9.6), $\bar{y}$ is a feasible one to dual problem (9.7) coming from (9.6), with $\bar{y}b = (m + \beta dn^{-1})x_0$, where $\bar{y} = y(\beta)$.

But, the fuzzy solution of (9.4) is given by the optimal solution of the parametric linear problem [Ver84], therefore, the theorem holds. □

Similar to a corollary in References [Man79], we can confirm the following.

Corollary 9.1.5 *The dual optimal solution y is unique to (9.2) associated with a primal optimal solution x_0 to (9.1), iff, for a CSM membership function $\phi : \Re \to [0,1]$, such that $\forall\beta \in [0,1]$, y remains a dual optimal parameter solution for the perturbed LP*

$$\begin{aligned} \min \quad & z = cx \\ \text{s.t.} \quad & Ax = b_1 + \beta d b_2^{-1}, \beta \in [0,1] \\ & x \geqslant 0 \end{aligned} \tag{9.11}$$

where b_1, b_2 are constant vectors, $d > 0$ an arbitrary real number. ◊

Theorem 9.1.3 *Let $\tilde{\phi} : \Re \to [0,1]$ be a CSM membership function. A solution x_0 is unique to LP (9.1) iff x_0 is still a fuzzy optimal solution to fuzzy LP (9.4).* ◊

Proof From Lemma 9.1.1, we know (9.4) $\Longleftrightarrow$ (9.6), so, from the result where Theorem 9.1.2 is applied to (9.6), x_0 is a unique solution to (9.1) iff x_0 remains a parameter optimal solution to (9.6). But the minimization in (9.4) is equivalent to that in (9.6), and x_0 is a parameter optimum solution to (9.6) iff x_0 is an fuzzy optimal solution to (9.4).

Hence, the theorem holds. □

Corollary 9.1.6 *The dual optimal solution y to (9.2) corresponding to the primal optimum solution x_0 to (9.1) is unique, iff, for a CSM membership function $\tilde{\phi} : \Re \to [0,1], \tilde{y}$ is still a dual optimum solution to the program (9.4).* ◊

Proof Let $\tilde{\phi}$ be formula (1.14) in Section 1.6.3. Then we have

$$\begin{aligned} (9.4) \Longleftrightarrow \min \quad & z = cx \\ \text{s.t.} \quad & Ax = b, \tilde{\phi}(c) \geqslant \beta, \beta \in [0,1] \\ & x \geqslant 0 \\ \Longleftrightarrow (9.11) & \end{aligned}$$

by Reference [Ver84]. Apply Corollary 9.1.5 to (9.11) and the conclusion holds. □

Definition 9.1.3 Let $\tilde{A}_0(x), \tilde{F}(x)$ be membership functions of fuzzy objection and fuzzy constraints. Then we call a fuzzy set $\tilde{D}$ satisfying $\tilde{D}(x) = \tilde{A}_0(x) \wedge \tilde{F}(x), x \geqslant 0$ a fuzzy decision for the problem

$$\begin{aligned} \min \quad & \tilde{z} = \tilde{c}x \\ \text{s.t.} \quad & Ax \lesssim b \\ & x \geqslant 0 \end{aligned} \tag{9.12}$$

while we call a point x satisfying $\tilde{D}(x) = \max_{x \geqslant 0}\{(1 - \tilde{A}_0(x)) \wedge \tilde{F}(x)\}$ an optimal solution to (9.12).

Theorem 9.1.4 *The maximization of $\tilde{D}(x)$ is equivalent to LP*

$$\begin{aligned} \min\ & (m + \beta M_0 n^{-1})x \\ \text{s.t. }\ & Ax \leqslant b_1 + B\beta b_2^{-1} + d\alpha, \alpha, \beta \in [0,1] \\ & x \geqslant 0 \end{aligned} \tag{9.13}$$

d denoting a flexible index; M_0 and B representing the length in intervals $[m,n], [b_1,b_2]$, respectively. ◊

Proof From formula (1.14) (1.15) and (1.16) in Section 1.6.3, we have

$$\begin{aligned} \tilde{D}(x) \Longleftrightarrow & \max\ (-\tilde{c})x \\ & \text{s.t. } Ax \lesssim \tilde{b} \\ & \qquad x \geqslant 0 \\ \Longleftrightarrow & \min\ \tilde{c}x \\ & \text{s.t. } \tilde{\phi}(c) \geqslant \beta \\ & \qquad Ax \leqslant b + d\alpha, \tilde{\phi}(b) \geqslant \beta \\ & \qquad c \in [m,n], b \in [b_1,b_2], \beta \in [0,1] \\ & \qquad x \geqslant 0 \\ \Longleftrightarrow & (9.14) \end{aligned}$$

where c and b can be freely fixed in the close value interval $[m,n]$ and $[b_1,b_2]$.

Overall, if antinomy is changed into formula (9.12) for solution, the antinomy of a fuzzy LP is not contrary. If the optimal solution in a primal LP problem is unique, then antinomy does not exist, which can be concluded as solutions to fuzzy LP (9.4). In the light of Theorem 9.1.4, an ordinary LP is only a particular example of fuzzy LP (9.12) for $d_i = 0$. Therefore, it can be changed into finding solutions by a fuzzy set method no matter whether it is antinomy of LP or fuzzy LP. □

9.1.4 Example

Example 9.2 The fuzzy LP corresponding to Example 9.1 in this section is

$$\begin{aligned} \min\ & z = 2x_1 + 3x_2 + x_3 + 2x_4 \\ \text{s.t. }\ & x_1 + x_2 + x_3 + x_4 \lesssim 9 \\ & 4x_2 + 2x_3 + 6x_4 \lesssim 12 \\ & 5x_1 + 6x_2 + 5x_3 + 4x_4 \lesssim 46 \\ & x_i \geqslant 0 (i = 1, \cdots, 4) \end{aligned} \tag{9.14}$$

Assume $d_1 = 1, d_2 = 6, d_3 = 4$ and we make a parameter program, then (9.14) is turned into

$$\begin{aligned}
\min\ \ & z = 2x_1 + 3x_2 + x_3 + 2x_4 \\
\text{s.t.}\ \ & x_1 + x_2 + x_3 + x_4 + x_5 = 10 \\
& 4x_2 + 2x_3 + 6x_4 + x_6 = 18 \\
& 5x_1 + 6x_2 + 5x_3 + 4x_4 + x_7 = 50 \\
& x_i \geqslant 0 (i = 1, \cdots, 7)
\end{aligned}$$

Under the unchangeable condition of basis matrix B, an optimal parameter solution denotes $x = B^{-1}b(\alpha) = (4 - 3\frac{1}{2}\alpha, 1 - 4\alpha, 4 + 8\frac{1}{2}\alpha)^{\mathrm{T}}, z = 15 - 10\frac{1}{2}\alpha$. When b is added from $b = (9, 12, 6)^{\mathrm{T}}$ to $b' = (10, 18, 50)^{\mathrm{T}}$, z decreases from $z = 15$ to 11. Therefore, antinomy comes into being because negative components exist in a solution vector for $\alpha > \frac{1}{4}$.

9.2 Antinomy in Fuzzy GP

First, we construct antinomy in a fuzzy PGP on the basis of an antinomy problem in a fuzzy LP. Second, we analyze a fuzzy GP for convexities, non-convex but quasi-convex or pseudo-convex. Finally, we discuss the condition in which antinomy arises.

9.2.1 Introduction

Consider a PGP to be

$$\begin{aligned} &\min\ g_0(x) \\ &\text{s.t. } g_i(x) \leqslant 1\ (1 \leqslant i \leqslant p) \\ &\qquad x > 0 \end{aligned} \tag{9.15}$$

and

$$\begin{aligned} &\min\ g_0(x) \\ &\text{s.t. } g_i(x) \leqslant 1 + \alpha d_i, \alpha \in [0,1] (1 \leqslant i \leqslant p) \\ &\qquad x > 0 \end{aligned} \tag{9.16}$$

where $g_i(x) = \sum\limits_{i=1}^{J_i} c_{ik} \prod\limits_{l=1}^{m} x_l^{\gamma_{ikl}} (0 \leqslant i \leqslant p)$ is posynomial; c_{ik} denotes a nonnegative constant coefficient; γ_{ikl} an exponent and d_i a flexible index which is a nonnegative real number. Obviously, (9.16) is obtained by generalizing a constraint constant in (9.15).

Definition 9.2.1 Suppose that an optimal solution exists in (9.15). If $\exists$: $d_1, d_2, \ldots, d_p > 0$ which are not all zero, an optimal solution exists in (9.16) $\forall \alpha \in [0,1]$. Again, if a minimal value in (9.15) is greater than that in (9.16), then there appears antinomy in (9.15).

In this section, we describe the antinomy problem above and detect the condition under which antinomy appears by using fuzzy sets.

9.2.2 Antinomy Phenomenon in Fuzzy PGP

Renew consideration of a PGP (9.15), whose fuzziness form is

$$\begin{aligned} &\widetilde{\min}\ g_0(x) \\ &\text{s.t. } g_i(x) \lesssim 1\ (1 \leqslant i \leqslant p) \\ &\qquad x > 0 \end{aligned} \tag{9.17}$$

where $x = (x_1, x_2, \cdots, x_m)^{\mathrm{T}}$ denotes a variable vector and

$$g_i(x) = \sum_{i=1}^{J_i} c_{ik} \prod_{l=1}^{m} x_l^{\gamma_{ikl}} (0 \leqslant i \leqslant p)$$

a fuzzy polynomial of x, $\widetilde{\min}$ means finding a fuzzy minimal value and "$\lesssim$" stands for "approximately less than or equal to".

Since any fuzzy PGP (9.17) can be turned into a fuzzy convex programming from Section 2.5, we know that any fuzzy PGP can be turned into a monomial fuzzy PGP. Therefore, we only consider the monomial fuzzy PGP problem

$$\begin{aligned} &\widetilde{\min}\ c_0 \prod_{l=1}^{m} x_l^{\gamma_{0l}} \\ &\text{s.t. } c_i \prod_{l=1}^{m} x_l^{\gamma_{il}} \lesssim b_i\ (1 \leqslant i \leqslant p) \\ &\qquad x > 0 \end{aligned} \tag{9.18}$$

whose dual form is

$$\begin{aligned} &\widetilde{\max}\ \prod_{i=1}^{p} \left(\frac{c_i}{b_i}\right)^{w_i} \\ &\text{s.t. } w_0 = 1 \\ &\qquad \sum_{i=0}^{p} \gamma_{il} w_i = 0\ (1 \leqslant l \leqslant m) \\ &\qquad w \geqslant 0 \end{aligned} \tag{9.19}$$

Given that the monomial fuzzy PGP can be converted into a fuzzy LP

$$\begin{aligned} &\widetilde{\min}\ \sum_{l=1}^{m} \gamma_{0l} z_l \\ &\text{s.t. } \sum_{l=1}^{m} \gamma_{il} z_l + \ln c_i \lesssim \log b_i\ (1 \leqslant i \leqslant p) \end{aligned} \tag{9.20}$$

(Theorem 3.3.1), it is easy to prove that (9.20) and (9.18) have the same fuzzy optimal value.

Again, by formula (1.15) (1.16) in Section 1.6.3, we nonfuzzify (9.17) into (9.16). Therefore, antinomy arises in (9.15) in Definition 9.2.1 (as it does in the fuzzy PGP in (9.17), actually).

Definition 9.2.2 Let an optimal solution exist in the fuzzy PGP (9.17). For formula (1.15) and nonnegative numbers d_i which are not all zero, such that we have an optimal parameter solution exists in program

$$\begin{aligned} &\max\ \alpha \\ &\text{s.t. } g_0(x) \leqslant z_0 + \alpha d_0, \\ &\qquad g_i(x) \leqslant 1 + \alpha d_i,\ 1 \leqslant i \leqslant p, \alpha \in [0,1], \\ &\qquad x > 0 \end{aligned} \tag{9.21}$$

for $\forall \alpha \in [0,1]$. Here z_0 denotes an expectation object, $d_0 = g_0^{(1)} - g_0^{(2)}, d_i > 0 (1 \leqslant i \leqslant p)$ represents flexible indexes in objective and constraint functions, respectively. If minimal value of $g_0^{(1)}$ in (9.15) is larger than that of $g_0^{(2)}$ in (9.16), then we call the fuzzy PGP (9.17) a phenomenon of antinomy appearance.

Generally, one of the solution methods most used in fuzzy GP is to obtain two optimal values $g_0^{(1)}$ and $g_0^{(2)}$ by turning (9.17) into two determinacy GP (9.15) and (9.16) respectively before we find an answer to GP (9.21).

We regard constant terms in constraint functions as tasks or resources and object $g_0(x)$ as cost functions. The other conditions in (9.15) and (9.16) are unchangeable but tasks or resources increase from 1 to $1 + \alpha d_i$ with cost decreasing instead. Antinomy in such a strange phenomenon as a fuzzy GP is described.

What is the reason for such a strange phenomenon?

Here are several results with respect to nondegeneration fuzzy PGP as follow:

Theorem 9.2.1 *Let fuzzy PGP be (9.17) with a nondegeneration fuzzy optimal solution. Then (9.17) with antinomy appearance is equivalent to a fact that a negative component exists in a dual basis solution $w = C_B B^{-1}$ in its corresponding fuzzy LP (9.20) when a basis solution $z^* = (z_B, z_N)$ corresponding to basis B denotes a nondegeneration optimal solution.* ◊

Proof From the discussion of this section mentioned above we know that any fuzzy PGP (9.17) can be changed into fuzzy LP (9.20), such that antinomy appearance in (9.17) is equivalent to antinomy appearance in (9.20). We have proved [Cao91] that when a basis solution is a nondegeneration optimal solution with respect to basis B, antinomy appearance in (9.20) means that a negative component exists in its dual basis solution. Therefore, the theorem holds. □

Corollary 9.2.1 *Let (9.17) contain a nondegeneration fuzzy optimal solution. If $\exists j_0$, such that there is a certain discriminate coefficient $\sigma_{j_0} < 0$ in (9.20) for nondegeneration basis optimal solution $z^* = (z_B, z_N)$ corresponding to fuzzy LP (9.20), then antinomy appears in (9.17).* ◊

Proof In fact, (9.17) can be turned into (9.20) and a negative component must exist in w in (9.20) from the knowledge of $\sigma_j = wP_j < 0$, so that this corollary holds from Theorem 9.2.1. □

Example 9.3 Suppose that a certain fuzzy PGP is changed into a fuzzy monomial PGP as follows:

$$\begin{aligned} &\widetilde{\min}\ x_1 x_2 x_3^6 x_4^3 \\ &\text{s.t. } x_1 x_2^{-1} x_3^{-1} x_4^2 \cong \tilde{b}_1 \\ &\quad x_1^2 x_2 x_3^2 x_4^{-1} \cong \tilde{b}_2 \\ &\quad x_1 x_2 x_3^2 x_4^3 \cong \tilde{b}_3 \\ &\quad x_1, x_2, x_3, x_4 > 0 \end{aligned} \tag{9.22}$$

where " $\cong$ " refers to the fuzzification of expectation level [AG01].

From (9.20), (9.22) can be changed into

$$\begin{aligned}&\widetilde{\min}\ (z_1 + z_2 + 6z_3 + 3z_4)\\&\text{s.t. } z_1 - z_2 - z_3 + 2z_4 \cong \log \tilde{b}_1\\&\quad 2z_1 + z_2 + 2z_3 - z_4 \cong \log \tilde{b}_2\\&\quad z_1 + z_2 + 2z_3 + 3z_4 \cong \log \tilde{b}_3\end{aligned}$$

where $z_l = \log x_l (l = 1,2,3,4), \tilde{b}_i (i = 1,2,3)$ is freely fixed in a certain closed value interval. If we choose z_1, z_2, z_3 as basis variables, a corresponding basis matrix B and basis reverse matrix B^{-1} arrive at

$$B = \begin{pmatrix} 1 & -1 & -1 \\ 2 & 1 & 2 \\ 1 & 1 & 2 \end{pmatrix}, \qquad B^{-1} = \begin{pmatrix} 0 & 1 & -1 \\ -2 & 3 & -4 \\ 1 & -2 & 3 \end{pmatrix}.$$

When we take $\tilde{b}_1$ as e, $\tilde{b}_2$ as e^6, $\tilde{b}_3$ as e^4, an optimal solution is $z_1 = 2, z_2 = 0, z_3 = 1$, and a minimum value is $g_0^1 = C_B z_B = 8$. When we select $\tilde{b}_1 = e^2$, $\tilde{b}_2 = e^7$, $\tilde{b}_3 = e^4$, an optimal solution is $z_1 = 3, z_2 = 1, z_3 = 0$ and minimum value is $g_0^2 = 4$. i.e., under the unchanging state of other conditions, constraint conditions (resp. task volume) in (9.22) is added from e^1, e^6, e^4 to e^2, e^7, e^4, while objective function (resp. cost) $g_0(x)$ decreases by $e^8 \cdot e^{-4} = e^4$ unit.

The reason is that negative component -8 exists in

$$w = C_B B^{-1} = (1,1,6) \begin{pmatrix} 0 & 1 & -1 \\ -2 & 3 & -4 \\ 1 & -2 & 3 \end{pmatrix} = (4, -8, 13).$$

In fact, if we take $d_0 = 4, d_1 = d_2 = 1, d_3 = 0$, the crisp programming corresponding to (9.22) is

$$\begin{aligned}&\max\ S = \alpha\\&\text{s.t. } z_1 - z_2 - z_3 + 2z_4 + \alpha \leqslant 2\\&\quad 2z_1 + z_2 + 2z_3 - z_4 + \alpha \leqslant 7\\&\quad z_1 + z_2 + 2z_3 + 3z_4 \leqslant 4\\&\quad -z_1 - z_2 - 6z_3 - 3z_4 - 4\alpha \geqslant -8\end{aligned}$$

Solve it by a simplex method through five repeaters and we have an optimal parameter solution to the problem

$$z^* = (z_1, z_2, z_3, z_4, \alpha) = \left(\frac{30}{13}, \frac{19}{13}, 0, \frac{1}{13}, 1\right).$$

And an optimal value is $g_0(z^*) = 4$, such that an optimal value e^4 is obtained for the primal fuzzy GP (9.22).

9.2.3 Extension

The above outcome has been revealed by turning a fuzzy GP into a fuzzy LP and on the assumption that the programming is convex. Now, extend convexity and we have the following results:

Theorem 9.2.2 *Let x^* be an optimal solution to fuzzy PGP (9.17), $g_0(x)$, $g_i(x)(1 \leqslant i \leqslant p)$ differentiable, $g_0(x)$ pseudoconvex and $g_i(x)$ quasiconvex at x^*, and $\nabla g_i(x^*)$ $(1 \leqslant i \leqslant p)$ be linear independence and $\tilde{A}_i(0 \leqslant i \leqslant p)$ be a CSM function. Then antinomy appears in (9.17)$\Longleftrightarrow$ the combination coefficients of $\nabla g_i(x)$ contains a negative component in the Kuhn-Tucker condition at x^* in (9.17), i.e., $\exists : \mu_i(1 \leqslant i \leqslant p)$, such that*

$$\nabla g_0(x^*) - \sum_{i=1}^{p} \mu_i \nabla g_i(x^*) = 0 \tag{9.23}$$

contains at least a negative component: $\mu_i < 0$. ◊

Proof Since (9.17) is equivalent to the determinacy GP (9.21) [Cao93a], and x^* is an optimal solution to (9.17), and again $g_i(x)(0 \leqslant i \leqslant p)$ is a convexity assumption, of course, x^* is an optimal solution to (9.21). Because (9.15) and (9.16) are obtained by taking $\alpha = 0$ and $\alpha = 1$ in (9.21), x^* also is an optimal solution to (9.15) and (9.16). Therefore, under assumption of the theorem and according to proof in Theorem 7 in Reference [YJ91], it is easy to prove that antinomy arises in (9.15) $\Longleftrightarrow$ at least a negative component μ_i exists in (9.15) in x^*, such that (9.23) holds.

Again, let optimal values in (9.15) and (9.16) be $g_0^{(1)}$ and $g_0^{(2)}$, respectively. When constraint constants in (9.16) are greater than those in (9.15), we have $g_0^{(1)} > g_0^{(2)}$, i.e., on the contrary, an objective value in (9.16) is smaller than that in (9.15) and antinomy arises. $\forall \alpha \in [0, 1)$, constraint constants in (9.21) are greater than those in (9.15), but their objective values are $g_{0_\alpha}^{(0)} < g_0^{(1)}$, instead. i.e, antinomy appears. Therefore, the theorem conclusion holds. □

Obviously, Theorem 9.2.2 is true for a fuzzy PGP since any fuzzy PGP can be deformed into a fuzzy convex programming [Cao93a].

Example 9.4 Consider Example 9.1. When e, e^6, e^4 is chosen in $\tilde{b}$, an optimal solution to (9.22) is $x^* = (e^2, 1, e, 0)^{\mathrm{T}}$; now,

$$\nabla g_0(x^*) = (1, 1, 6e^5, 0)^{\mathrm{T}}, \ \nabla g_1(x^*) = (1, -1, -e^{-2}, 0)^{\mathrm{T}},$$
$$\nabla g_2(x^*) = (2e^2, 1, 2e, 0)^{\mathrm{T}}, \ \nabla g_3(x^*) = (1, 1, 2e, 0)^{\mathrm{T}}.$$

Let

$$\nabla g_0(x^*) - \mu_1 \nabla g_1(x^*) - \mu_2 \nabla g_2(x^*) - \mu_3 \nabla g_3(x^*) = 0.$$

Then,

$$(1, 1, 6e^5, 0)^{\mathrm{T}} = \mu_1(1, -1, -e^{-2}, 0)^{\mathrm{T}} + \mu_2(2e^2, 1, 2e, 0)^{\mathrm{T}} + \mu_3(1, 1, 2e, 0)^{\mathrm{T}}.$$

We have the following:

$$\mu_2 = \frac{2-6e^4}{-2+e^{-3}} \frac{2}{1-2e^2}$$
$$\approx -24.23 < 0$$

such that at least $\mu_2 < 0$ satisfies the sufficient condition in Theorem 9.2.2. Therefore, antinomy arises in the above fuzzy GP.

Definition 9.2.3 Change min in (9.20) into inf, writing $(\tilde{P})$, and its dual programming is

$$(\tilde{D}) \quad \sup\ \tilde{d}(w) = \prod_{i=1}^{p}\prod_{k=1}^{J_i}\left(\frac{\tilde{c}_{ik}}{w_{ik}}\right)^{w_{ik}} \prod_{i=1}^{p} w_{i0}^{w_{i0}}$$
$$\text{s.t. } w_{00} = 1$$
$$\Gamma^{\mathrm{T}} w = 0$$
$$w \geqslant 0$$

Based on a strongly dual theory, there exist the results as follows:

Theorem 9.2.3 [Cao93a] *Let $\tilde{A}_i (0 \leqslant i \leqslant p)$ be a CSM function and $(\tilde{P})$ be fuzzy super-consistent, with $M_{\tilde{P}} > 0$. Then a dual program $(\tilde{D})$ must be fuzzy consistent and we have*

1^0 *An optimal solution;*

2^0 $M_{\tilde{D}} = \tilde{d}(w^*) = M_{\tilde{P}}$,

where $M_{\tilde{P}}$ and $M_{\tilde{D}}$ represent fuzzy constraints infimum of $(\tilde{P})$ and supremum of $(\tilde{D})$, respectively. ◊

Theorem 9.2.4 *On the assumption of Theorem 9.2.2, antinomy appears in $(\tilde{P})$ if the fuzzy strong duality holds iff there exists at least a negative component in solution $w^* = (w_1^*, w_2^*, \ldots, w_p^*)^T$ to $(\tilde{D})$ or iff the contradictory equation exists in the constraint to $(\tilde{D})$.* ◊

Example 9.5 Consider antinomy like Example 9.1, whose dual programming is

$$\max\ (1)^{w_1} \left(\frac{1}{\tilde{b}_1}\right)^{w_2} \left(\frac{1}{\tilde{b}_2}\right)^{w_3} \left(\frac{1}{\tilde{b}_3}\right)^{w_4}$$
$$\text{s.t. } w_0 = w_1 + w_2 + w_3 + w_4 = 1$$
$$w_1 + w_2 + 2w_3 + w_4 = 0$$
$$w_1 - w_2 + w_3 + w_4 = 0$$
$$6w_1 - w_2 + 2w_3 + 2w_4 = 0$$
$$3w_1 + 2w_2 - w_3 + 3w_4 = 0$$
$$w_1, w_2, w_3, w_4 \geqslant 0.$$

Calculate a constraint equation group and a contradictory equation $0 = d_4$ appears, i.e., we know no solution in the dual programming.

Example 9.6 Find an antinomy problem

$$\begin{aligned}
&\min\ x_1^2 x_2^3 x_3 x_4^2 \\
&\text{s.t. } x_1 x_2 x_3 x_4 \cong \tilde{b}_1 \\
&\qquad x_2^4 x_3^2 x_4^6 \cong \tilde{b}_2 \\
&\qquad x_1^5 x_2^6 x_3^5 x_4^4 \cong \tilde{b}_3 \\
&\qquad x_1, x_2, x_3, x_4 > 0
\end{aligned} \tag{9.24}$$

whose dual programming is

$$\begin{aligned}
&\max\ (1)^{w_1} \left(\frac{1}{\tilde{b}_1}\right)^{w_2} \left(\frac{1}{\tilde{b}_2}\right)^{w_3} \left(\frac{1}{\tilde{b}_3}\right)^{w_4} \\
&\text{s.t. } w_1 + w_2 + w_3 + w_4 = 1 \\
&\qquad 2w_1 + w_2 + 5w_4 = 0 \\
&\qquad 3w_1 + w_2 + 4w_3 + 6w_4 = 0 \\
&\qquad w_1 + w_2 + 2w_3 + 5w_4 = 0 \\
&\qquad w_1, w_2, w_3, w_4 \geqslant 0.
\end{aligned}$$

Solve the constraint equation group and we get w_4 to be negative. Therefore, there is no feasible solution in the dual programming.

Obviously, (9.24) can be changed into the following LP to solve

$$\begin{aligned}
&\min\ z = 2y_1 + 3y_2 + y_3 + 2y_4 \\
&\text{s.t. } y_1 + y_2 + y_3 + y_4 \cong \log \tilde{b}_1 \\
&\qquad 4y_2 + 2y_3 + 6y_4 \cong \log \tilde{b}_2 \\
&\qquad 5y_1 + 6y_2 + 5y_3 + 4y_4 \cong \log \tilde{b}_3 \\
&\qquad y_1, y_2, y_3, y_4 \geqslant 0
\end{aligned}$$

It is an antinomy problem of LP in Reference [Lin86] when $\log \tilde{b}_1, \log \tilde{b}_2, \log \tilde{b}_3$ are 9,12,46, respectively. It is easy to determine whether it is an antinomy problem by Example 9.3.

9.2.4 Conclusion Remarks

The antinomy problem helps us not only to make optimal use of resources, but also to diagnose a variety of systems in order to better the systems.

9.3 Directions in Fuzzy GP Research

We have obtained a series of results in fuzzy GP since we began to study it 14 years ago. It looks like a mine containing rich resources which remains to be excavated. Meanwhile, it can be generalized into investigation in many other fields.

9.3.1 Study of Fuzzy GP Model and Its Method

Fuzzy GP will attract us to further research because many aspects remain untouched. In the basic field, we shall consider the following topics.

1. Fuzzy reverse GP, including a GP problem with mixed sign-terms, is much more complex than the fuzzy convex (resp. concave) GP and fuzzy PGP, which differ a lot in their properties.

2. Fuzzy fractional GP model and its method remain to be researched.

3. The discussion of the GP with fuzzy coefficients and fuzzy variables is yet to be further perfected as well as extended.

4. It is very interesting to solve antinomy in the realistic world by using fuzzy GP further.

5. It will be a creative job to combine fuzzy GP and other fuzzy programming with Data Mining before dealing with their data and classification.

6. We have developed the concepts of extension convex function in extendable-valued sets based on the extendable sets [Cai83] and its convex sets [Cao90b], built a matter-element model and a mathematical model in extendable GP, and discussed solution properties in an extendable mathematical model and given an algorithm of the programming which can be changed into an extendable LP by the solution of "contrary theory" [Cao01d]. This model plays a practical role in solving the fuzziness at the border as well as contradictory conflicts in building a fuzzy non-compatible GP model and in turning a non-compatible problem into a compatible one.

9.3.2 Further Application of Fuzzy GP

In the real world, the problems, which can be expressed once by an exponent polynomial functions in the relation under fuzzy environment, are concluded as a fuzzy GP. We cite only practical examples in electrical power and postal service. In practice, fuzzy GP bears broad prospects in application in fields such as optimal design, management, electronics, chemical industry, biology and automation control.

Such research directions as mentioned above are not limited to those fields. Take an application of stochastic factors to fuzzy GP for example; it has captivating prospects.

It is not totally confirmed whether the application of fuzzy GP is successful since this branch is like a newborn baby. We contribute this book to readers in the belief that fuzzy GP will become an important branch of fuzzy mathematics after scholars over the world make an effort to study it, and that it will add luster to the development of science in the world and bring benefit to humanity.

Chapter 10

BIBLIOGRAPHY

[AG01] F. Arikan and Z. Güngör. An application of fuzzy goal programming to a multiobjective project network problem. *International J. of Fuzzy Sets and Systems,* 2001, 119: 49-58.

[AMA93] P. Aliev, G. Mamedova, and R. Aliev. *Fuzzy Sets Theory and Its Application.* Talriz University Press, 1993.

[Asa82] K. Asai Writing, R.H.Zhao Translation. *An Introduction to the Theory of Fuzzy Systems.* Peking: Peking Norm. University Press, 1982.

[Avr76] M. Avriel. *Nonlinear programming Analysis and Methods.* Englewood Cliffs, New Jersey: Prentice Hall Co. Inc., 1976.

[AW70] M. Avriel and A.C. Williams. Complementary geometric programming. *SIAM J. Appl. Math.,* 1970, 19: 125-141.

[Biw92] M.P. Biwal. Fuzzy programming technique to solve multi-objective geometric programming problems. *International J. of Fuzzy Sets and Systems,* 1992, 51: 67-71.

[BP76] C.S. Beightler and D.T. Phillips. *Applied Geometric Programming.* New York: John Wiley and Sons, 1976.

[BS79] M.S. Bazaraa and C.M. Shetty. *Nonlinear Programming—Theory and Algorithms.* New York: John Wiley and Sons, 1979.

[Cai83] W. Cai. The extension set and incompatible problem. *J. of Scientific Exploration,* 1983, 1: 81-93.

[Cao87a] B.Y. Cao. Solution and theory of question for a kind of fuzzy positive geometric program. *Proc. of 2nd IFSA Congress,* 1987 July 20-July 25, Vol.I, Tokyo, 205-208.

[Cao87b] B.Y. Cao. Solution and theorem of question for a kind of fuzzy positive geometric program. *J. of Changsha Norm. Univ. of Water Resources and Electric Power (Natural Sci. Ed.)*, 1987, 2(4): 51-61).

[Cao89a] B.Y. Cao. Study of fuzzy positive geometric programming dual form. *Proc. 3rd IFSA Congress*, 1989, August 6-August 11; Seattle. 775-778.

[Cao89b] B.Y.Cao. Study on non-distinct self-regression forecast model. *Kexue Tongbao*, 1989, 34(17): 1291-1294.

[Cao89c] B.Y.Cao. Study for a kind of regression forecasting model with fuzzy datums. *J. of Mathematical Statistics and Applied Probability*, 1989, 4(2): 182-189.

[Cao90a] B.Y. Cao. Study on non-distinct self-regression forecast model. *Chinese Sci. Bull.*, 1990, 35(13): 1057-1062.

[Cao90b] B.Y. Cao. Extension convex set. *Acta Science Naturalium Univ. Norm Hunanenist*, 1990, 13(1): 18-24.

[Cao91] B.Y. Cao. A method of fuzzy set for studying linear programming "Contrary Theory". *J. Hunan Educational Institute*, 1991, 9(2): 17-22.

[Cao92a] B.Y.Cao. Further study of posynomial geometric programming with fuzzy coefficients. *Mathematics Applicata*, 1992, 5(4): 119-120.

[Cao92b] B.Y.Cao. Another proof of fuzzy posynomial geometric programming dual theorem, *BUSEFAL*, 1996, Vol.66: 43-47.

[Cao92c] B.Y.Cao, Chief Editor. *Proceedings of the Results Congress on Fuzzy Sets and Systems*. Changsha: Hunan Science Technology Press, 1992.

[Cao93a] B.Y. Cao. Fuzzy geometric programming(I). *International J. of Fuzzy Sets and Systems*, 1993, 53: 135-153.

[Cao93b] B. Y. Cao. Input-output mathematical model with T-fuzzy data. *International J. of Fuzzy Sets and Systems*, 1993, 59: 15-23.

[Cao93c] B.Y. Cao. Extended fuzzy geometric programming. *J. of Fuzzy Mathematics*, 1993, 2: 285-293.

[Cao93d] B.Y. Cao. Fuzzy strong dual results for fuzzy posynomial geometric programming. *Proc.of 5th IFSA Congress*, 1993, July 4-July 9, Seoul. 588-591.

[Cao94a] B.Y.Cao. Posynomial geometric programming with L-R fuzzy coefficients. *International Fuzzy Sets and Systems*, 1994, 64: 267-276.

[Cao94b] B.Y.Cao. *Lecture in Economic Mathematics—Linear Programming and Fuzzy Mathematics*. Tiangjing: Tiangjing Translating Press of Science and Technology, 1994.

[Cao95a] B.Y.Cao. The study of geometric programming with $(\cdot, c)$-fuzzy parameters. *J. of Changsha Univ. of Electric Power (Natural Sci. Ed.)*, 1995, 1: 15-21.

[Cao95b] B.Y.Cao. Fuzzy geometric programming optimum seeking of scheme for waste water disposal in power plant. *Proc. of FUZZY-IEEE/IFES' 95*, 1995 August 22-August 25; Yokohama. 793-798.

[Cao95c] B.Y.Cao. B.Y.Cao, Types of non-distinct multi-objective geometric programming. *Hunan Annals of Mathematics*, 1995, 15(1): 99-106.

[Cao95d] B.Y.Cao. Study of fuzzy positive geometric programming dual form. *J. of Changsha Univ. of Electric Power (Natural Sci. Ed.)*, 1995, 10(4): 343-351).

[Cao95e] B.Y.Cao. Classification of fuzzy posynomial geometric programming and corresponding class properties. *J. of Fuzzy Systems and Mathematics*, 1995, 9(4): 60-64.

[Cao96a] B.Y.Cao. New model with T-fuzzy variable in linear programming. *International Fuzzy Sets and Systems*, 1996; 78: 289-292.

[Cao96b] B.Y.Cao Fuzzy geometric programming (II)—Fuzzy strong dual results for fuzzy posynomial geometric programming. *J. of Fuzzy Mathematics*, 1996, 4(1): 119-129.

[Cao97a] B.Y.Cao. Research for a geometric programming model with T-fuzzy variable. *J. of Fuzzy Mathematics*, 1997, 5(3): 625-632.

[Cao97b] B.Y.Cao. Fuzzy geometric programming optimum seeking of scheme for wastewater disposal in power plant. *Systems Engineering—Theory & Practice*, 1997, 5: 140-144).

[Cao98a] B.Y.Cao. Further research of solution to fuzzy posynomial geometric programming.*Academic Periodical Abstracts of China*,1998, 4(12): 1435-1437.

[Cao98b] B.Y. Cao. "Advance in fuzzy geometric programming". In: *Popular Works by Centuries' World Celebrities*, Z. X. Mah ed., *U. S. World Celebrity Books LLC*, California: World Science Press, 1998; 15-20.

[Cao99] B.Y.Cao. Fuzzy geometric programming optimum seeking in power supply radius of transformer substation. *1999 IEEE Int. Fuzzy Systems Conference Proceedings*, 1999 July 25-July 29; Korea. 3: III–1749-1753.

[Cao00] B.Y.Cao. Research of posynomial geometric programming with flat fuzzy coefficients. *J. of Shantou University (Natural Sci. Ed.)*, 2000, 15(1): 13-19.

[Cao01a] B.Y.Cao. Primal algorithm of fuzzy posynomial geometric programming. *Joint 9th IFSA World Congress and 20th NAFIPS International Conference Proceedings*, July 25-July 28; Vancouver. 31-34, 2001.

[Cao01b] B.Y.Cao. Model of fuzzy geometric programming in economical power supply radius and optimum seeking method. *Engineer Sciences*, 2001, 3: 52-55.

[Cao01c] B.Y.Cao. Application of geometric programming and fuzzy geometric one with fuzzy coefficients in seeking power supply radius transformer substation. *Systems Engineering—Theory & Practice*, 2001, 21(7): 92-95.

[Cao01d] B.Y.Cao. Extension posynomial geometric programming. *J. of Guangdong University of Technology,* 2001, 18(1): 61-64.

[Cao01e] B.Y.Cao. Direct algorithm of fuzzy posynomial geometric programming. *J. of Fuzzy Systems and Mathematics,* 2001, 15(4): 81-86.

[Caoupd] B.Y.Cao. Fuzzy reversed posynomial geometric programming and its dual form, update.

[CDR87] A.Charnes, S.Duffuaa, and M. Ryan. The more-for-less paradox in linear programming. *European Journal of Operational Research,* 1987, 31: 194-197.

[Cha83] S. Chants. The use of parametric programming in fuzzy linear programming. *International J. of Fuzzy Sets and Systems,* 1983, 11: 243-251.

[CK71] A. Charnes and D. Klingman. The more-for-less paradox in the distribution model. *Cahiers de Center d'Etudes Recharche Operationelle,* 1971, 13(1): 11-22.

[Dan63] G.B.Dantzing. *Linear Programming and Extensions.* New Jersey: Princeton U.P., Princeton,1963.

[Dia87] P.Diamond. "Least squares fitting of several fuzzy data". *Proc. of IFSA Congress,* 1987, July 20-July 25, Vol.I, Tokyo, pp. 329-332.

[DP66] R.J.Duffin and E.L.Peterson. Duality theory for geometric programming. *SIAM J. Appl. Math.,* 1966, 14: 1307-1349.

[DP72] R.J.Duffin and E.L.Peterson. Reversed geometric programming treated by harmonic means. *Indiana Univ. Math. J.,* 1972; 22: 531-550.

[DP73] R.J.Duffin and E.L.Peterson. Geometric programming with signomials. *J. Optimization Theory Appl.,* 1973; 11: 3-35.

[DP80] D.Dubois and H.Prade. *Fuzzy Sets and Systems—Theory and Applications.* New York: Academic Press, 1980.

[DPZ67] R.J.Duffin, E.L. Peterson, and C. Zener. *Geometric Programming: Theory and Applications.* New York: John Wiley and Sons, 1967.

[Duf62a] R.J.Duffin. Dual programs and minimum cost. *SIAM J.Appl. Math.,* 1962, 10: 119-123.

[Duf62b] R.J.Duffin. Cost minimization problems treated by geometric means. *Operations Res.,* 1962, 10: 668-675.

[Duf70] R.J.Duffin. Linearizing geometric programs. *SIAM. Review,* 1970, 12: 211-227.

[Eck80] J.G. Ecker. Geometric programming. *SIAM Review,* 1980, 22(3): 338-362.

[EK87] J.G. Eckerand and M. Kupferschmid. An ellipsoid algorithm for non-linear programming. *Mathematical Programming,* 1987, 27: 83-106.

[Fin77] G. Finke. Aunified approach to reshipment, overshipmemt and post-optimization problems. *Proceedings of the 8th IFIP Conference on Optimization Techniques,* 1977, part 2: 201-208.

[Gog67] J.A.Goguen. L—fuzzy sets. *J. Math. Anal. Appl.,* 1967; 18: 145-174.

[GZ83] M.G.Guan and H.D.Zheng. *Linear Programming.* Jinan: Shandong Science Technology Press, 1983.

[JS78] T.R. Jefferson and C.H. Scott. Avenues of geometric programming. *New Zealand Operational Res.,* 1978, 6: 109-136.

[KM73] J.C. Kuester and J.H. Mize. *Optimization Techniques with Fortran.* New York: McGraw-Hill, 1973.

[Lin86] Y.Lin. On "Contrary Theory" in general linear programming. *Chinese J. of Operations Research,* 1986, 5(1): 79-81.

[LL01] T. León and V. Liern. A fuzzy method to repair infeasibility in linearly constrained problems. *International J. of Fuzzy Sets and Systems,* 2001, 122: 237-243.

[Luo84a] C.Z.Luo. The extension principle and fuzzy numbers (I). *Fuzzy Mathematics,* 1984, 4(3): 109-116.

[Luo84b] C.Z.Luo. The extension principle and fuzzy numbers (II). *Fuzzy Mathematics,* 1984, 4(4): 105-114.

[LW92] Lu Minggen, Wu Wangming. Interval value and derivate of fuzzy-valued function. *J. Fuzzy Systems and Mathematics,* 1992, 6(Special Issue): 182-184.

[LZ98] Liu B.D., Zhao R.Q., *Stochastic Programming and Fuzzy Programming.* Peking: Tsinghua University Press, 1998.

[Man79] O.L. Mangasarian. Uniqueness of solution in linear programming. *Linear Algebra and Its Applications,* 1979, 25: 151-162.

[MTM00] H.R. Maleki, M. Tata, M. Mashinchi. Linear programming with fuzzy variable. *International J. of Fuzzy Sets and Systems,* 2000, 109: 21-33.

[NS76] C.V. Negoita and M. Sularia. Fuzzy linear programming and tolerance in planning. *Econ. Comp. Econ. Cybernetic Stud. Res.,* 1976, 1: 613-615.

[Obr86] M.M. Obrad. Mathematical dynamic model for long-term distribution system planning. *IEEE Transaction on Power Systems,* 1986, 1: 34-41.

[PB71] L.D. Pascual and A. Ben-Israel. Vector-valued criteria in geometric programming. *Operations Res.,* 1971, 19: 98-104.

[Pet78] E.L. Peterson. Geometric programming. *SIAM Review,* 1978, 18: 1-51.

[Rij74] M.J. Rijckaert. Survey of programs in geometric programming. *C.C.E.R.O.,* 1974, 16: 369-382.

[Rou91] M. Roubens. "Inequality constraints between fuzzy numbers and their use in mathematical programming." In *Stochastic Versus Fuzzy Approaches to Multiobjective Mathematical Programming Under Uncertainty,* Slowinski R. and Teghem J., eds, Dordrecht: Kluwer Academic Publishers, 1991, 321-330.

[RT91] M. Roubens and J.J. Teghem. Comparison of methodologies for fuzzy and stochastic multi-objective programming. *International J. of Fuzzy Sets and Systems,* 1991, 42: 119-132.

[Shi81] G.Y.Shi. Algorithm and convergence about a general geometric programming. *J. Dalian Institute Technol.,* 1981, 20: 19-25.

[TA84] H. Tanaka and K. Asai. Fuzzy linear programming problem with fuzzy number. *International J. of Fuzzy Sets and Systems,* 1984, 13: 1-10.

[TOA73] H. Tanaka, T. Okuda, and K. Asai. On fuzzy mathematical programming. *J. Cybern.,* 1973, 3(4): 37-46.

[Ver84] J.L. Verdegay. A dual approach to solve the fuzzy linear programming problem. *International J. of Fuzzy Sets and Systems,* 1984, 14: 131-140.

[Ver90] R.K. Verma. Fuzzy geometric programming with several objective functions. *International Fuzzy Sets and Systems,* 1990, 35: 115-120.

[WB67] D.J. Wilde and C.S. Beightler. *Foundations of Optimization.* Englewood Cliffs, New Jersey: Prentice Hall Co. Inc., 1967, 76-109.

[WX99] W.X.Xing and J.X.Xie. *Modern Optimization for Calculation Method.* Peking: Tsinghua Press, 1999.

[WY82] F.Wu and Y.Y.Yuan. Geometric programming. *Math. in Practice and Theory,* 1982, 1-2: 46-63, 61-72, 60-80, 68-81.

[YCL95] Y.Y. Yu, B.Y. Cao, and X.R. Li. "The application of geometric and fuzzy geometric programming in option of economic supply radius of transformer substations." In: *Proceedings of Int. Conference on Inform and Knowledge Engineering*, K.Q.Zhou ed., Dalian, China: Dalian Maritime University Publishing House, August 21-August 25, 245-249, 1995. (or in *Information Engineering*, K.Q.Zhou ed., Harbin, China: Harbin Engineering University Press, 69-74, 1996.)

[YJ91] C.E.Yang, D.Y.Jin. The more-for-less paradox in linear programming and nonlinear programming. *Systems Engineering,* 1991, 9(2): 62-68.

[YWY91] Y.Y.Yu, X.Z.Wang, Y.W. Yang. Optimizational selection for substation feel economic radius. *J. of Changsha Normal University of Water Resources and Electric Power,* 1991, 6(1): 118-124.

[YZY87] Q.Y.Yang, Z.L.Zhang, M.Z.Yang. Transformer substation capacity dynamic optimizing in city power network planning. *Proc. of Colleges and Univ. Speciality of Power System and Its Automation. The Third a Academic Annual Conference,* Xian: Xian Jiao Tong Univ. Press, 1987, 7-11.

[Zad65a] L.A. Zadeh. Fuzzy Sets. *Inform. and Control.*, 1965, 8: 338-353.

[Zad65b] L.A. Zadeh. Fuzzy sets and systems. *Proc. of the Symposium on Systems Theory,* NY: Polytechnic Press of Polytechnic Institute of Brooklyn, 1965.

[Zen61] C.Zener. A mathematical aid in optimizing engineering design. *Proc. Nat. Acad. Sci. U. S. A.*, 1961; 47: 537-539.

[Zim76] H.-J. Zimmermann. Description and optimization of fuzzy systems. *Internat. J. General Systems,* 1976, 2: 209-215.

[Zim78] H.-J. Zimmermann. Fuzzy programming and linear programming with several objective functions. *International J. of Fuzzy Sets and Systems,* 1978, 1: 45-55.

[Zim91] H.-J. Zimmermann. *Fuzzy Sets Theory and Its Application.* Boston: Kluwer Academic Publishers, 1991.

Chapter 11

INDEX

Applied Optimization

72. A.H. Siddiqi and M. Kočvara (eds.): *Trends in Industrial and Applied Mathematics.* Proceedings of the 1st International Conference on Industrial and Applied Mathematics of the Indian Subcontinent. 2002 ISBN 1-4020-0751-5
73. M. Doumpos and C. Zopounidis: *Multicriteria Decision Aid Classification Methods.* 2002 ISBN 1-4020-0805-8
74. E.J. Kontoghiorghes, B. Rustem and S. Siokos (eds.): *Computational Methods in Decision-Making, Economics and Finance.* 2002 ISBN 1-4020-0839-2
75. J. Dupačová, J. Hurt and J. Štěpán: *Stochastic Modeling in Economics and Finance.* 2002 ISBN 1-4020-0840-6
76. Bing-Yuan Cao: *Fuzzy Geometric Programming.* 2002 ISBN 1-4020-0876-7

KLUWER ACADEMIC PUBLISHERS – DORDRECHT / BOSTON / LONDON

MIX
Papier aus verantwortungsvollen Quellen
Paper from responsible sources
FSC® C105338

If you have any concerns about our products,
you can contact us on
ProductSafety@springernature.com

In case Publisher is established outside the EU,
the EU authorized representative is:
Springer Nature Customer Service Center GmbH
Europaplatz 3, 69115 Heidelberg, Germany

Printed by Libri Plureos GmbH
in Hamburg, Germany